AF466116

Advanced Information and Knowledge Processing

Springer-Verlag London Ltd.

Also in this series

Gregoris Mentzas, Dimitris Apostolou, Andreas Abecker and Ron Young
Knowledge Asset Management
1-85233-583-1

Michalis Vazirgiannis, Maria Halkidi and Dimitrios Gunopulos
Uncertainty Handling and Quality Assessment in Data Mining
1-85233-655-2

Asunción Gómez-Pérez, Mariano Fernández-López and Oscar Corcho
Ontological Engineering
1-85233-551-3

Arno Scharl (Ed.)

Environmental Online Communication

With 64 Figures

ECOresearch.net

Arno Scharl, DDr.
Professor of Information Systems
The University of Western Australia, Business School,
35 Stirling Highway, Crawley, WA 6009, Australia

Series Editors
Xindong Wu
Lakhmi Jain

British Library Cataloguing in Publication Data
Environmental online communication. - (Advanced information and knowledge processing)
1. Environment protection - Computer network resources
2. Sustainable development - Computer network resources
3. Web site development - Computer network resources
I. Scharl, Arno
025'.0633372

DOI 10.1007/978-1-4471-3798-6

Library of Congress Cataloging-in-Publication Data
A catalogue record for this book is available from the Library of Congress

AI&KP ISSN 1610-3947

springeronline.com

Originally published by Springer-Verlag London in 2004
MyCopy version of the original edition 2004

Typesetting: Electronic files prepared by editor
34/3830-543210 Printed on acid-free paper SPIN 10959756
www.springer.com/mycopy

Preface

This book brings together articles exploring the design, implementation, management, funding, promotion, and evaluation of networked information systems that advocate sustainability and the protection of natural ecosystems. Case studies of deployed and planned information systems complement theoretical work on the methodological, technological, and organizational foundations of environmental online communication.

Studying communicative strategies and processes that function between disciplines and worldviews, this book helps unearth hidden assumptions and misconceptions about environmental issues, contributes to a mutual understanding of existing problems, and suggests priorities for research and policy development. The 25 chapters contained in this volume analyze environmental online communication from four interrelated perspectives: *Raising Environmental Awareness*, *Environmental Science*, *Corporate Sustainability*, and *Networks and Virtual Communities*. The chapters document social and technological challenges and allow readers to appreciate the diversity of approaches and projects.

The initiative to compile this edited volume stems from the activities of the ECOresearch Network [696], an interdisciplinary collaboration that encourages scientific discourse and critical debate to establish a shared understanding of environmental online communication. The network also hosts the book's official Web site at

www.ecoresearch.net/springer [701],

which provides supplemental material including text samples, bibliographic resources, project descriptions, and detailed author profiles including current affiliations and research interests.

Emerging Technologies

Advances in information technology are transforming the way society handles the explosive growth and dwindling half-life of knowledge in general [46; 218; 607], and of environmentally relevant information in particular [435]. Interactive media such as the World Wide Web, electronic mail, or text messaging via wireless devices revolutionize the reach and efficiency of individuals and organizations alike – from simple electronic mailing lists to video activists broadcasting protests live via palm-top computers and mobile phones [587], or attempts to run an organization remotely while climbing K2, the second highest mountain in the world [600; 643; 791].

Broadband connectivity, global ubiquity, portability, 24/7 availability, and adaptivity transcend historical modes of production. They trigger social change and catalyze advanced economic systems, creating both opportunities and constraints [91; 610; 612]. Besides improving productivity and reducing transaction costs, new technologies enable people to participate in decision-making [135] and increasingly align networked information systems with the visions of their pioneers and early proponents [81; 153; 386]. The proliferation of these technologies ers the transition to a knowledge-based economy, connects the world's poor to entrepreneurial and educational opportunities, and helps balance power by enabling businesses, government, and civil society to scrutinize each other in collaborative, consensus-building processes [135].

Online communication arguably has the potential to support democratic decision structures and decrease inequality as outlined above. Since stakeholders are not equally well-positioned to express their views in online environments, however, the Internet may "just as easily undermine sound decision-making and public participation" [636, p41]. It may exacerbate existing inequalities if steps are not taken to ensure widespread access [135]. Barriers of distance, language, literacy, and connectivity hinder participation [560]. They limit the Internet in its reach, which is helpful but not yet adequate to ensure access in all countries.

Particularly unlikely to serve citizens in poor countries or rural areas, affordability and usability of the Internet vary greatly between and within countries [636]. Not surprisingly, the virtual reflects power differentials of the real. Advanced systems cater to the needs of preferential clients. Their usage remains differentiated in territorial terms, following the uneven distribution of infrastructure, wealth, and education [94]. Emphasizing this argument, Tehranian describes McLuhan's global village [364] as a "neofeudal manor with highly fortified and opulent castles (centers of industrial, financial, and media power) surrounded by vast hinterlands of working peasants clamoring for survival and recognition" [557, p55f.].

Such conflicting claims regarding the social and environmental cost-benefit ratios of information technology often arise when either side prescribes a one-way causal relationship between technology and society [47]. Yet technology only conditions society, but does not determine it. Assigning univocal meaning to information technology or attempting to calculate its impacts deterministically neglect the real world's ambivalence and multiplicity [334]. Information technology is often used in ways never intended or foreseen by its inventors. Neither good nor bad (depending on context, use, and point of view), information technology cannot even be considered neutral since it "conditions or constrains, exposes or closes off, the range of possibilities" [334, p8].

With regard to environmental communication, emerging technologies facilitate transitions [397] from broadcast to interactive communication, cognitive to experiential learning, product to process orientation, conflict positions to shared meaning, distributive to integrative negotiations, and from isolated, compartmentalized knowledge [599] to the ability to deal effectively with complexity, uncertainty, and risk. One-dimensional, linear ecological moralization is replaced by a discourse on value judgments from competing social, economic, and ethical points of view [397].

Although environmental online communication has an immediate positive effect on agenda-setting within the target group, its long-term impacts are hard to evaluate. To a large extent, the long-term impacts depend on the quality, professional representation, and credibility of communicated content. Constant progress monitoring and participatory peer review among network members and external experts improve quality and representation [5]. Transparent processes that respect professional norms and procedural fairness help produce credible information [295].

Part I: Raising Environmental Awareness

Transcending geographic boundaries and exploring new symbiotic relations between human society and nature [216], sustainability calls for participatory decision-making and cooperative approaches to environmental governance [245; 599]. Public awareness, a cornerstone of such participatory strategies, contributes to informed personal choices, protection of the environment, and improved environmental performance by the corporate sector [636]. With its emphasis on public access to environmentally relevant information, online advocacy, and environmental education, the first part of the book covers three important aspects of raising environmental awareness.

Adopted by 178 nations in 1992, Principle 10 of the *Rio Declaration on Environment and Development* [567] articulated access to environmental information, the opportunity to participate in decision-making, and access to justice as cornerstones of environmental governance. The principle was elaborated by the targets of the *Aarhus Convention* [566], adopted in 1998, and entered into force in 2001. Governments reaffirmed these targets during the *Johannesburg Summit 2002* [730] with the launch of the *Partnership for Principle 10* [758], a cooperation of government agencies and civil society groups to translate the principle into action at the national level.

At least three interrelated forces drive worldwide demand for public access to environmental information: the urgency and scope of environmental problems, the increasingly active character of civil society, and the revolution in information and communication technologies [432]. Disseminating environmental information via the Internet, directly or via online media as intermediaries, enhances public knowledge and builds awareness about the interdependency of ecological, economic, and social issues. Web-based or wireless services, for example, provide instant feedback on diverse issues such as weather, air quality, speed of traffic, or location of accidents [259].

Online advocacy also benefits from Internet technology in its aims to raise environmental awareness. While sharing environmental concerns and responsibilities, individuals and organizations often lack coordination. The competition for budget, jurisdiction, and influence increases their insularity [636]. Environmental online communication addresses this problem by encouraging networking and facilitating "environmental discourse that recognizes and incorporates the social construction of meaning in communicative processes" [366, p392]. Such a discourse improves the quality of decisions, builds trust in institutions, and helps resolve conflict among competing interest [636].

The expansion of formal and non-formal environmental education in terms of both geography and curriculum through traditional educational institutions, advocacy organizations, and the media are indispensable for achieving sustainability. Environmental education requires accurate and credible information [287]. Underpinned by science and technology as providers of such information, environmental education is a powerful tool for understanding natural and social processes and their complex interrelationships [458].

Part II: Environmental Science

The rapid expansion of science results in differentiation and the creation of new "niches" succumbing to scientific scrutiny. Environmental science is no exception. Ecology expanded into a wide range of specialized fields such as risk assessment, environmental ethics, resource economics, ecological psychology, environmental education, and environmental communication [607]. The second part of the book addresses three problems intrinsically linked to this differentiation and the increasing complexity of environmental science. Complex environmental models are difficult to handle computationally, require advanced methods to represent model structure and results, and introduce communicative challenges both within research communities and between researchers and external stakeholders.

Science is an important source of environmental indicators [365] and essential for promoting sustainability [90]. Online databases such as the World Resource Institute's *EarthTrends* [691] and the United Nations Environment Programme's *GEO Data Portal* [708], for example, synthesize information gathered from a variety of institutions, geographical regions, and economic sectors [432]. Interactive visualizations allow exploring such multidimensional and often geographically referenced data [562].

Despite advances in representing and sharing complex data, some environmental simulations are either not feasible economically or still beyond the reach of supercomputers such as the *Earth Simulator*, the currently most powerful machine [689]. Distributed Internet computing overcomes these limitations by exploiting spare computer cycles and breaking down the calculations into parallel units, which are then processed by networks of globally distributed volunteers.

Environmental scientific disciplines have superior expertise in their focal activities and specific means of disseminating information. In many cases, however, breaking away from "the mono-logical habits of entrenched and specialized disciplines" [366, p390] remains a challenge. Lack of awareness regarding scientific expertise continues to jeopardize good intentions. Thus it is a shared responsibility of researchers and the media to choose accessible and intuitively understandable formats for describing complex phenomena [624].

Many failures to raise awareness stem from environmentalists' beliefs that scientific evidence and related ecological concerns are convincing on their own, a phenomenon commonly referred to as "comprehension gap" [397]. To improve mutual understanding and overcome comprehension gaps, Meppem and Bourke suggest communicative strategies for developing and implementing environmental policy that explicitly deal with the "assumptions and motivations of contested positions in the sustainability debate" [366, p389]. This can increase transparency and "give coherence and credibility to masses of scientific information" [295, p92]. Good communicative strategies accept diverging perspectives, provide rules of conduct, and establish objective criteria for decision-making [90].

Part III: Corporate Sustainability

Environmental degradation is intimately linked to global modes of production and accumulation [279]. Economic activities that degrade the environment generally yield winners and losers with regard to the distribution of costs and bene

[67; 636]. Environmental communication can spark interest in redistributing environmental costs more fairly throughout society. Fuelled by the World Summit on Sustainable Development in Johannesburg [730] and the numbers and magnitude of accounting failures on Wall Street, there is a growing trend towards accountability, transparency, and stakeholder engagement across all levels, functions, and operations [230; 605]. This trend promotes the concept of sustainability and fuels stakeholders' interest in the allocation and acquisition of corporate resources.

Sustainable resource allocation requires the reporting of comprehensive and reliable indicators based on integrated analysis of the environmental, social, and economic dimensions of corporate activities [112; 437]. Rather than directly influencing decisions, environmental reporting often creates conditions under which decisions crystallize [258]. It represents an additional stimulus to consider social and environmental issues [138] and sheds light on an organization's resource allocation and corporate communication policy.

Socially responsible investing affects the acquisition of corporate resources by integrating social and environmental values into the decision-making process. Evaluating the environmental performance of an organization based on its reports and external information sources, socially responsible investing strives to deploy capital consistent with the needs of society and the natural limits of ecosystems [70]. External information sources include government reports on company emissions and pollution registers such as *Pollution Watch* [760] and the *Toxics Release Inventory* [783]. By addressing the stakeholders' interest, these sources affect media coverage, public opinion, and stock market fluctuations [636].

Part IV: Networks and Virtual Communities

The final part of the book pays attention to communicating, managing, and applying knowledge within virtual communities, the roles trust and credibility play in these processes, and how networked information systems catalyze the emergence and growth of virtual communities.

Simultaneously arousing both hope and fear among observers [37; 353], networks and virtual communities have been associated with conflicting claims about their impact on social interaction. On the one hand, online communication is seen as re-enchanting community (perceived as eroded in "real life"), substituting territorially bound forms of human interaction by more selective patterns of social relations. On the other hand, online communication is accused of damaging community by encouraging domination, dogmatism, social isolation, and a withdrawal from the real world [47; 94; 353; 610; 612].

With the Internet diffusing into the mainstream of society, the dichotomy of these contested positions is becoming less pronounced. Independent of geographic proximity and institutional affiliation, networks and virtual communities emerge through processes of cooperation and social exchange [334; 454]. They depend on and benefit from a synergy of skills, resources, and projects, the constitution and dynamic maintenance of shared knowledge, flexible and nonhierarchical modes of cooperation, and distributed decision-making [334].

Built on common interests, virtual communities often establish social bonds that extend beyond the narrow focus of those interests [47]. People in networked societies live and work in overlapping relationships, typically cycling through interactions with multiple sets of others [610; 612]. Such social networks have always existed, but new technologies have supported and enabled their emergence as a dominant form of social organization [611].

Acknowledgements

This book would not have been possible without the help and contributions of many colleagues. My first word of appreciation goes to the authors for their excellent chapters and active participation in the peer-review process. Each chapter was evaluated by two or three referees and revised at least once on the basis of their comments and criticism. Apologies in advance to the people I have failed to acknowledge, and thanks to the following colleagues who generously provided additional reviews and feedback for the authors: Mitra Arami, Christian Bauer, Astrid Dickinger, Nicolas Knotzer, Robert Krimmer, Renée Lertzman, Sarah Lumley, Ken Morgan, Jamie Murphy, Barbara Pedersen, Richard Thomas, and Dave Webb. I would also like to recognize the financial support of the University of Western Australia's School of Economics and Commerce to help establish and promote the ECOresearch Network. Annette McNamara is to be commended for her valuable assistance in the editorial process. At Springer, I would like to thank Stephen Bailey, Joanne Cooling, Catherine Drury, Beverley Ford, Lyn Imeson, Melanie Jackson, Rebecca Mowat, and Jenny Wolkowicki for their support and help in the materialization of this book.

Arno Scharl
Perth, January 2004

Table of Contents

Geographically Referenced Data

PART III: CORPORATE SUSTAINABILITY

Sustainability and Social Responsibility

Environmental Reporting

PART IV: NETWORKS AND VIRTUAL COMMUNITIES

Online Collaboration, Trust and Credibility

Knowledge Management

List of Authors

The book's Web site at www.ecoresearch.net/springer [701] provides additional author information including contact details, research interests, current projects, and a short biography.

BSc MSc
Tolu Aina
Researcher
University of Oxford, Atmospheric, Oceanic and Planetary Physics
Oxford, UK

MSEE PhD
Gary Alexander
Senior Lecturer in Telematics
The Open University, Faculty of Technology
Milton Keynes, UK

BA Hons (Oxon) DPhil
Myles R. Allen
Lecturer in Physics
University of Oxford, Atmospheric, Oceanic and Planetary Physics
Oxford, UK

PhD
Wolfgang Amann
Senior Researcher
University of St. Gallen, Research Institute for International Management
St. Gallen, Switzerland

MSc
Andri Baltensweiler
Senior Scientist
Swiss Federal Institute for Forest, Snow and Landscape Research
Birmensdorf, Switzerland

Anastasios Bassoukos
Software Developer
Aristotle University, Mechanical Engineering Department
Thessaloniki, Greece

Dr
Martin Brändli
Senior Scientist
Swiss Federal Institute for Forest, ow and Landscape Research
ensdorf, Switzerland

Dipl oec
Rudolph Braun
Research Assistant
University of Hohenheim, Chair of Environmental Management
Stuttgart, Germany

PhD
Steve Cabaniss
Professor
University of New Mexico, Department of Chemistry
Albuquerque, USA

BNE MSc
Carl Christensen
Researcher
University of Oxford, Computing Laboratory
Oxford, UK

BSc (Hons), PhD
Matthew Collins
Climate Scientist
Hadley Centre for Climate Prediction and Research, The Met Office
Exeter, UK

Renate Ebel
State Institute for Environmental Protection Baden-Wuerttemberg
Karlsruhe, Germany

Diplôme ESSEC, PhD
Corinne Faure
Professor of eCommerce/Marketing
European Business School
Oestrich-Winkel, Germany

BSc MSc (Hons) PhD
David Frame
Researcher
University of Oxford, Atmospheric Oceanic and Planetary Physics
Oxford, UK

PhD
Michael Haase
Division Head
Environmental Information Systems, Research Institute for Applied Knowledge Processing; Ulm, Germany

Benjamin Hermann
System Engineer
Informations- und Technologiemanagement Beratungsgesellschaft mbH
Stuttgart, Germany

Dipl Ing (FH)
Thomas Hillenbrand
Project Manager
Fraunhofer Institute for Systems and Innovation Research, Department of Environmental Technology and Environmental Economics; Karlsruhe, Germany

RNDr CSc
Jiri Hrebicek
Professor of Informatics
Masaryk Univerzity Brno,
Faculty of Informatics
Brno, Czech Republic

MS
Yingping Huang
Research Assistant
University of Notre Dame, Department of Computer Science and Engineering
Notre Dame, USA

Dr
Ralf Isenmann
Senior Researcher
Kaiserslautern University of Technology, Department of Business Information Systems and Operations Research
Kaiserslautern, Germany

Dipl Eng MSc
Athina Kaprara
Electrical and Computer Engineer
Aristotle University, Mechanical Engineering Department
Thessaloniki, Greece

Dr Eng
Kostas Karatzas
Assistant Professor of Informatics Systems and Applications; Aristotle University, Mechanical Engineering Department
Thessaloniki, Greece

BSc PGDFM
Prakash Kashwan
Program Associate
The Ford Foundation
New Delhi, India

BSc (Hons) PhD
Jamie Kettleborough
Researcher Associate
Rutherford Appleton Laboratory
Oxfordshire, UK

MES
Shiban Khan
Researcher
University of Pennsylvania
Philadelphia, USA

Ir
Rolf Kleef
Consultant
AIDEnvironment
Amsterdam, The Netherlands

Henning Kolb
Student, Academic Staff
University of Hohenheim,
Chair of Environmental Management
Stuttgart, Germany

Dr
Helmut Krcmar
Professor of Information Systems
Technical University of Munich,
Department of Informatics
Munich, Germany

Dipl oec
Martin Kreeb
Research Assistant
University of Hohenheim,
Chair of Environmental Management
Stuttgart, Germany

Mgr
Miroslav Kubasek
PhD Student
Masaryk Univerzity Brno,
Faculty of Informatics
Brno, Czech Republic

Dieter Lehne
State Office for Environment and Geology Hesse
Wiesbaden, Germany

Dr
Wolfgang Loibl
GIS Modeling Scientist
ARC Systems Research GmbH
Seibersdorf, Austria

MS PhD
Gregory Madey
Associate Professor
University of Notre Dame, Department of Computer Science and Engineering
Notre Dame, USA

Dr. Ing.
Jorge Marx-Gómez
Assistant Professor of Business Informatics
Otto-von-Guericke Universität
Magdeburg, Germany

Dipl Math
Asteris Masouras
Software Developer
Aristotle University, Mechanical Engineering Department
Thessaloniki, Greece

Sebastian Maute
System Administrator
Informations- und Technologiemanagement Beratungsgesellschaft mbH
Stuttgart, Germany

PhD MPhil
Fiona M. Haslam McKenzie
Lecturer
University of Western Australia, School of Earth and Geographical Sciences
Perth, Australia

Dipl Ing
Roman Mesicek
Researcher
Sustainable Europe Research Institute
Vienna, Austria

Dr
Aldo de Moor
Assistant Professor
Tilburg University, Department of Information Systems and Management
Tilburg, The Netherlands

Dr Ing Habil
Nicolas Moussiopoulos
Professor of Heat Transfer and Environmental Engineering; Aristotle University, Mechanical Engineering Department
Thessaloniki, Greece

Dipl oec
Valentin Nicolescu
Research Assistant
Technical University of Munich, Chair of Information Systems
Munich, Germany

Andreas Niederl
Research Assistant
Sustainable Europe Research Institute
Vienna, Austria

Dr
Rudolf Orthofer
Environmental Assessment Scientist
ARC Systems Research GmbH
Seibersdorf, Austria

Ioannis Papaioannou
Software Developer
Aristotle University, Mechanical Engineering Department
Thessaloniki, Greece

Dr
A. Townsend Peterson
Professor and Curator
University of Kansas, Department of Ecology and Evolutionary Biology; Biodiversity Research Center
Lawrence, USA

Ing Mag
Michael Pieber
Assistant Professor
Vienna University of Economics and Business Administration, Department of Technology and Commodity Science
Vienna, Austria

MSIS
Ingrid Pohl
Information Analyst
Campbell Alliance
Raleigh, USA

Dr
Irene Pollach
Assistant Professor
Vienna University of Economics and Business Administration, Department of English Business Communication
Vienna, Austria

Dipl Kffr
Susan Röver
Account Representative
The Dow Chemical Company
Schwalbach, Germany

Markus Russ
Student, Academic Staff
University of Hohenheim,
Chair of Environmental Management
Stuttgart, Germany

DDr
Arno Scharl
Professor of Information Systems
University of Western Australia
UWA Business School
Perth, Australia

Dr
Werner F. Schulz
Professor of Environmental Management
University of Hohenheim,
Chair of Environmental Management
Stuttgart, Germany

Dr
Heather Sehmel
Assistant Professor of Writing
Richard Stockton College of New Jersey
Pomona, USA

Dipl Math
Theodora Slini
Mathematician
Aristotle University, Mechanical Engineering Department
Thessaloniki, Greece

MA PhD
George R. Smith
Associate Professor
Ball State University, Department of Landscape Architecture
Muncie, USA

MA
Lars T. Soeftestad
Coordinator and Manager
Community-Based Natural Resource Management Network (CBNRM Net) and Supras Consult
Kristiansand, Norway

BA Hons (Oxon) MSc
David A. Stainforth
Research Fellow
University of Oxford
Oxford, UK

Dr
Horst Treiblmaier
Assistant Professor
Vienna University of Economics and Business Administration, Information Systems Department; Vienna, Austria

MLIS PhD
Nancy A. Van House
Professor
University of California at Berkeley, School of Information Management and Systems; Berkeley, USA

PhD
Jörg Westbomke
Deputy Division Head
Environmental Information Systems, Research Institute for Applied Knowledge Processing; Ulm, Germany

Dr
Edward O. Wiley
Professor and Curator
University of Kansas, Department of Ecology and Evolutionary Biology; Biodiversity Research Center
Lawrence, USA

BSc(Env)(Hons)
Emma Woodward
PhD Candidate
University of Western Australia, School of Earth and Geographical Sciences
Perth, Australia

MS
Xiaorong Xiang
Research Assistant
University of Notre Dame, Department of Computer Science and Engineering
Notre Dame, USA

Part I
Raising Environmental Awareness

Public Access to Environmental Information • Online Advocacy • Environmental Education

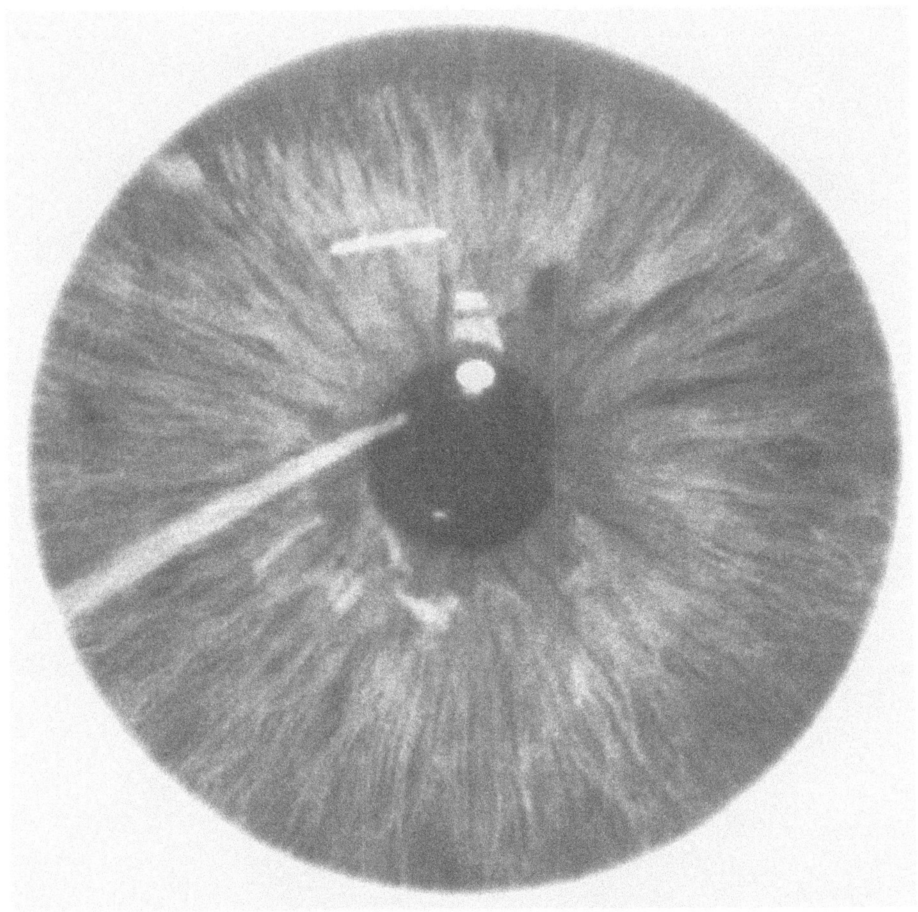

Chapter 1
Environmental Information Systems and the Concept of Environmental Informatics

Kostas Karatzas • Asteris Masouras • Athina Kaprara • Anastasios Bassoukos
Ioannis Papaioannou • Theodora Slini • Nicolas Moussiopoulos

Abstract. *Environmental Information Systems are the core of contemporary urban environment management systems, and a prerequisite for the proper, timely information of the public. Technological advances require systems that can make optimum use of informatics and telecommunications infrastructures to address environmental management needs, yet which are open, flexible, modular and inexpensive to implement and operate. For this reason, the use of Open Source software resources is presented as an appealing solution. On this basis, a number of existing environmental information components for wireless and Web-based applications are presented. Overall, the right of access to information related to the quality of the environment citizens live in via personalized electronic information services, appears to suit well the concept of open source, and may lead to a better quality of life and more effective citizen involvement and participation in environmental decision-making.*

1.1 Introduction

Environmental information was always at the top of the priority list for societies and individuals from ancient times, as it was considered to be crucial for survival and a prerequisite of prosperity. As environmental data (the raw material of information) result from "instrumentation" (i.e., measurements compiled via hardware devices, computations, or human estimations), it is evident that in ancient times environmental estimations were qualitative rather than quantitative, as such "instruments" did not exist. The situation is of course completely the opposite today, and the production of environmental data has increased dramatically, while in addition the communication channels available for their dissemination have increased with the aid of information and communication technologies, offering the potential to provide citizens with reliable, comprehensive and timely environmental information, especially in cases where environmental pressure-related problems are pronounced, as it is often the case with contemporary urban agglomerations. Improved access to environmental information is the basis for a higher degree of involvement of citizens and stakeholders in environmental decision-making. Following the development of new types of personalized electronic service technologies, especially in the mobile device industry, a new, service-oriented concept is being formulated: that of the citizen-centered environmental information services. The availability of these applications has also created a conceptual change in urban environment management, in stating that the urban environment should be treated as a whole, and that managing environmental problems in a city may only be effective if an integrated, multi-domain, holistic and systematic approach is followed. In this way, environmental management techniques and methods developed for dealing with domain-specific environmental problems (like air quality) foment the creation of a citywide environmental conscience from the citizen's point of view. Additionally, the near real-time availability of information on the quality of the urban environment to the citizen helps to bolster citizen involvement in civic affairs, as well as the feeling of community affluence.

1.2 Public Access to Environmental Information in the European Union

A robust legal body has been developed to support public access to environmental information within the European Union. The core of this legal body consists of the Council Directive 90/313/EEC [162] for Public Access to environmental information as repealed by Council Directive 2003/4/EC [164]. The latter defines environmental information as "information in any form on the state of the environment, on factors, measures or activities affecting or likely to affect the environment or designed to protect it", and recognizes that increased access to environmental information and its dissemination may contribute to a greater awareness of environmental matters, a free exchange of views, more effective participation by the public in environmental decision-making and, eventually, to a better environment. The Directive moves one step further in the way that environmental information systems should be built for the public, by specifically stating that environmental information should be disseminated via available computer telecommunications and/or electronic technology, an aspect in the heart of Environmental Informatics, as discussed hereafter.

1.3 Environmental Informatics as the Backbone of Environmental Information Services

Environmental Informatics is a special sub-discipline of Applied Informatics that utilizes the methods, techniques and tools of Computer Science for analyzing, supporting and instituting those information processing procedures which contribute to the investigation, removal, avoidance and minimization of environmental burdens and damages [238; 404]. Environmental Informatics should be considered a new knowledge-paradigm towards supporting environmental well-being on an international, national, regional, community or personal level; therefore Environmental Informatics is an integration of scientific domains and not just the result of using software engineering resources to serve environmental engineering needs [284].

The infusion of Environmental Informatics principles and knowledge into the everyday life of citizens is expected to lead to a new paradigm for quality of life within the urban web, with citizen-centered environmental information services that will support societal sustainability while promoting personal well-being [285].

Examples of such services are the Urban Air Quality Management and Information Systems [285] presented in this chapter, that is, operational applications that were funded by past research projects. These are entities that cover the life cycle of environmental information: from production (monitoring, modeling) to consumption (decision-making, public information, strategic planning). This holistic approach provides a better model of the dynamics involved when urban air quality is being addressed. Such an Urban Air Quality Management and Information System considers the dynamics of a city, i.e. the interactions, dependencies and limitations that go well beyond specific air quality management actions but certainly influence the quality of the atmospheric environment. As a result, systems like these pose the need for a modular, flexible, scalable and open architecture, which will make optimum use of existing information and communication technologies in order to provide similar functionalities to different organizations and authorities, thus making compliance with environmental regulations and legislation more effective.

An additional parameter affecting the design and development of such systems relates to the cost and resources that have to be invested by the organizations wishing to operate them, and their capacity to be integrated with existing infrastructures and decision-making chains. The previous constraints fit well with the use of Open Source software resources for the development of environmental information systems and their components.

1.4 Using Open Source Software Resources

A new software development paradigm has emerged in the last decade; contrasting previously established practices [451], the Free Software [192] or Open Source movement [416] relies directly on the volunteer efforts of geographically dispersed developers of varying affiliations and proficiencies, and recognizes a number of "freedoms" granted to the user regarding his ability to interact with the software and propagate its use (mention of the terms Open Source and Free software and their corresponding movements should be taken to be interchangeable throughout this chapter, for purposes of simplicity).

Unrestricted access to the software source code is a precondition for most of these freedoms, and it is implied that the usefulness and potential for reuse of such software is dependent on the continual revision and adaptation of its source code. It follows that the "life expectancy" of software developed in this way is a direct outcome of its popularity with developers, who will choose to devote time to improve functionality, and with users, who will provide constant feedback to developers on needed improvements and modifications.

The utilization of Open Source software towards building environmental information systems hinges on three points [484] that provide benefits to users, developers and operators of the software: economy, quality and philosophy.

Minimizing redundancy in development efforts by reusing and adapting freely available best practice software, instead of resorting to monolithic proprietary solutions or developing everything from scratch, leads to concentrating investment on innovation and obtaining the best value for the citizens' money; while relying on the community to spark developer interest in the software and provide feedback reduces maintenance costs and prolongs its life cycle. A corollary of this is that the functionality and maintainability of the software is not impaired by artificial limitations, such as expiring licenses and financial plights affecting the producing company.

The Total Cost of Ownership for solutions based on Open Source software is alleviated [168; 293], since service and consulting fees are the only recurring expenditures, providing value to operators, in contrast with licensing fees which serve as instruments of amortization to developers, this being in line with the principle for providing access to Environmental Information under a low-fee, low-cost framework which provides opportunities for open access for all. As Environmental Information Systems development is usually supported by public funding, no such amortization must burden the customers of these systems. Moreover, the service-oriented model of Open Source allows for a broad range of contractors providing support and maintenance [332], thus further minimizing operating costs.

Reliance on proprietary data formats for science results in vendor lock-in, making it difficult to pursue common protocols for data interchange and storage, for instance, as required by modern systems dealing with the problem of environmental data heterogeneity [589]. In contrast, Open Source developers promote the use of open standards as a means of consolidating researcher efforts and increasing technical interoperability [165] to minimize the cost and dependencies of technical innovation.

Although not a panacea, the free dissemination, in terms of cost as well as availability, of technological advances in information and telecommunication services will eventually help eclipse the "digital divide" [476] by allowing developing countries to catch up [107]. Finally, the source code itself can be seen as literature [211; 301], and therefore the product of a human endeavor worthy of preservation as a cultural heritage for future generations.

It is these characteristics of Open Source software that render it appropriate for use in developing publicly funded information and telecommunication projects [51], especially those aiming at the dissemination of information to citizens, such as online environmental portals.

1.5 Examples of Projects Implementing Environmental Information Services

This section presents a number of application modules for environmental information services that were developed based on Open Source software tools to improve interoperability and serviceability, and minimize operating and use fees (an issue of major importance for public authorities which usually face problems in financing information systems). In this way, it is demonstrated that efficiency can be combined with quality development and a low cost of ownership, in the spirit of European Community recommendations [484] for public administration systems. It should also be underlined that all modules presented here share some common characteristics: they operate in combination or independently of a central environmental information service, are capable of collaborating with a variety of technical platforms (i.e., operating systems and database-, mail-, or Web servers) and employ resources available over the Internet.

1.5.1 Web-based Environmental Information Services

Environmental Information Systems often have to resolve problems related to collecting data in a central server from remote monitoring stations. Optimally, the data must then be formatted for dissemination over the Web, applying user-friendly presentation techniques, like graphs or maps [286]. Moving one step further, an information portal may be envisaged that allows for electronic, personalized, environmental information dissemination (both online and early warnings), via multiple information channels. This was the main aim of the Air Pollution Network for Early Warning and Online Information Exchange in Europe (APNEE) and its follow-up APNEE-TU [652], resulting in the creation of an information portal, the APNEE Regional Server (ARS), implemented in various application sites across Europe, based on Open Source technologies, supporting a multichannel bouquet of pull and push environmental information services. In the rest of this chapter, examples are presented of the services developed and implemented in the frame of APNEE-TU for the city of Thessaloniki, Greece [652].

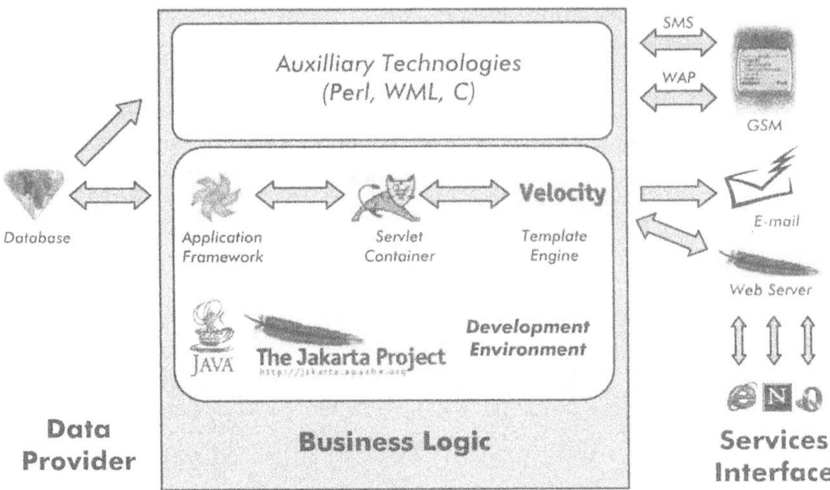

Figure 1.1: Architecture of the APNEE platform on the basis of the Thessaloniki regional server

The regional server application (developed by the APNEE project consortium and schematically presented in Figure 1.1) is the centerpiece of the APNEE system, directing the information flow to and from the output and input modules and providing an open-ended interface for resource handling. The APNEE Consortium elected to base the ARS architecture on Jakarta Turbine [726], a modular Web application framework in conjunction with the Jakarta Velocity [727] macro-enabled HTML template engine, both extensions of the Java language. Turbine provides a large array of service modules that handle all aspects of functionality common to Web applications (from interfacing with the template engine, to user authentication and caching) and follows the Model-View-Controller design pattern of obligation separation, leaving Velocity to handle content presentation. The APNEE system comprises a collection of "servlets", server-side Java applications. During its pilot operation, the system was hosted on a Debian Linux server running the Tomcat servlet container [651], in conjunction with the Apache Web server, while PostgreSQL [761] provided the database backend. The decision to base APNEE development and deployment on Open Source components and tools was made not only with cost considerations in mind, but also for compatibility, availability and performance reasons. APNEE was designed to be platform-independent and can also perform in a variety of enterprise environments.

1.5.2 Wireless Environmental Information Services

In the frame of the APNEE-TU project (the follow-up of APNEE), a set of environmental information services were developed, capitalizing on the spread of mobile devices and making use of available technologies, to provide continuous and affordable personalized information to citizens on the quality of the atmospheric environment.

One such service is the lightweight APNEE regional server version for Web-enabled personal digital assistants (PDAs) and handhelds. This application uses a trimmed layout, navigation menus and images to provide access to the same content and bundle of information services, including administration functionalities, as the Thessaloniki regional server [653], thus allowing for the smaller display size of the average PDA and limited bandwidth of wireless communication protocols (e.g. GPRS). This was made possible by adapting ARS templates to be "user-agent sensitive", i.e. to detect the client browser version and serve the appropriate layout (light graphical content, less menu options, sparser text), if a PDA-based browser is detected (Figure 1.2, left).

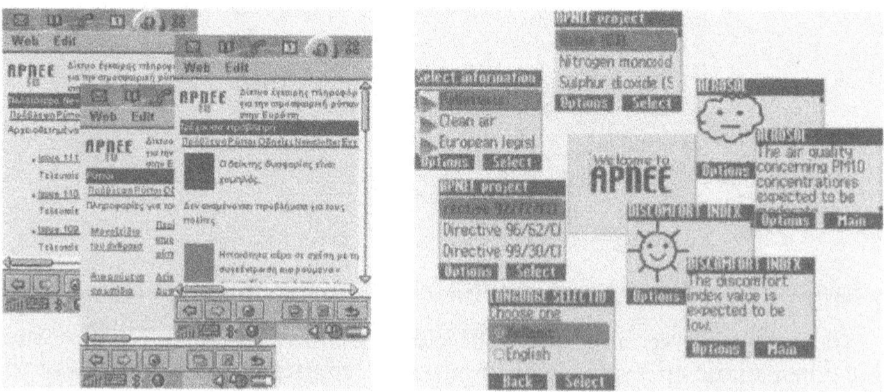

Figure 1.2: The APNEE-TU application for PDA (left) and mobile phones (right)

Another such service was developed by Siemens Hellas (a partner in the project), on the basis of specifications provided by the authors, making use of a Java 2 Mobile Edition [729] application for mobile devices, an overview of which is provided in Figure 1.2 (right). The application enables the mobile device to periodically connect to the APNEE-TU server via an HTTP connection over GPRS, access dynamic environmental information and display to the mobile screen the corresponding PNG image and text available in the mobile device. Thus, the user ma obtain, instead of a static screen saver image, a dynamic image that represents current, real time, environmental information, updated regularly. In addition, the application provides access to an electronic book on the urban atmospheric environment that corresponds to the content of the Thessaloniki ARS. Currently, the application provides static information regarding legislation, pollutants, hospitals in the area of Thessaloniki, and dynamic information such as the current level of the Discomfort Index, the forecasted PM_{10} concentration levels for the city centre and current and forecasted weather conditions parsed by a publicly available site.

The APNEE-TU J2ME application provides a good example of a flexible and informative mobile environmental information service as it is portable, easy to use, and inexpensive both to operate and access. Furthermore, the content of the application is mainly static, providing useful information to the user even when the device is offline, while the dynamic part is optimized for limited bandwidth consumption.

1.6 Conclusions

Personal well-being is directly linked to the quality of the environment that we live in, the latter being heavily dependent for its improvement on our knowledge and willingness to participate in decision-making. In addition, the right of access to environmental information has been established in contemporary European legislation. In order to design and develop efficient information systems that will help us revise our way of thinking and acting towards resolving environmental problems, a synergy of informatics and environmental engineering methods and tools is required. Environmental Informatics is a comparatively new scientific domain that aims to address this requirement by introducing service-oriented systems that encompass the life cycle of environmental information.

The dynamic and holistic nature of such systems, requiring developers to produce modular, flexible, scalable and open architectures, is best served by making use of Open Source software resources. This approach allows reusing components to enhance harmonization, in compliance with regulations, and decreasing the cost of implementation. It requires resorting to the Open Source community to ensure continued technical support and minimize operating costs. The use of Open Source (publicly available free software tools) to distribute environmental information (a public good) also suggests the concept of an open community of developers and scientists that contributes to the creation and maintenance of a collaborative, participative citizen-centered society.

Examples provided on the basis of existing applications support the feasibility of the approach while underlining the importance of incorporating contemporary informatics and telecommunications technologies to construct electronic services that address the information needs of both the general public and individuals.

Acknowledgements

The authors greatly acknowledge the European Commission for supporting research projects APNEE (IST 1999-11517) and APNEE-TU (IST 2001-34154), and their partners within these projects for their collaboration and contribution towards implementing the common APNEE vision.

Chapter 2
Mobile Access to Environmental Information

Jörg Westbomke • Michael Haase • Renate Ebel • Dieter Lehne

Abstract. *Mobile phones have almost replaced conventional telephones for public use in Western European countries. Modern WAP telephones offer mobile access to the Internet, anytime and anyplace. This technique therefore appears to be very attractive for transferring up-to-date information about the environmental situation to the public, i.e. air and water quality measurements or weather conditions as well as forecasted values of these processes. On this account WAP services complete existing information sources for public use like the Internet, videotext or print media. Location-based services add a new dimension to these services. By automatically detecting the current location of the calling telephone the information offered to the user may thus be focused to this location. Due to the relatively small size of the telephone displays and the low resolution of these devices new concepts for designing such WAP-based environmental information services are essential.*

2.1 Introduction

In the early 1990s digital mobile communication networks were set up all over Western Europe. These networks are based upon the GSM (Global System for Mobile Communication) standard, which has evolved to become the most used communication platform nowadays. For instance, mobile telephones outnumbered standard telephones in the year 2000 in Germany. Presently more than 75% of the population of Germany (aged from 14 to 64) owns a mobile telephone [648]. The use of these mobile devices has been subject to a drastic change within recent years. The first generation of mobile telephones were only designed for voice communication, but with the new mobile devices available (second and third generation of mobile telephones) the use of mobile data services like SMS (Short Message Service), MMS (Multimedia Messaging Services), or WAP (Wireless Application Protocol) increased substantially. For example, in the year 2000 mobile data services contributed only 2% to the business volume of the network providers, whereas in 2003 an amount of up to 25% is forecasted in this field [400]. This trend clearly demonstrates that there exists a strong increasing demand for mobile data services. These services also include WAP-based Internet access.

For these reasons mobile devices are assigned a key role in retrieving up-to-date environmental information [191], i.e. present local air quality measurements for asthmatic people, present water quality measurements of lakes and rivers for bathers, current depths of rivers for skippers or short-term weather forecasts for avelers. WAP services in this context offer new means of enhancing information from existing media channels like the Internet, videotext or print media as well as gaining new information related to environmental processes [272; 413]. These new services will help public authorities to fulfill their duty of supplying the public with up-to-date environmental information.

WAP-based services might be linked with location-based services (LBS), which integrate the present location of the mobile device into the information retrieval and presentation process, i.e. by offering air quality measurement from stations that are close to a user's present location. This is an especially interesting feature for environmental information services, since environmental data are spatially attributed in general.

WAP services offer great potential for realizing modern information services which are user centered. On the other hand it is a demanding task to realize such services. The strongly limited size and resolution of the displays call for new approaches to information presentation. Concepts and architectures which were developed for the Internet cannot be easily transferred to WAP-based systems. New ways of designing WAP-based services with a special focus on the appropriate presentation of environmental information are needed.

2.2 Design of an Environmental WAP Service

In order to utilize the full potential of WAP services, developers have to consider the limitations of the micro-browsers embedded in mobile devices. Thus adequate concepts for the information presentation and corresponding navigation strategies need to be designed and realized.

2.2.1 Data Visualization

The mediation of complex environmental data (i.e. air pollutant concentration) via mobile devices calls for user-centered visualization concepts which take the special demands and needs of these users into account. Public electronic information services should offer intuitive means for comprehending the contents which are presented to the public. This for instance includes intuitive visualization techniques for classifying critical values, i.e. critical ozone concentrations at a certain gauging station. Just simply displaying numerical data as well as extensive tables is not an appropriate way to handle this challenge. Only relevant data should be presented to the user in order to avoid "visual overload", and to reduce the amount of data which needs to be transmitted to the mobile device. The process of navigating in these systems should be designed intuitively as well. This includes the requirement that the user should be able to access the required information via short paths. Taking these issues into account WAP-based environmental information services can offer real value to users.

Currently most mobile telephones are equipped with monochrome displays [57] with a resolution of 100 × 60 pixels. This screen resolution is only suitable for displaying four or five lines of text with up to 26 characters. These hardware properties very much confine the ways of presenting information on these displays which has to be taken into account when a WAP service is designed. Besides the limited size of the display the page markup language WML (Wireless Mark Language) only offers a few possibilities to define the layout of a page. Ther it is necessary to reduce the information offered to a WAP compliant measure in the process of designing a WML page. The information which is to be presented should be split into just a few pages in order to facilitate the user prompt and to avoid frequent loading of new pages.

For graphical representations on screened displays it is very important to follow certain design rules to ensure that the graphics will be readable on the display [73]. These rules apply to the sizes of graphical objects and the distances between them. Due to the limited resolution of mobile telephone displays small fonts and structures are subject to distortion when they are displayed on the screen. Depending on the location of the graphical objects relative to the picture matrix an image disruption may occur. Due to the rectangular shape of a pixel, only structures adjusted to the picture matrix are displayed undistortedly. For this reason rectangular shapes are preferred when designing applications for low-resolution displays. Figure 2.1 illustrates some effects occuring on low-resolution displays. The left display depicts lines and circles which do not fit into the screen matrix and therefore are presented in a distorted manner. Fonts for displaying text ought to be chosen with care. The readability of text highly depends on the font being used. The right display depicts examples of different fonts being displayed on a low-resolution display. Sans-serif fonts (i.e. Arial, Verdana) should preferably be used for displaying text on low-resolution displays.

Figure 2.1: Picture distortion on non-vertical or non-horizontal graphical elements (left); Different font types – Small Fonts, Arial, Verdana, Times – shown on a small resolution raster display (right)

2.2.2 Navigation

Billing rates for mobile phones depend on either login time or data transfer volume. Thus an information service based on WAP technology should offer efficient and user-friendly access to information. It should provide a combination of small and ergonomic WML pages, which may be accessed by just a few navigational steps.

Geographic selection mechanisms form the main task of environmental information services, i.e. selecting a gauging station from a flood prediction service or selecting an air pollutant gauging station which is located closest to the user's present location. By utilizing location-based services this selection process may be automated. The present location of the user may be detected automatically (i.e. Global Positioning System, Cell ID) and transferred to a Geographic Information System (GIS) which in turn determines the gauging station closest to the present location of the user. This gauging station may then be automatically selected for further analysis by the user. Unfortunately this technology is presently not accessible for all content providers. Due to legal restrictions and data privacy regulations location information is presently only accessible to the network providers (see Section 2.4). Therefore other ways of selecting the area of interest need to be ought of. Topograms offer a solution to this problem. Topograms are maps that not show the topography on an exact scale, but conserve all topological relaips [24; 508]. They can be used in WAP services to offer the user the possibility of selecting a region of interest without being forced to know the exact name of that region or having to choose the region from a long alphabetical list. Figure 2.2 shows some examples of topograms which have been realized for a WAP service. This WAP service will be subject to a detailed analysis in the following sections of this contribution.

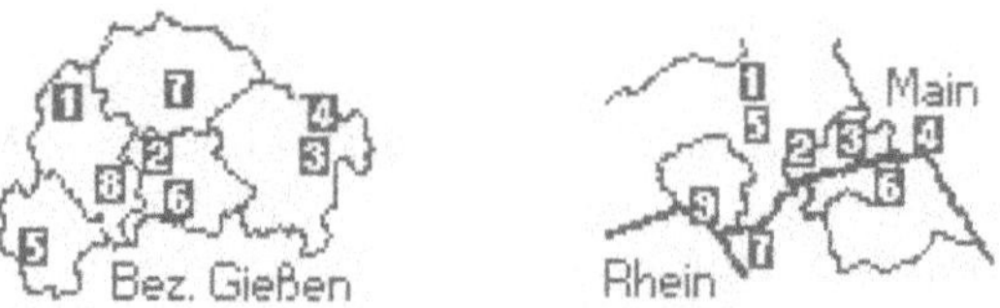

Figure 2.2: Topograms depicting air pollutant gauging stations from the district of Giessen and the Rhine-Main area

The numbers in the topograms refer to air pollutant gauging stations in the area covered by the topogram. To select a particular gauging station the user just has to enter the number assigned to this gauging station using the keyboard of the mobile telephone. This eases the selection process, because the user does not need to know the names of the gauging stations. The user of course has to be aware of his present location to be able to locate himself on the map.

2.3 WAP-based Environmental Information Services

Based upon the concepts which were presented in the previous section, two environmental WAP services were realized [359]. The first one, commissioned by the State Office for Environment and Geology Hesse, focuses on air pollutant ozone and on meteorological parameters. The second service, which was realized for the State Office for Environmental Protection Baden-Württemberg, aims to inform the public about the up-to-date local gamma dosage rate of the nuclear power plants in the state of Baden-Württemberg as well as those power plants in Switzerland and France which are located close to the border of Baden-Württemberg.

2.3.1 O_3-WAP

The O_3-WAP service [751], which can be called on WAP devices, offers access to ozone concentration and meteorological parameters like temperature and wind velocity to the public of Hesse. All these environmental data are measured and recorded by more than 30 automatic gauging stations throughout Hesse. The recorded data are transmitted to the central office in Wiesbaden and are stored periodically in an Oracle database. The WAP service is realized as a Java Servlet. It consists of static WML pages which are stored on the local file system of the server, and dynamic WML pages which are generated on-the-fly from the recorded values stored in the database (Figure 2.3).

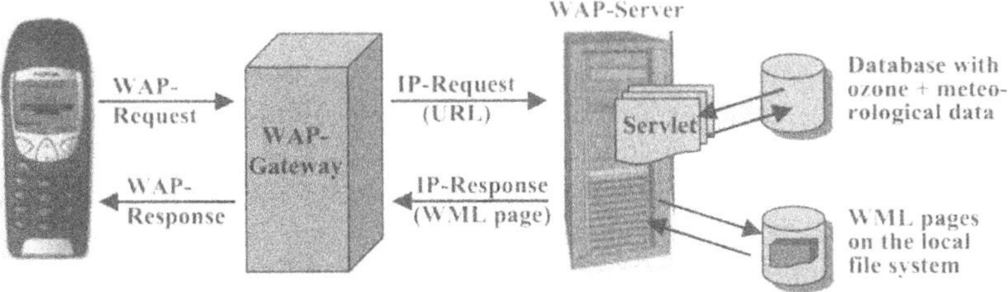

Figure 2.3: System architecture of the O_3-WAP service

The content of the O_3-WAP service is structured hierarchically. Firstly users have to decide what kind of information (ozone / weather conditions) they are interested in. They then have to select a particular gauging station. This task can be accomplished by selecting the requested station from an alphabetical list of all gauging stations, from a regionally ordered list, or by using a topogram visualizing the distribution of the gauging stations in the state of Hesse.

Figure 2.4: Menu structure of the O_3-WAP service (German)

Figure 2.4 depicts a sequence of screenshots showing the selection process of the O_3-WAP service. The reception of the measured values is supported (Figure 2.5) by icons and corresponding diagrams. The icons are designed in order to support the user in the process of interpreting the recorded values. The latter will be illustrated by two examples: ozone values above 180 µg O_3/m^3 are marked with a warning sign; the wind direction is visualized by an arrow, which is embedded into an icon.

Figure 2.5: Information offered by the O_3-WAP service (German)

For every gauging station the present ozone concentration as well as the ozone concentration for the past 24 hours may be retrieved. The latter is presented to the user by means of a bar chart. Additionally an ozone forecast service has been integrated in order to inform the user about estimated maximum values of ozone oncentration for the present day for four major areas of Hesse. This service also rs support on how to personally deal with high ozone concentrations.

2.3.2 KFUE-WAP

The KFUE-WAP service [732] was realized for the State Institute for Environmental Protection Baden-Württemberg. KFUE-WAP is an information service that is predominantly designed for the employees of the agency for nuclear power plant observation in Baden-Württemberg in order to supply them with important measurement data in case of an emergency. Thus they can obtain a first impression of the situation even when they are not in their offices. Some of these data are accessible to the public.

KFUE-WAP offers access to the local gamma dosage rate recorded at the sites of the five nuclear power plants which are either located in the state or close to the border of the state. The dosage rate is updated every 24 hours for all three nuclear power plants in Baden-Württemberg. Measured dosage rates from the foreign power plants are updated on an hourly basis.

The selection process implemented in the KFUE-WAP service is much simpler than in the case of the O_3-WAP service, because there are just a few regions of interest in this service. However, a topogram was implemented to visualize the distribution of the nuclear power plants, because the user might not know the name of the different power plants and their location locations. The topogram offers the user easy access to the region of interest. LBS could be utilized here as well to simplify the selection process and to point the user directly to the measurements of the nearest power plant.

Figure 2.6 depicts some screenshots from the publicly accessible part of the KFUE-WAP service. The interpretation of the present local gamma dosage rate is supported by two values calculated from measurements of the previous year. Users are supplied with the current dosage rate, and the maximal and average dosage rates of the previous year. They may compare the up-to-date value with the values from the previous year, which helps to interpret the present values. For casual users and users who are not familiar with measured dosage rates, there is a short explanation how to interpret measured local gamma dosage rates.

The KFUE-WAP service was predominantly designed for the employees of the agency for nuclear power plant observation to offer them direct and mobile access to the up-to-date measurements. These people are familiar with the interpretation of local gamma dosage rates. In case of emergency they can get a first impression of the current situation without having to access their office computers.

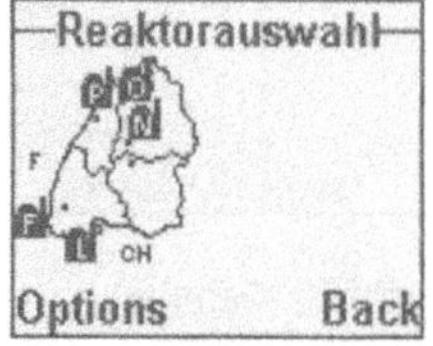

Neckarwesthein
28.08.2003,05 Uhr
Dosisleistung in µSv/h:
0.14 (aktuell)
0.11 (Mittel 2002)
0.22 (Maximum 2002)
Options Back

Erläuterungen
Die
Gamma-Ortsdosisleistung ist
ein Maß für die ionisierende
Strahlung.
Sie wird beeinflusst durch:
Options Back

Figure 2.6: Screenshots from the KFUE-WAP service (German)

The system architecture of the KFUE-WAP service is roughly the same as presented for the O_3-WAP service in Section 2.3.1. There is a slight difference due to firewall limitations and organizational restrictions. In this case it is not possible to connect directly to the database with the KFUE data. Therefore the current loc gamma dosage rates need to be extracted from a HTML page, which is part of the Internet homepage provided by the State Institute for Environmental Protection Baden-Württemberg.

2.4 Location-based Services

Geographic selection mechanisms are important to environmental information services (see Section 2.2.2) which of course also holds true for WAP services in this field. WAP services for retrieving environmental information should therefore offer easy and efficient ways for users to define their region of interest. The O_3-WAP and KFUE-WAP services make use of topograms to help users find the gauging stations they are interested in. This approach assumes that the users are always aware of their present location in order to identify the gauging stations on the topogram which are closest to them. Such services may be implemented with the technologies which are currently at hand. It would be, however, much more convenient for users if their locations could be auto-detected by the system. In turn, the information being presented to them could be automatically bounded to the region where they are currently located. This would positive influence the attractiveness of WAP services.

2.4.1 Methods for Locating Mobile Phones

From the technical viewpoint two technologies may be used at present to detect the location of a WAP user: localization by identification of the radio cell (Cell ID), and localization by utilizing the Global Positioning System (GPS).

In the first case a characteristic of a mobile network is used to determine the position of a mobile phone. Every radio cell is characterized by its identification number (Cell ID). This ID is used by the mobile network to identify a radio cell where a mobile phone is logged on. The coordinates of the center of this cell (i.e. the cell's center of gravity) are well determined. Depending on the size of the cell the present location of the mobile phone may be identified with a certain tolerance. This technique of locating a mobile phone may be applied to urban areas with an accuracy of 100 to 150 meters, which is adequate for the services we are investigating in this article. In rural areas, radio cells may expand up to 35 kilometers in diameter. The localization of mobile phones may be improved by taking detailed information about the structure of the radio cell into account. A radio cell is often composed of three antennas, each of them ideally covering a sector of 120 degrees. If the antenna that a telephone is connected to is known, the location of the telephone may then be identified to be in a certain sector of the radio cell [231]. Other techniques have been introduced in order to determine a unique set of coordinates for the current location of a calling mobile telephone by analyzing the structural information of the radio cell [453]. All these improvements cannot overcome the fact that the determination of the location of a mobile phone in rural areas still remains very inaccurate. However, these methods may be applied to identify the location of a calling mobile phone without having to modify any hardware components. These services could be offered by the network providers right away.

Integrating a GPS (Global Positioning System) receiver and a GPS processor into a mobile phone would offer more precise means for determining the location of this device. The GPS processor software could continuously determine the present location of the device on the basis of the satellite-based GPS signals being received. This technique allows a very precise determination of the location (the average tolerance is about ten meters) and is independent of radio cell structures. But again there is a drawback to this technique: GPS signals cannot be received in tunnels and especially in urban areas difficulties exist receiving these signals on narrow streets which are lined by high buildings. This technique is based on special mobile telephone hardware which up to now has not been introduced to the mobile communication market. Therefore this technique does not at present constitute an alternative that should be taken into account when WAP services are designed for public use.

2.4.2 LBS Enriched WAP Services

Environmental information services may be substantially enhanced by introducing LBS as an integral part. In this way new environmental information services could be offered to users, automatically focusing on their location and thus saving them time and money. Figure 2.7 depicts the system architecture of an environmental information service which utilizes the GPS-assisted localization of mobile telephones.

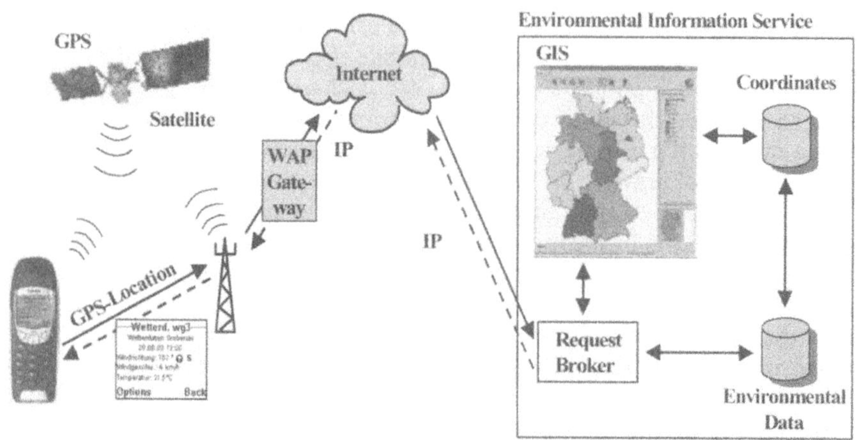

Figure 2.7: System architecture of an environmental information service with GPS-assisted localization of mobile telephones

The built-in GPS receiver of the mobile telephone receives the GPS signals from the satellites and determines its present location on this basis. These coordinates are transmitted via WAP to the request broker of the environmental service. The request broker will submit these coordinates to a Geographic Information System (GIS) which in turn will either determine the gauging stations in the neighborhood of this location or the gauging station closest to this location depending o the service the user requested. The request broker then will generate a WML r with the up-to-date measurement data of these stations and send it back to the mobile telephone, thus allowing users to get the desired information related to their present surroundings without requiring any manual input.

This procedure is applicable to a broad variety of environmental information services that provide information about spatially distributed environmental processes (i.e. rainfall, ozone concentrations, water quality of running waters). Such a system could also be used to build much more sophisticated services. For example, a visitor information system for national parks could be set up on this basis which would guide visitors to locations of special interest within the park and supply them with important information about these places once they have reached them. Identifying objects which are closest to a given location is only one basic service from a broad set of spatial analysis services which could be provided by Geographic Information Systems. These other services could very well enhance WAP-based environmental information systems. For instance, all ozone gauging stations in a predefined neighborhood of the user's present location could be identified in order to supply the user with an overview of the current ozone conditions within this area.

2.4.3 Application Barriers

In spite of all the interesting new information services that could be build on top of the LBS technology only a few of these services exist at present. One example of such an LBS-based service is the StreetGuide service of the provider Vodafone [590] in Germany which generates maps of the neighborhood of the user's present location. This service thus helps users to obtain orientation on where they are located at the present time. They may also request a route description from this service which will guide them to a selected address starting from their present location. PASSO is another LBS service offered by Vodafone to its customers in Germany [590]. This service takes the current location of a mobile telephone into account and reports all traffic jams, construction sites and other traffic information in the area.

At present it does not make sense for content providers to offer services which are based upon GPS-assisted localization of the users (see Section 2.4.2). Only the network providers are able to localize mobile phones by Cell IDs in their network. Thus content providers would have to enter a liaison with network providers in order to be able to offer such services. A major drawback of this procedure would be that these services would be exclusive to the customers of a particular network provider. In order to make these services available to the broad public content providers would have to enter liaisons with all network providers. Taking the present situation in this market into account it seems very unlikely that the network providers will be entering such agreements since they would then lose sole attributes. But on the other hand public authorities cannot rely on certain network providers in order to inform the public about environmental issues.

Location-based services with respect to the field of environmental information systems can be characterized as a very interesting technology from today's point of view. Unfortunately up to now sustainable business models are missing. "Killer applications" that potential users are willing to pay for have not yet been identified.

2.5 Conclusions

The problems which arise when complex information needs to be presented on screens with low resolutions are discussed in detail in this chapter. Design rules for WAP services which have been derived in order to meet these challenges are presented. These design rules take into account the low screen resolutions of the mobile telephones which are at present dominating the market. Optimized navigation strategies for environmental information services based on the Wireless Application Protocol (WAP) are part of these design rules. In this context topograms (iconized maps) are assigned a key role in the design of the user interface. The topography is not represented on a correct scale on these topograms. Only the relevant topological relationships between the objects are conserved on these iconized maps.

Two environmental information services are introduced, which are realized on the basis of these design concepts. O_3-WAP presents ozone and meteorological data to the public, while the KFUE-WAP service is designed to be used mainly by the employees of the agency for nuclear power plant observation in Baden-Württemberg to give them access to important gamma dosage rates in case of an accident. Furthermore the improvements that could be achieved by combining environmental information services with location-based services are pointed out.

Despite the mentioned difficulties, WAP technology will play a key role in future mobile accessible information systems. These third generation mobile networks and telephones will offer higher transmitting rates and better means of information processing. The introduction of UMTS (Universal Mobile Telecommunication System) will extend the transmission rates up to 2 Mbit/s for mobile communication systems [598]. In addition the mobile telephones are continuously being equipped with more powerful processors and colored screens with higher resolutions.

WAP has a high potential for future services in mobile networks. Its role may then be characterized similar to the role of the Short Message Service (SMS) in our present mobile networks. At present there is no alternative to mobile handsets available, although the size and the resolution of the display is too small for most advanced applications. TabletPCs with WLAN (Wireless Local Area Network) or PocketPCs with GSM (Global System for Mobile Communication) adapters may serve as a front-end for specialists but they will not be able to replace mobile phones as a public gateway to environmental information services.

Chapter 3
EnviWeb and Environmental Web Services: Case Study of an Environmental Web Portal

Jiri Hrebicek • Miroslav Kubasek

Abstract. *The EnviWeb Environmental Web Portal on the Czech Internet was launched in October 2001 and has become the most visited portal in this sector (www.enviweb.cz). The portal is divided into the following sections: Water, Air, Waste, Chemicals, Noise, Nature, Environmental Impact Assessment, Environmental Management Systems, and Miscellaneous. EnviWeb provides information for experts, professionals, companies, governments, and citizens. It contains a directory of companies in the Czech Republic, directories of inspections, ministries, regional offices, department offices, branch associations, health departments, business support organizations, and so forth. This chapter presents the current status and planned development of EnviWeb based on Web Services technology.*

3.1 Introduction

Development can only be sustainable when communities, planners, policy makers, and all stakeholders of civil society participate actively. This can only be achieved through constant sharing of knowledge. The Rio Declaration [567], Agenda 21 [569], and the Earth Charter [690] emphasize access to and deliberate sharing of information and expertise between all nations and agents to ensure a sustainable future for humankind. Environmental information is one of the most fundamental resources for environmental protection. Environmental information should be available not only for awareness, education, understanding, empowerment and participation, but also for sound decision-making because it influences the values, attitudes and behaviors necessary for solving environmental problems and sustaining natural resources. The need for information arises at every level, from senior decision-makers at the national and international levels to the grass roots and individual levels. In this regard, information and communication technology (ICT) plays a significant role [270].

Modern ICTs have been presented to the world and help a variety of sectors including environmental management. ICTs involve and reach more people and achieve deeper geographic penetration than is the case with traditional means and modalities. ICTs allow global access to information and represent a novel and effective tool, helping to enhance sustainable human development.

Since 1994, students of the Faculty of Informatics of Masaryk University, Brno have applied modern ICTs to reduce the gap in public access to environmental information in the Czech Republic [247]. This chapter presents a case study of the EnviWeb environmental Web portal, the outcome of a PhD thesis and related research [318; 320].

3.2 EnviWeb Project

3.2.1 History and Target Group

In August 1998 it was decided to create a waste Web portal. In line with the Waste Act of the Czech Republic, this would ensure a supply of waste for future utilization. In May 1999, the waste server "Skladka" [769] was launched to ensure the supply and demand of waste, and to offer a complex service in waste management (e.g. a list of companies, legal issues, technical articles, waste software, literature, advertising, etc.). By 2001, 100 visitors per day accessed the site.

After two years' experience with managing the portal "Skladka", we decided to build a new environmental Web portal to cover all aspects of the environment and increase the number of visitors. It was named EnviWeb and launched in October 2001 [769]. The EnviWeb portal was built on standard architectures and technologies for creating Web sites [143]. It runs under Linux with an Apache Web server, with PHP (PHP: Hypertext Preprocessor; recursive acronym) as a server-side HTML-embedded scripting language.

EnviWeb is primarily intended for experts, professionals, firms, government and citizens. The main purpose of EnviWeb is to provide free access to public environmental information. The main target groups of EnviWeb are environment professionals, enterprise environmentalists, and employees of companies involved in technology for environmental protection and labor safety. Other visitors are representatives and students of high schools and universities, representatives and members of professional unions, employees of public administration and local government, organizers and participants of special events, and editors and readers of professional publications. Nowdays EnviWeb has approximately 500 visitors per day [319].

EnviWeb serves as a crossroads to environmental resources in the Czech Republic. For this reason, a huge catalogue of links to other Web sites is maintained. For foreign users, who do not understand the Czech language, English, German, Russian, and Italian versions of EnviWeb are available. Accessing EnviWeb by hand-held vendors (e.g. PalmPC) is possible.

3.2.2 Structure of the Portal

The EnviWeb portal contains a large and varied amount of environmental information, which is collected and verified by a group of environmental experts. Information is divided into several sections to cover particular elements of the environment (see Figure 3.1): *Water* (water management and its aspects), *Air* (air protection problems, monitoring and reducing pollution), *Waste* (waste management, demeaning production of waste and waste recycling), *Chemicals* (environmental aspects of chemical treatment and their influence on the living environment), *Disasters* (prevention of serious industrial accidents, reactions to accidents and removing their consequences), *Noise* (noise problems, noise measurement, noise reduction, modeling and abnormal noise prevention), *Nature* (problems concerning nature protection), *Soil* (soil conservation and ecological agronomy), *Forest* (forest management and forest protection), *Geology* (environmental aspects of geologic prospecting, extraction and rehabilitation), *Environmental pact Assessment* (EIA), *Environmental Management Systems* (EMS) and the Eco Management and Audit Scheme (EMAS) of the European Union, *Integrated Prevention and Pollution Control*, and *Miscellaneous* (articles and links, which cannot be placed into specific sections or are universal in scope).

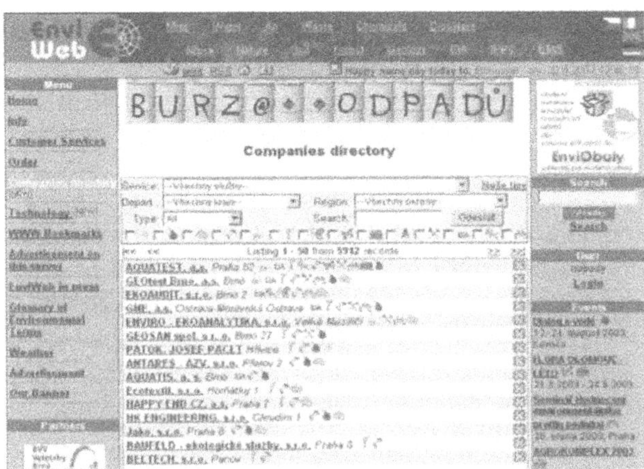

Figure 3.1: Directory of companies

3.3 EnviWeb Components

EnviWeb as a live system is developed, improved, and updated on a daily basis. This section describes fundamental services offered by EnviWeb. *Press Monitoring* collects articles for EnviWeb databases by monitoring dailies, regional press, and technical journals. Date, source, and key words mark every article placed in its relevant section. This update is carried out daily by editors who are experts in particular areas of the environment. Articles about public orders are very popular. Therefore, EnviWeb monitors important information sources and presents public orders that are connected with the environment.

The *Events Calendar* contains information about upcoming seminars, conferences, workshops, and symposiums relevant to the environment. These events are held not only in the Czech Republic, but also in Europe and around the world. An archive of past events is also available.

Waste Stock is the most popular service. Companies who registered for an account can offer and demand waste. They can specify the waste including related details, filter the offers by waste category, and view information about other organizations. The *Interactive Waste Catalogue* lets users browse and search the Czech Waste Catalogue (compatible with the European Waste Catalogue).

A very important part of environmental information is *environmental legislation*. There are two ways the visitor can access environmental legislation: either via Web browser after logging in to the portal, or through the software "Envi-Paragraf". The user can install the client on his computer and then obtain national environmental legislation with an option of full-text searching. This program is automatically updated from the Internet after any change is made in the ch environmental legislation.

In the *Advertisement Services* section, users can enter non-commercial advertisements for free. The most frequent types of advertisement are job offers and education in the field of environmental protection services.

The *Directory of Companies* contains about 6,000 companies from the Czech Republic (Figure 3.1). It is the only complete and up-to-date directory in the Czech environmental sector. Companies are classified by the services and products provided. There are approximately 150 categories. Every company has an opportunity to add to or edit its own information directly on the Web.

3.4 Future Development

EnviWeb components are currently implemented as scripts included in the portal structure. To construct a complex services system or cooperate among services, it is necessary to use components with compatible, standardized interfaces. Web Services technology based on the Extensible Markup Language (XML) with the HTTP protocol acting as a transport layer can be used for this purpose [86]. A new version of EnviWeb will use Web Services for Remote Portlets (WSRP), briefly described in the following.

Web services are self-contained, self-describing modular applications describing collections of operations that are network-accessible through standardized XML messaging. They are packaged into a single entity and publicly available via the network for use by other programs. Web services are a building block for creating open distributed systems and allow companies and individuals to quickly and cheaply make their digital assets available worldwide. A Web service can aggregate with other Web services to provide a higher-level set of features [346]. Web Services for Remote Portlets (WSRP) is a specification that will allow for "plug-and-play" of portals, intermediary content aggregation applications, and integration with applications from disparate sources [789]. WSRP will enable applications to either consume Web services, produce Web services, or both. These Web services will allow portal administrators to select and display portlets that originate from virtually anywhere on the Web without writing any integration code. WSRP producers and consumers may be implemented on different platforms, such as a J2EE Web service, .NET Web service, or JSR 168 Portlet [148].

3.5 Conclusions

EnviWeb was initiated and developed as an environmental Web portal during the PhD research of one of the authors at the Faculty of Informatics, Masaryk University Brno. EnviWeb is dedicated to providing comprehensive and up-to-date environmental information and news to the public. After two years it turned into a daily environmental information source for hundreds of visitors who are professionals, enterprise environmentalists, and company managers.

EnviWeb helps high school and university students interested in environmental protection; informs representatives and members of professional unions, employees of public administration, organizers and participants in special events, editors and readers of professional publications. It acts as an information broker for environmental information in the Czech Republic.

Chapter 4
Web Coverage of Renewable Energy

Arno Scharl

Abstract. *The Web assessment and media monitoring methodology presented in this chapter provides empirical Web data by automatically mirroring and processing large samples of Web information systems. Fifty international media sites have been monitored in monthly intervals. The wealth of information contained in these sites was converted into aggregated representations through structural and textual analysis. This process yielded detailed profiles for each site and roughly one billion words of textual raw data. The resulting indicators and visualizations describe complex relationships and developments in the Web coverage of renewable energy that can hardly be captured and analyzed in traditional ways.*

4.1 Introduction

The commercial success of renewable energies and their potential to substitute fossil fuels depend on improved communication channels, greater participation by non-adopters, and sufficient funds for research and development [283]. The transition to renewable energies will boost demand for resources to carry out theoretical and applied research, to construct prototypes, and to increase the installed capacity of solar power plants and wind farms. To efficiently allocate these resources, analysts and decision-makers need adequate information about emerging technologies, available products, market trends, and public opinion.

Monitoring online media provides a rich source of such information, as news media have actively embraced the diffusion of networked information technology. As a fundamental source of environmental information, they contribute to social change and help limit corporate and government abuses [636]. While the study of traditional media has moved beyond the point where further descriptions of content are needed, the situation differs when investigating electronic sources: "We have no real idea what Web content consists of, and it is arguably important ... to obtain that knowledge before we chase after the content trying to superimpose theories upon it" [442]. Textual Web data not only contains valuable information about sector trends and online strategies, but also hidden assumptions and misconceptions about environmental issues. Concentrating on the structure and technical implementation of Web sites often neglects opportunities to capture contextual meaning.

To reduce this shortcoming, the media monitoring architecture introduced in this chapter illustrates how to discover regularities in textual Web data. It aims to identify and visualize environmental trends, thereby creating a well-informed public that is more likely to support sustainable policies, and more capable of holding decision-makers accountable [636].

4.2 Literature Review

When investigating large collections of empirical data, the term "mining" has become increasingly popular. Text and data mining are part of the general field of information mining, reflecting the variety of forms information may take [102]. Mining draws upon the inductive discovery and visualization of patterns and non-obvious relationships between data elements that might be useful in a particular context [298; 447]. With regard to the World Wide Web, mining tasks fall into three categories [102]: *Web content mining* analyzes a site's textual or audio-visual content; *Web structure mining* performs a structural analysis of the hypertext and its navigational mechanisms; *Web usage mining* focuses on behavioral data such as hits, visits, and clickstreams derived from server log-files or client state cookies. Some authors treat Web structures as part of Web content and only distinguish two aspects, content and usage [226; 373]. Excluding the usage aspect [41; 474], this chapter concentrates on analyzing textual content gathered from the Web sites of international media.

Automated Web assessment differs from traditional approaches that rely on the judgments of individuals, often on the basis of a list of weighted attributes [398]. Many of these attributes tend to be approximations for which exact data cannot be measured [74]. Thus expert evaluations take place with varying degrees of subjectivity regarding the process itself and the observed data structures. Surveys that integrate results from user questionnaires also suffer from respondent inaccuracy due to differences between reported and actual behavior [628].

Scalability, speed, consistency, rigorous structure, and abundance of data are assured if information is gathered and extracted automatically. Software tools alleviate methodological limitations of subjective impressions and anecdotal evidence [42; 267; 474]. They process large samples of Web sites simultaneously and generate abundant longitudinal data. Judged against human evaluation, automated approaches are more efficient at handling dynamically changing Web data and immune to inter- and intra-personal variances. Thus the regular sampling of thousands of systems becomes feasible compared with samples limited to tens or hundreds in previous studies [76; 268; 496; 626]. These advantages come at the expense of sacrificing recipient-dependent attributes, which are difficult to quantify. Domain knowledge and expert opinions, therefore, continue to play an important role when interpreting and applying the results.

4.3 Methodology

The nonprobability sample underlying this research contains 50 international media sites from five English-speaking countries: United States (20), United Kingdom (11), Australia (10), Canada (5), and New Zealand (4). The sites were selected via judgment sampling from the *Kidon* [733], *ABYZ* [644], and *NewsLink* [749] directories. The left diagram of Figure 4.1 summarizes the methodology with an emphasis on textual analysis. *Sample specification* includes uploading new sites, updating external rankings, and regularly checking the validity of primary and secondary (redirected) addresses. Between May 1999 and August 2003, the overall sample size of the webLyzard project [790] grew to more than 4,000 Web sites, which are monitored on a monthly basis (Figure 4.1; right diagram).

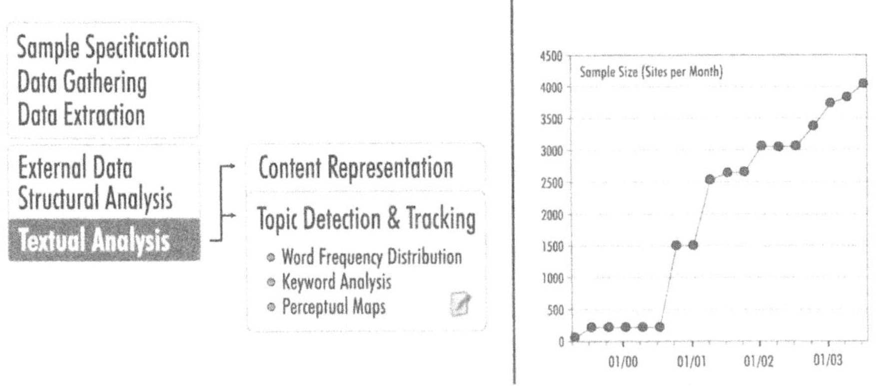

Figure 4.1: Methodology (left) and sample size (right) of the Web monitoring project

As the volume and constantly changing character of Web data entails ongoing analysis, a software tool automatically mirrors Web sites, captures their characteristics, and stores the resulting profiles in a relational database. Between February and June 2003, 40 gigabytes of textual Web data were mirrored and processed, comprising about two million documents with almost one billion words.

Ignoring graphics and multimedia files, the software tool follows a site's hierarchical structure until reaching 10 megabytes for site analysis, or 50 megabytes for media monitoring. Only textual information is considered, which includes both visible (raw text including headings, menus, or link descriptions) and invisible text (embedded markup tags, scripting elements, and so forth). This focused strategy helps compare systems of heterogeneous size and manage local storage capacity. Documents found in lower hierarchical levels do not contribute to the main user interface and can be disregarded for most analytical objectives.

Data extraction yields actionable knowledge by processing and automatically coding the mirrored data. Automated coding attenuates subjective interpretations and many of the questionable aspects of manual coding. It renders the problems of time-consuming and expensive coder training or intra- and intercoder reliability obsolete, as the process measures variables directly and does not involve human coders who could disagree on particular attribute values.

After extracting the data, *structural analysis* relies upon single- and multivariate statistical techniques to search for underlying relationships [206]. This analysis may incorporate *external data* such as network statistics, server log-files, or questionnaire data to determine success factors of Web information systems. The computation of ratios, means and variances reduces data redundancy and provides aggregated parameters that are easier to interpret. These parameters fall into four categories: (1) *navigational mechanisms:* structure and accessibility of internal links, external links, or anchor links within and between documents; (2) variables that characterize *interactive features* such as forms, scripts, applets, and so forth; (3) *layout and multimedia characteristics:* information on frames, embedded pictures, fonts and styles, etc.; (4) *linguistic descriptives:* average lengths of linguistic units such as words, sentences, or documents. For most variables, total and distinctive values are stored in absolute terms and relative to the number of mirrored documents.

Textual analysis, which is the focus of the remainder of this chapter, converts the raw data into adequate representations, assigns languages to each document, and eliminates non-relevant information in order to provide linguistic site metrics and identify focal points of Web coverage with regard to renewable energy. The following sections apply and extend a variety of methods for analyzing raw textual data, most of which originate from corpus linguistics and textual statistics [55; 69; 362].

4.4 Content Representation

Given the raw text from a Web site, the first step is segmenting the textual chain into minimal units, removing coding ambiguities such as punctuation marks, the case of letters, hyphens, or dots occurring within abbreviations. Most regular texts contain paralinguistic items such as word delimiters (blanks), sequence delimiters (weak punctuation marks), and sentence separators (strong punctuation marks). All other characters represent non-delimiting characters. Two or more identical series of non-delimiters constitute occurrences of the same word [326]. Exhaustive indices of these occurrences (word lists) typically use decreasing frequency of occurrence as primary sorting criterion and lexicographic order as secondary criterion.

4.4.1 Type Token Ratio

The creation of word lists yields several important variables. *Tokens* represent the number of all running words encountered when processing the text file. Each word is counted as a string of valid letters with a separator at each end. *Types*, the distinct words identified when first processing the text, represent the system's complete vocabulary. The mirror of an average media site as of June 2003, for example, contained about 1.3 million running words, but less than 35,000 distinct occurrences.

The type token ratio (TTR) is a syntactical index that divides the number of distinct words (types) by the total number of words (tokens). To accommodate the varying sizes of Web sites, a running average based on consecutive 1,000-word chunks of text is computed. This standardized TTR indicates the quality of a text in terms of richness of its vocabulary and has been identified as an important success factor in previous Web evaluation projects [474; 627]. Percy and Rossiter [415] also emphasize the importance of linguistic complexity. Higher TTRs may indicate extensive use of words with technical meanings, more exact lexical choice, or greater information density [189]. Conversely, lower TTRs may imply a high degree of repetition and reduced complexity of a text [565]. Although evidence exists that linguistic complexity improves readership [101], it should not be confused with copy length. As reading from a computer monitor is more demanding on the eye than reading in print, most Web design guidelines suggest short copy [388; 585]. Whether high or low TTRs are more desirable depends on the target audience, which explains conflicting guidelines found in the literature.

Topical diversity and frequent updates represent core competencies of international media sites that positively influence the TTR. The longitudinal diagram i Figure 4.2 plots TTR values for selected international media sites from Janu

2002 to July 2003: BBC [661], CNN [671], USA Today [786], Fortune [706], Los Angeles Times [738], Scientific American [767], and Environmental News Network [700]. General scientific publications tend to use more heterogeneous vocabularies, reflecting a detailed coverage of a broad spectrum of topics. Lower TTRs reflect sites that focus on groups that prefer standardized terminology. A comparison of June 2003 data reveals that the media's textual richness with a mean TTR of 0.434 (sd = 0.077) exceeds the corporate sector. The Web sites of Fortune 1000 companies achieved a significantly lower ratio ($M = 0.358$, $sd = 0.053$, $t(55) = 7.01$, $p < .001$).

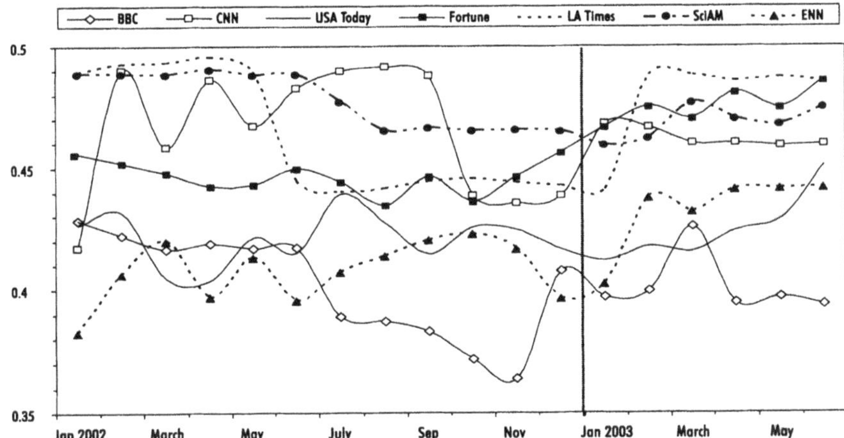

Figure 4.2: Longitudinal comparison of the standardized type token ratio (TTRs) of selected international media sites between January 2002 and June 2003

4.4.2 Language Identification

The study of effective communication requires explicit attention to the role of language [366]. Automatic treatment of textual content, therefore, necessitates an identification of the language(s) used. The identification algorithm should classify both multilingual sites (i.e., localized versions of the same content) and multilingual content within one particular document [249]. For this classification problem, techniques based on trigrams and common short words are the most popular.

The *trigram* technique compares frequencies of three letter sequences within a document with a typical distribution for a particular language. Similarly, common *short words* such as determiners, conjunctions or prepositions are good clues for guessing a language. As the results of the two methods are nearly identical for textual segments comprising more than ten words [214], the computationally less intensive technique based on short words has been chosen for this research.

4.4.3 Exclusion of Non-significant Words

In sharp contrast to stylometry [243] – i.e., analyzing literary style – the relevance function words for content-oriented statistical analyses is low. The nature of e texts like those contained in Web sites generally limits the role of these words and suggests eliminating them. Lebart et al. compare them to noise that is introduced into the signal – i.e., the document's primary content [326]. After text segmentation and identification of occurrences, two simple techniques are applied to remove function words and obtain a list of semantically more meaningful terms: specifying enumerative stop lists and requiring a minimum word length. Stop lists are lists of words to be excluded from the analysis. Function words such as articles, prepositions, or conjunctions are natural candidates for such a stop list [34]: *the, and, for, with, that, are, this, from, all, can, will, its, have, has, other, not, about, one, also, which, may, these, any, than, was, you, your, our, more, their, they, when, how, based, first, only, such, through, into, over, what, most, end.* Short words can also be deleted with minor impact on the result. For the media sample, all words with a length of less than three letters were disregarded.

4.4.4 Lemmatization (Word Stemming)

Plurals, gerund forms, and past tense suffixes are syntactical variations that complicate the process of interpreting word lists. Lemmatizing the text partially overcomes this problem, putting verb forms into the infinitive, nouns into the singular, and removing elisions. The research presented in this chapter uses a lemma list containing 40,569 words in 14,762 lemma groups [521]. More sophisticated approaches for eliminating ambiguities concerning the syntactic function or semantic nature of words require grammatical analysis of the sentence structures. While being indispensable for socio-linguistic or morpho-syntactic studies, such methods are less essential for Web assessment.

Lemmatization reduces the number of words as compared with the initial text. While the vocabulary of a text consists of all *distinct* occurrences, the size (or length) of a text is determined by the total number of occurrences. Both lemmatizing texts and applying a frequency threshold for words considerably reduces the vocabulary by grouping words of similar meaning together and discarding occurrences that only appear once (hapaxes) or rarely used expressions. In many cases, the reduced vocabulary contributes to more significant patterns [326].

4.5 Topic Detection and Tracking

In the context of media monitoring, topic detection identifies semantic clusters in continuous streams of raw data that correspond to previously unidentified events [102]. To detect these clusters, a keyword analysis compares the relative term frequencies of different corpora. The dynamic characteristics of such agglomerations can be visualized using temporal histograms or perceptual maps based on correspondence analyses.

4.5.1 Keyword Analysis

Keyword analysis locates and identifies characteristic words in a given text. The frequency of words in the text is compared with a reference distribution usually taken from a larger corpus of text. Such a reference corpus can be obtained by integrating the textual content of a large sample of sites, for example, or fror external sources such as the *British National Corpus* or Collins' *Bank of Eng*

[657; 663]. Any word that is outstanding in its frequency in comparison with the reference corpus is considered characteristic. The keywords are then presented in order of significance, which is commonly computed via Dunning's *log likelihood test* [142], or the classic *chi-square test* of significance with Yates' correction for continuity. Dunning's test was chosen because it gives a better estimate when contrasting long texts against a reference corpus [492].

Analyzing the morphological structure of electronic content via keywords yields thematically unified clusters; alternative approaches to determine such clusters are based on graph theory [133]. Lists of keywords also allow the assessment of how closely the information contained in published documents matches the user's informational needs (as expressed by the most popular terms that have been entered into the organization's search engine). Besides its obvious value for analyzing and comparing the textual content of online media, therefore, the discriminative potential of keyword analysis can also be leveraged for information retrieval tasks of a more general nature. Search engines rely on link topologies (e.g., Google [710]) or document-oriented metrics (e.g., AltaVista [649]) for determining how precisely a given document matches the user's query. By neglecting site-oriented parameters such as the "keyness" of terms, most rankings do not reflect the organization's level of expertise in a certain area.

4.5.2 Correspondence Analysis

While simple descriptive representations highlight the relative importance of topics for media segments, they fail to reflect trends or complex semantic relationships. Alternative visualization techniques such as dendrograms or perceptual maps based on correspondence analyses help address this shortcoming.

Web analysts often have to encode multidimensional data comprising so many variables that an orthogonal visual structure is not sufficient. Subdividing a given problem or computationally reducing the dimensionality often leads to results that are faster to interpret, convey more information, and lead to fewer interpretive problems [326]. Correspondence analysis exemplifies this class of computational solutions. It represents the rows and columns of a contingency table in a joint plot, referred to as a perceptual map [221].

Correspondence analysis belongs to the family of principal axes methods. Trying to uncover major structural traits, it identifies lower-dimensional sub-spaces that approximate the original distribution of data points. Correspondence analysis portrays general patterns contained in textual Web data. Its easily interpretable displays facilitate heuristic exploration of these patterns, without the need to specify a priori assumptions about the distribution that yielded the values [58; 573]. Correspondence analysis emphasizes the importance of models that fit the data, rather than the rejection of hypotheses based on statistical significance tests. Although not requiring a priori models itself, it is often a useful preliminary to a more structured multivariate modeling of categorical data [434; 533].

4.5.2.1 Data Preparation

Before applying correspondence analysis to visualizing the emergence and decay of topics, the word lists have to be integrated into a contingency table by cross-tabulating word occurrences and sites in a two-dimensional matrix. Characteristic terms related to renewable energy were selected from the glossaries of the *Energy Information Association* [147], the *US Department of Energy* [134], and the *California Energy Commission* [100]. The cumulated frequencies for groups of these terms such as "*hydro*" (hydro, hydropower, hydroelectricity, hydroenergy, hydro-energy) or "*efficiency*" (energy efficiency, conversion efficiency, fuel efficiency, power efficiency, engine efficiency) were then compared among the sample of 50 media sites. Between February and June 2003, the number of words for the individual media sites contained in the sample ranged from 0.03 to 6.48 million. The number of absolute occurrences was corrected for these differences proportionally by converting the absolute values into relative values per one million words.

Correspondence analysis can go beyond static snapshots based on particular archives or values aggregated over time. Instead of presenting separate diagrams for monthly data, a combined sub-space for the time series addresses the dynamic character of Web content [372] and facilitates the interpretation of results.

4.5.2.2 Perceptual Map

Subsequent computations using the *CORRESPONDENCE* command of SPSS [558; 573] resulted in the perceptual map of Figure 4.3. While the rectangular markers represent the concepts of interest, the round markers denote the position of a particular media site in the computationally created two-dimensional space. For the sub-sample of media sites used for the type token ratio analysis in Section 4.4.1 *(CNN, BBC, LA Times, Fortune, Scientific American, Environment Daily, Environmental News Network)*, the arrows demonstrate the monthly changes between February and June 2003.

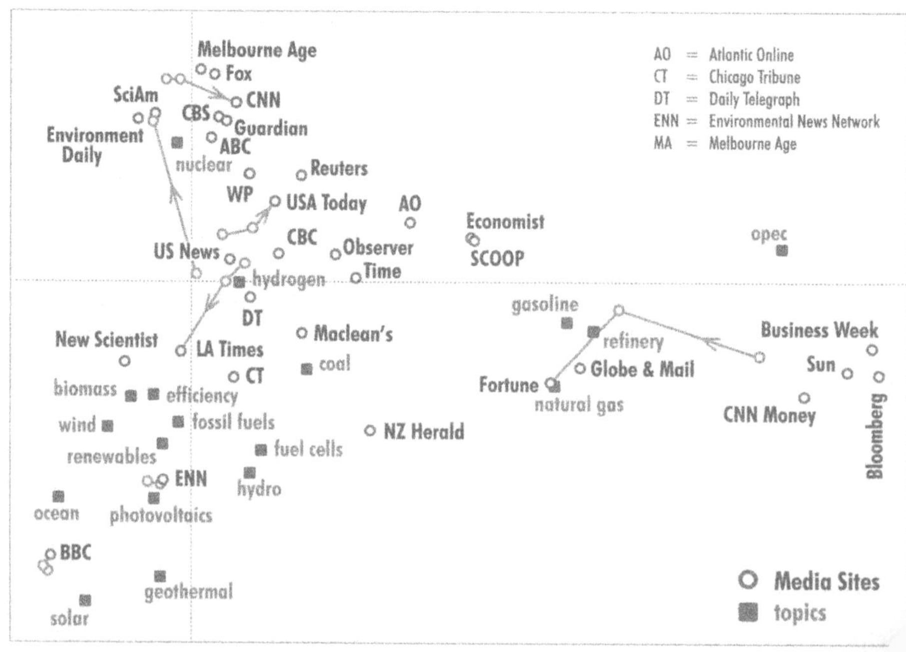

Figure 4.3: Perceptual mapping via correspondence analysis (02-06/2003)

It is often tempting but generally not permitted to interpret cross-proximities between row- and column-points of such displays (i.e., terms and media sites), because they originate from different initial spaces. Nevertheless, a joint interpretation of the row- and column-points with respect to the principal axes of the map is possible. As transition relationships link the coordinates of one point to those of all the points of the other space, the position of this single point may be interpreted with respect to the set of points of the other dimension.

The distribution of data points shows that most renewable energies congregate in the bottom left corner of the diagram, while specific fossil fuels and nuclear technologies are located in the right and upper left part of the diagram, respectively. By contrast, general concepts such as "fossil fuels", "renewable energies", and "energy efficiency" occupy the center of the diagram, indicating comparative analyses on a strategic level rather than focused coverage on particular energy sources or carriers.

While business media such as Fortune [706], CNN Money [672], or Business Week [664] continued to emphasize articles investigating fossil fuels including related organizations and market trends, the spatial proximity of data points in Figure 4.3 shows that two topics dominated the mainstream coverage: nuclear technologies and hydrogen. Querying a number of press release databases and newspaper archives for articles published during the first six months of 2003 clarified the changes within those sites.

The war in Iraq and the hunt for weapons of mass destruction, rising tensions between the Unites States and North Korea, discussions about tactical nuclear weapons, and renewed interest in nuclear power as a means to secure energy supplies are primarily responsible for this dominance. This reflects a tendency of mass media to cover the environment only occasionally, focusing on accidents and controversy rather than substantive issues or policies required to address them [552; 636]. The observed trends were not restricted to news media. The data points of *Scientific American* for February and April 2003 document similar changes, which are primarily due to an exclusive online issue on the "Science of War: Nuclear History" [491] and numerous links to this issue from various other parts of the site.

Hydrogen traditionally holds a pre-eminent position among alternative fuel candidates because of its comparably high energy content and environmental compatibility [377]. The hydrogen debate intensified early 2003 after the "State of the Union" speech of George W. Bush, who demanded US-$ 1.2 billion to speed development of hydrogen-powered automobiles in a program called *Freedom Fuel* [80]. Despite some initial praise for the effort, critics soon emphasized the following limitations [32; 98; 99; 349; 394; 490; 586]:

- technical difficulties of producing, storing, and transporting hydrogen as a chemical energy carrier;
- the US administration's emphasis on nuclear power (through electrolysis, which splits water into hydrogen and oxygen) and fossil fuels (e.g., through catalyzed steam reformation of natural gas) as sources of hydrogen, largely ignoring pollution-free renewable alternatives;
- the limited scale of committed resources, which is inadequate as an incentive to disentangle current business models from crude oil.

Boosted by this initial publicity and ongoing discussions about the monetary and environmental costs of oil dependence, increased coverage on hydrogen, hydrogen-powered cars and related technologies such as fuel cells continued throughout the first six months of 2003 [96; 97; 438].

The obvious outlier in Figure 4.3 is the site of BBC. The Energy White Paper, published in February 2003, committed the British government to providing 10% of energy from renewable resources by 2010. A big step towards meeting this commitment is the new licensing round for offshore windfarms announced by the British government in July 2003 [44; 240]. The initiative plans to attract 6,000 megawatts of offshore capacity, enough power for almost nine million people. This capacity approaches the 6,500 megawatts of new wind energy that came online worldwide in 2002 [349]. The ongoing discussion of renewable energy policies in the United Kingdom [195; 586] helps explain the isolated location of the BBC site. Another factor is the heavily linked and well-promoted climate change and alternative energy section of the BBC Weather Centre [586].

4.6 Conclusions and Outlook

The automated media monitoring methodology described in this chapter can process thousands of Web sites without human intervention. Fifty international media sites were chosen from a sample of more than 4,000 Web sites to investigate the coverage of renewable energies. Hydrogen as an emerging energy carrier and nuclear technology were identified as two dominant themes, reflecting current events in the first six months of 2003.

Future research should use a higher frequency of data gathering for certain types of media (mirroring in weekly or daily intervals, for example, depending on the media's dynamic characteristics). In addition, the presented methodology neglects the usage context of a particular term. Determining not only the term frequency, but also the semantic orientation of a site towards specific concepts, would effectively address this problem. The semantic orientation can be computed from the co-occurrence of a term with positive and negative terms taken from previous classifications such as Harvard's *General Inquirer* [545; 546]. This approach, which incorporates a limited amount of syntactic information contained in the textual raw data, is referred to as shallow or partial parsing [225].

Refining the list of query terms would increase the method's discriminative power, for example distinguishing between civilian and military use of nuclear technology. Online tools such as Princeton's WordNet [792], inspired by psycholinguistic theories of human lexical memory, could support this task by suggesting synonym sets in order to better capture underlying lexical concepts [179; 446].

Future research should extend and leverage the existing architecture to derive theoretical knowledge and advance existing theories about the nature and dynamics of Web content in general and of specific focus areas in particular. Similar to traditional media research [543], the aim is to build theories concerning organizational influences on the production, their connections to the final content, and on inter-relationships between content and target audience. Longitudinal studies of electronic content on the macro-level (comparative analyses of specific media segments) and the micro-level (trends in the coverage of certain topics, Web benchmarks, etc.) would facilitate and complement this theory-building process.

Chapter 5
Online Communication for Environmental Fundraising

Wolfgang Amann • Shiban Khan

Abstract. *Non-profit organizations avail the Internet to disseminate information to the public and to build community support for their causes. This chapter explores the status quo of Internet use for fundraising activities within environmental non-profits and proposes necessary improvements. A quantitative study was conducted via interviews with key fundraising personnel of multiple organizations. Non-profit environmental organizations from the state of Pennsylvania, USA, served as the sample. The results clearly indicate the overall effectiveness of Internet use for fundraising purposes to be quite disheartening. Due to resource constraints and a lack of technical knowledge, compounded by organizational reservations about its advanced use, online fundraising emerged as a low priority for the targeted organizations. None of the organizations seemed content with their level of online fundraising, despite acknowledging its vast potentials. Two strategy roadmaps are recommended to overcome the status quo.*

5.1 Introduction to Online Fundraising

Online communication boasts of advantages in speed, accessibility, customization, interactivity, and pervasive information over other media channels. This explains why this rather new and unique way of communication has the potential to offer major benefits to environmental organizations, especially with regard to fundraising activities. The advantages of online fundraising activities over other media channels are manifold. If implemented correctly, this media can prove to be the most cost and time effective way of reaching a virtually inestimable audience. Non-profit organizations already avail themselves of the Internet to disseminate information to the public and to build a support community for their causes. Customized information can be directed at various segments, some of which have not been as accessible before. The Internet enables interactions with key stakeholder and target groups, or for that matter, among members of target groups, at amazingly low cost levels. Therefore, the Internet as a media for fundraising has vast untapped potentials for non-profits. As for the technical aspects, technology-shy organizations have the advantage of being able to rely on professional Application Service Providers (ASPs) which provide excellent support in Web site design and maintenance. This should ensure that the technological side represents no fundamental problem. As for the knowledge on fundraising in the Internet age, a plethora of "how to" guides for online fundraising has also emerged recently [54; 88; 118; 199; 205; 273; 602]. Furthermore, a personal concern for the environment among the primary target groups can be deemed a significant plus in creating traffic to the online media and simultaneously facilitate the process of achieving the goals of online communication.

However, very few empirical studies embark on the identification of the actual status quo of online fundraising. Studies have shown that although corporations have been quite successful in these processes, making the Internet work for environmental organizations with regard to fundraising seems to take a backseat [468]. This initial result holds true for other societies as well, interestingly enough also for the Germanic culture cluster, where the environmental concern is more strongly rooted [63; 242]. Beyond normative and heuristic guidelines, further research is needed to explore the current situation of online fundraising. Online communication and online fundraising are likely to follow the overall development of the Internet as depicted in Figure 5.1, although with a time lag.

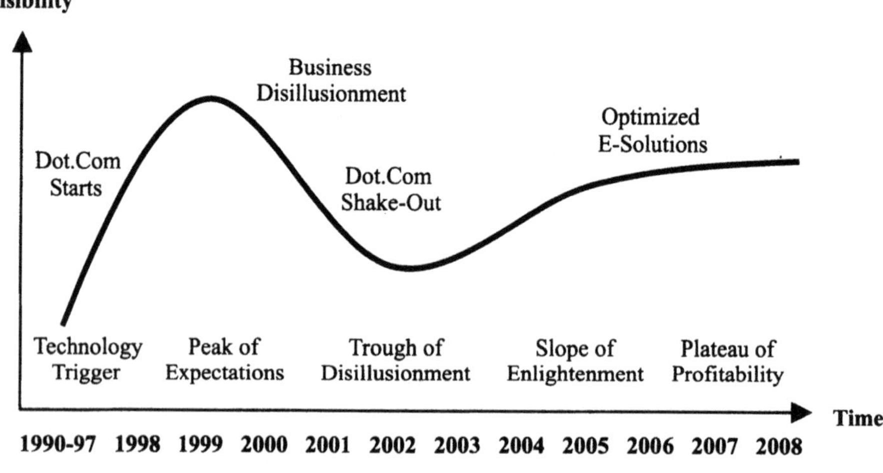

Figure 5.1: Visibility of the Internet over time

A phase of disillusionment has followed the hype of the early years, but the plateau of optimized e-solutions with their respective financial success has not yet been reached. This explorative study pursues the goal of addressing the identified gap with extended empirical work on the actual situation of online fundraising within the environmental non-profits in the state of Pennsylvania, USA, and of proposing improvements, if necessary, that could be made to enable progress towards satisfaction and financial success. This chapter is structured as follows: the next section outlines the methodology, and the subsequent section presents the empirical results, followed by a discussion. The chapter closes with specific recommendations, limitations, and the final conclusions.

5.2 Methodology

Due to a dearth of existing empirical information on environmental online fundraising, an interview schedule with half-standardized questions was favored above other methods in order to achieve a proper balance between intra-organizatio depth and the identification of cross-organization patterns. The sample was lected based on the organizations' purpose, geographical location, size, and the nature of their online activity. The study sample therefore consisted of environmentally oriented organizations within the state of Pennsylvania, having an income range of roughly $1 million and above and currently maintaining a Web site with an online option for donation and fundraising. Following GuideStar [715], a list of 36 such environmental non-profit organizations emerged. For the purpose of this study, the head of fundraising (the development personnel) within the organizations was identified as the key informant [459]. Thus individuals who have extensive knowledge on the subject matter within the organization were chosen for in-depth interviews in March and April of 2003.

As suggested in the literature [503], a preliminary e-mail informed the targeted personnel of the study, asking them to participate, ensuring confidentiality of answers and offering to share the study findings. The interview schedule was sent out prior to the interview session to provide an opportunity to prepare answers.

Given the newness of the topic and the exploratory nature of the study, no previously proposed standard instruments were available. The final interview schedule is thus based on an extensive literature review and contained 23 items categorized in three parts.

The first part dealt with the organization's online communication and its Web site. The second part focused on fundraising activities in general, while the third part investigated online activities pertaining to fundraising. The format of the interview schedule and its implementation via interviews with key informants enabled responses that were of high richness and validity. The method thus suits the early stages of research in this field and promotes the understanding of the phenomenon. Consequently, this method was deemed suitable for the study's research question [269].

5.3 Results

5.3.1 General Insights into the Online Communication Patterns of Environmental Organizations

Non-profit environmental organizations seem to have added online communication only rather recently, since more than half of the observed organizations' Web sites are younger than five years. Twenty percent of the organizations only established a Web site within the last three years. The traditional control of Web site performance and traffic, e.g. in hits per month [274], is largely absent. The major reasons for this lack of measurement are to be found in technical constraints, specifically a lack of technical proficiency, according to the participants of the study. When asked to comment on the primary objective of the organizational Web site, 75% stated that it was to increase awareness of the organization's activities. The Web site thus primarily serves as a public relations tool. Only 25% of the responding organizations mention that online communication through Web sites aims at providing information on environmental issues. Rating the importance of a battery of goals for an online presence, online fundraising was actually rated very low n priority. The specific insights into online fundraising are further explored in following section.

5.3.2 Online Fundraising

In order to assess online fundraising and its success, a comparison with overall fundraising is useful. Asked about the targeted donors, 60% identified them to be environmentally oriented, local, and individual community members, followed by grant-giving entities, including foundations, governmental organizations and corporations. Sixty-five percent of the organizations mentioned that they had registered donors, or, in other words, individuals who are listed in the organization's database as contributing on a regular basis. Twenty percent did not have any registered donors.

Ninety percent of the organizations had options available on their Web site through which to donate to specific projects, stating that they increase accountability in respect of the donors and hence enhance the organization's credibility in the eyes of the donors. Sixty percent stated that options to give to specific projects yield more donations. Simultaneously, donations to specific projects entail the frequently cited disadvantage of reduced flexibility in allocating funds where this is most needed in the bigger scheme of fund allocation. This last aspect is especially important for these organizations with regard to their development towards becoming a self-sufficient entity.

While 90% of the researched organizations have online donation options, the percentage of revenue earned from online fundraising in relation to total fundraising proved to be very low, regardless of the total amount of money raised. As Figure 5.2 illustrates, an astonishing 80% of participating organizations in this study fail to raise more than 2% of their total funds per year online, and 95% of the organizations do not arrive at 5% of the total funds being raised online. Traditional means of raising funds, e.g. via postal mailings, door-to-door canvassing and especially lobbying and applying for grants, continue to prevail.

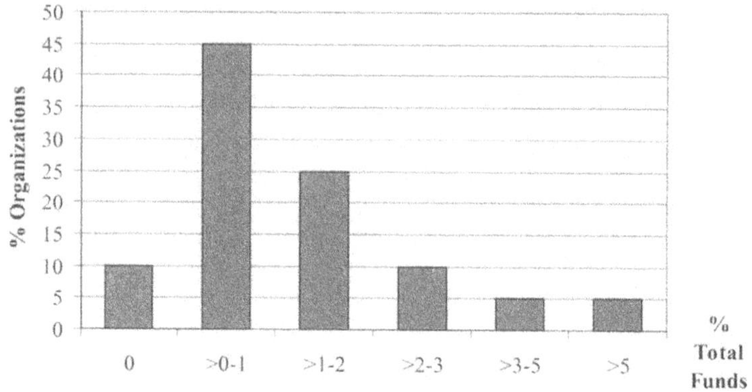

Figure 5.2: Success of online fundraising

Not surprisingly, the satisfaction rates reported in Figure 5.3 are low across organizations; only 5% deem their online fundraising efforts effective. The study participants were asked to provide an explanation for either the failure or success of their online fundraising activities. A striking three-quarters jointly suggest th following two reasons for the disillusioning results of online fundraising: insu cient promotion thereof in respect of the target groups and the low prior online fundraising within the organizations. A lack of technical knowledge and the reservations of some of the donor segments were also mentioned as possible and quite common reasons.

Overall, the way online fundraising is managed appears to be rather unprofessional. This is revealed in multiple aspects. Although 90% of the organizations mentioned collecting donor e-mail addresses, 70% admitted not using them to actively solicit funds at all. They also mentioned the lack of a comprehensive effort in gathering the addresses, or using them only for information dissemination. The source of the problem does not lie in environmental organizations not recognizing the potential of the Internet. Most organizations mention the access to a broader audience base (75%), the advantages for donors in respect of convenience (70%), the accessibility (55%) of provided information and an online donation option that is available around the clock.

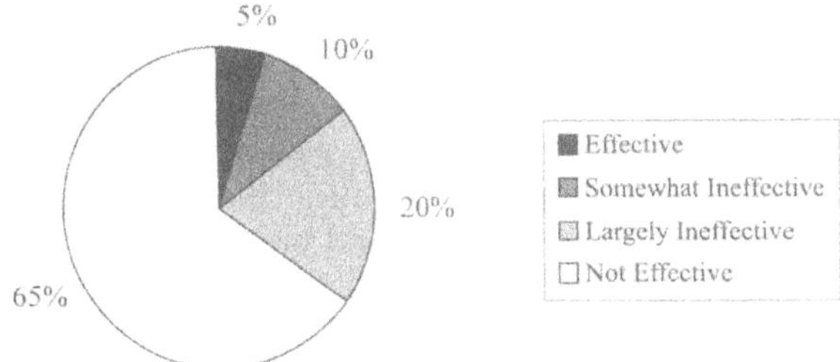

Figure 5.3: Satisfaction rates for online fundraising

In our interviews, the major causes clarifying the status quo among environmental organizations are primarily found in staff and time constraints or a lack of technical knowledge, and the cost and difficulty associated with the maintenance of a good Web site. A discussion and suggestions for improvement are outlined in the following section.

5.4 Discussion and Recommendations

The in-depth interviews clearly revealed that environmental organizations in Pennsylvania have not yet crossed the digital divide despite its significant potential. While their for-profit counterparts seem to leave the trough of disillusionment in order to embark on the slope of enlightenment in Figure 5.1, the non-profit environmental organizations have yet to fully commit to the digital world. To begin with, the organizations have reservations about adapting to new technology, and the widespread apprehension of the Internet is quite apparent. In many instances the organizations only maintain a halfhearted 'online brochure' of a Web site with limited opportunities for interactivity, or have out-of-date Web sites. This is further reflected in the lackluster attitude towards monitoring Web site traffic. This often stems from restricted resources on a tight budget, characteristic of most non-profits. Consequently, they do not actively promote their Web site for fundraising purposes.

Thus, the study results show that online fundraising is not particularly prevalent among the organizations studied, and consequently their efforts have not been effective. They have yielded less than 2% of the total funding for most organizations, and the benefits noticed have been minuscule. On the whole, online fundraising needs to be made an integrated part of the overall fundraising endeavors, as it clearly has the potential to make a difference. Therefore, online fundraising should be granted more focus and importance. It seems that organizations realize the importance of online fundraising, but are at a loss on how to go about it. The key findings from this study call for two major approaches for which practical steps towards implementation can be recommended - prioritizing online fundraising and professionalizing the online fundraising process. These are described below.

5.4.1 Prioritizing Online Fundraising

The results from the study confirm Boeder's [62] thoughts on non-profit Internet usage being dominated by communication, advocacy and community building, followed by fundraising. Despite its overall importance within the organization, online fundraising only has a low priority in the range of online communication objectives. As actions are guided by objectives, the low rank that fundraising enjoys as an objective of the Web site can be a primary reason for the general disheartening picture of online fundraising within the organizations. The respondents also mentioned the lack of priority and the promotion of online giving as the two most important reasons for unsuccessful outcomes. In addition, their often-limited budgets do not allow them to invest heavily in their Web sites. The situation is further compounded by the upper management's reservation about using e-mail technology and their presumptions that donor groups will not use it. However, trends show that more senior citizens are using the Internet and are developing into a significant Internet user group. The Internet represents a new channel to communicate with this already existing donor pool. Moreover, the members of the younger generation are more technologically savvy, and adept with shopping portals and online communication channels. Their chances of using the Internet for donation purposes are, thus, significantly high. Therefore, a precondition to prioritization is a shift in the mindsets of the upper management to become more adaptable and opportune in their approach.

Therefore, in the prioritization process, steps need to be taken to institutionalize online fundraising as part of the strategic planning, and integrate it into the mainstream fundraising program. For example, redesigning the Web site and orienting it more towards fundraising, as well as having prominent links to and reminders of the donation link on every page is highly recommended. Concurrent active promotion of the Web site and of online giving capabilities in all print and other outreach materials should also prove highly beneficial. This would require the allocation of substantial staff time and resources.

5.4.2 Professionalizing Online Fundraising

Prioritizing online fundraising is the first step towards success and changing the status quo, but the benefits will not become apparent unless suitable actions are taken. At this point, professionalizing the process becomes important. Essentially, this process has two sides, (1) relationship management with donors, and (2) dealing with the technical aspects of running a Web site.

5.4.2.1 Relationship Management with Donors

Building and maintaining a harmonious relationship with donors has been emphasized time and again by fundraising professionals as a precursor to amassing donations [344]. The first step in this regard is the active use of donor e-mail addresses, followed by high levels of personalization. The results indicate that in the interviewed organizations the database for donors' e-mail addresses is not used in an orderly and effective manner. None of the organizations maintained that the use of e-mails leads to successful fundraising. However, electronic newsletters are much cheaper to produce and to distribute than print newsletters, and thus boast a very high rate of return on the marketing and cultivation dollars invested [604]. It is also vital for driving traffic to an organization's Web site. Moreover, despite the generic broad appeal of a Web site, it also offers a unique opportunity for a high degree of customization according to personal preference. In order to offer what donors want, the organization needs to know who they are. Building a donor database through the Web site ranked as the lowest priority in the organizations studied. However, the Web site can be a wonderful tool with which to research interest groups (potential donors) by monitoring the visitors.

For-profit organizations have substantially increased the resources allocated to their Web-based marketing, and this can be easily replicated in the non-profit world to reap the benefits of online fundraising. In order to enable successful e-mail appeals for funds, non-profit organizations need to actively collect e-mail addresses from interest groups, donors, supporters and members and implement the structured and extensive use of a donor e-mail database to actively communicate information leading to fund solicitation. Other suggested steps include:

- Treat all Web site visitors as prospective donors, and build relationships based on a high level of interactivity. Provide an easy to fill out comment section and the contact information of all staff, thus encouraging visitors to reply. A quick response to general e-mails can be an effective method in personal relationship building and can help to overcome the disadvantage of the impersonal mass appeal of Internet fundraising.
- Provide options for the personalization of newsletters, so that subscribers only receive news and information on material that they have requested.
- Provide links to articles, essays or other materials of interest available on the Web site within the body of electronic newsletters.
- Provide numerous opportunities for donors to get involved in volunteering and other preferred activities.
- Promote all Web site content. Send alert e-mails about any new additions or features on the Web site.
- Every donation must be followed up by a 'thank you' e-mail. The e-mail should contain options for signing up for newsletters or e-mail alerts.
- Direct the potential donors to the Web site for more information, and provide the relevant guidelines for donation on the Web. A good Web site is also an effective tool for buttressing grant proposals.

5.4.2.2 Technical Aspects of Running a Web Site

Instant donation capability over the Internet is undoubtedly the most controversial issue in the context of online fundraising, with superlative advantages being counteracted by perilous risks. Therefore, to make online fundraising really effective, it is crucial for organizations to offer the most advanced security solutions in respect of financial information on the Web. The process of making donations online should be quick and easy and other options provided for those donors who are still not comfortable with providing their credit card information online, e.g. a telephone number, downloadable forms to be mailed and postal addresses to which they can send checks and bank drafts.

Secondly, as Carter [88] mentions, the Internet has leveled the playing field for organizations of all sizes to approach a vast base of donors. The results of this study also indicate that the organizations regard the Internet's universal reach, along with convenience, to be the most attractive benefit. However, there are millions of Web sites out there, and more are being added. In order to be conspicuous as an exceptional Web site and attract donor attention, the Web site should be well designed, easily navigable, interactive, quick loading, and updated at regular intervals. The attractiveness of the Web site should be maintained throughout by offering unique features, so that visitors keep returning to the site, which may ultimately lead to increased donation. A substantial initial investment is recommended to build the Web site, even outsourcing it for better quality if deemed necessary. Another important step is including as many key words (e.g., environment, nature, earth, as well as water, air, pollution, etc.) as possible on the Web site, so they come up more often in general search engines.

On a final note: the use of the Internet for environmental fundraising is a new playing field, and there is a substantial gap in experience and tried and tested methods to attain success. This study is the first of its kind in venturing into assessing the status quo. It offers a current situation analysis and covers the very basic questions on the scenario. In order to delve more deeply into the matter and analyze specific issues in more detail, further research on individual aspects of the study would be highly beneficial. These are discussed in the next section.

5.5 Limitations and Suggestions for Future Research

The limitations experienced by this study can essentially be divided into two subcategories, conceptual limitations on one hand and methodological or empirical limitations on the other. As for the first, one study can only address certain elements of any phenomenon. Future studies need to be carried out to illuminate additional research aspects in order to gain a better understanding of the phenomenon. The conceptual limitations can be overcome by including the donors' side and a respective gap analysis in future studies. Interviewing donors and contrasting the findings with actual online offers has a great potential to offer valuable insights. An interim step may be found in profiling the emerging online donor community better. One of the next steps from this largely descriptive research at hand is into normative research – investigating how the organizations should address donor expectations and how the status quo of environmental online fundraising can be improved for more effective results in future.

From the methodological point of view, the study is constrained in generalization potentials due to the sampling process. Firstly, the study is conducted only in one state, and while Pennsylvania has its fair share of ecological problems, the results may vary for regions with more or less acute cases of environmental degradation. The study regards organizational size as a decisive criterion on the assumption that more established organizations have a more structured fundraising process. This omits smaller organizations, although they may equally benefit from the study. Consequently, this also excludes the possibility of learning more about the innovative ideas that these smaller organizations may have to offer. Therefore, future research should include smaller organizations in the sampling frame and explore whether the results of this study hold true regardless of organizational size. While larger sample sizes enable generalizations, there is an inherent trade-off with within-company depth. Nonetheless, larger samples and sophisticated multivariate statistical analysis can be possible in combination with an enhanced generalization potential.

This study overviewed the nature of the phenomenon, interesting trends and peculiarities that have emerged from it. For a better understanding of these idiosyncratic traits in organizations, in-depth case studies are likely to provide an even better understanding of the phenomenon along with richer descriptions of different contexts. This would more aptly allow for a better means to investigate the questions on the hows and whys of, for example, organizational attitude towards online fundraising, and would produce better insights into the successes as well as the failures of various strategies. It would increase the depth of the knowledge of the phenomenon in the organizations.

5.6 Conclusions

The advancement of new technologies has always earmarked the advent of significant leaps in human civilization. Growth and progress have been experienced when these technologies are understood and their benefits optimized. Few other new technologies have revolutionized and infiltrated our day-to-day living as rapidly as the Internet. It has changed the face of communication and information dissemination, and no entity involved in these fields can afford to ignore it.

This definitely holds true for non-profit organizations. They excel in developing personal relationships with people [536], and the Internet offers by far the best opportunity to reach the largest number of people – theoretically, without limits. Better relationship building results in an increased supporter base, and ultimately leads to bigger donor groups. Thus, the Internet can be highly influential in boosting the donor database.

Results from this study indicate that, despite acknowledging the great potential of online communication, the organizations' utilization of the Internet is still in its infancy, especially in respect of fundraising activities. Organizational reluctance, stemming from resource constraints and a lack of prior experience in the field, is a major cause of this phenomenon. However, there are solutions to these drawbacks, and effective strategies that have yielded satisfactory outcomes. This study delineates these strategies and mentions how they may be helpful to the , nizations.

In a nutshell, most promising directions foresee prioritizing online fundraising on one hand, and professionalizing it in regard to relationship management with donors and technological implementation on the other. After proper implementation and adaptation of these approaches to the specific organizational context, non-profits are likely to enhance their success in fundraising. Successful utilization of the Internet for fundraising by non-profit organizations should not be too far off in the future.

Chapter 6
How a Small Environmental Group Uses the Web to Inform and Promote Action: A Content Analysis

Heather Sehmel

Abstract. *As small environmental advocacy groups increasingly create Web sites, we need to better understand how they use them. This chapter reports on research combining interview, observation, and textual analysis of a small environmental group's Web sites to see how the group uses Web sites to meet various goals, such as informing the public and encouraging citizens to take action for the environment. An initial quantitative content analysis demonstrates that more space on the Web sites is used to inform citizens than to persuade them or help them take action. However, more thorough consideration of content placement and combination of visual and verbal elements shows how the Web sites may actually do more to facilitate action than the initial analysis demonstrates. These and other results help us better understand how advocacy groups may be using Web sites as rhetorical and political tools.*

6.1 Introduction

The Internet is a significant tool for environmental advocacy groups, allowing them to meet their goals through textual and visual messages and electronic-supported actions like sending e-mail to politicians. However, the Web is not a panacea. Advocacy Web sites seldom take full advantage of hyped Web opportunities like creating dialogue or providing personalized information [296], despite a correlation between user satisfaction and Web site interactivity [241]. Therefore, we must wonder whether Web sites indeed help groups improve access to information or support action.

Current research inadequately answers most questions about Web site use, especially about environmental advocacy Web sites, which too seldom receive attention [522]. Furthermore, small advocacy Web sites have different issues than large ones, largely because of their differing resources. The group I discuss here benefits from six to seven hours of Web site design expertise a month, while the Environmental Defense Fund has at least twenty full-time Web site design employees [534]; the group's budgets for Web site development are also dissimilar. Groups are also constrained by whether their employees have the diverse skills needed for Web design [180]: the smaller the group and its budget, the more serious these constraints. Identifying the goals small environmental advocacy groups have and comparing them with their Web sites may help us determine whether they are able to do what they want to with their Web sites and why any discrepancies exist between what they want and what they create. This information can guide Web site design for small environmental advocacy groups.

This chapter investigates the Sustainable Energy and Economic Development (SEED) Coalition, a small environmental advocacy group based in Austin, Texas. The not-for-profit non-lobbying organization (with five full-time employees and a Web site designer contracted for about six to seven hours a month in 2000-2002) works with other Texas groups to raise awareness of and organize citizens against global warming and pollution. This study considers the main Web site and one of the group's many campaign-specific Web sites, the Toxic Two Step Web site. Both Web sites use textual and visual elements to support the SEED Coalition's goals. All of the group's full-time employees except the Office Manager contribute to site content, and their 2002 Web site designer primarily programmed and published this information (I am unsure of the scope of the role of the group's new designer). The main Web site is still (as of October 2003) based upon – and its updates and additions are constrained by – a pre-2000 design. The smaller, campaign-specific Web sites the group uses are, in contrast, made just in time and reflect newer Web site design characteristics. Many of the Web sites the group has made since receiving the results of this research in summer 2002 also reflect my recommendations, some of which answered the research questions to which this chapter responds: How do Web sites meet the goals and reflect the values of their developers? How do Web site characteristics promote action or inform visitors on environmental issues?

6.2 Methods

My methodology builds on systematic assessments of political Web sites. Like McAndrews [360], I consider the purpose for Web site material. Like Klotz [299], I look at visual, textual, and interactive elements and study use of electronic space. While Klotz and McAndrews' analyses are valuable, both researchers deduce purpose and audience based upon textual analysis. By combining elements of their textual analyses with interview data and observation, I provide a richer picture of the applications of the Web sites.

Therefore, between May 2001 and May 2002 I interviewed the Director, Office Manager, Clean Air Coordinator and her assistant, Climate Coordinator, and the Web Designer; I also observed the group's full-time employees for four full workdays in August 2001. My content analysis of the SEED Coalition's Web sites determined to what extent the group's Web sites reflect the group's goals for the Web sites, as identified through the interviews and observations. I coded units on the SEED Coalition's main Web site [779] and campaign-specific Toxic Two Step Web site [763] and analyzed them to learn if the group's Web sites reflected its stated goals and how the Web sites communicate the priority of each goal. This information informs us about the functions of environmental advocacy Web sites.

My unit of analysis was a discourse bloc: the material accessed by a given hyperlink; a broken hyperlink, itself, which indicates planning for material or material no longer there; or a text section with unified content and organization. I counted press releases, reports, photos, logos, and grouped photos as discourse blocs. I considered each discourse bloc on the main Web site separately, b counted reports, press releases, and other linked documents each as one course bloc. I did not code discourse blocs within linked reports, but treated a report as one discourse bloc; I did not identify separate blocs on external Web sites. I coded the Web sites for characteristics from word count to use of an *ethos, pathos*, or *logos* related appeal. The following coding characteristics relate to results reported here:

- The advocacy campaign to which the discourse bloc relates most (e.g., mercury, power plants, pollution, or global warming).
- The bloc's location on the Web site (e.g., on the homepage).
- Presence or absence of visual elements in the bloc.
- The purpose of the bloc relative to the SEED Coalition's professed goals: (1) Provide a means for taking political or personal action, (2) Provide tools, (3) Ask for donations, (4) Ask for time, e.g. invite to a meeting or rally, (5) Provide a fax-action opportunity, (6) Provide means for taking political or personal action, (7) Inform/Build awareness of the environment, (8) Inform/Build awareness of the group, (9) Persuade/Build support.

A sample coding of an online anti-mercury pollution petition informs us that the petition relates to power plant pollution campaigns and is accessed from the homepage and the mercury pollution pages on the Web site. Also, it has no visual elements and it asks for political action (encourages people to complete it online). I coded units for multiple characteristics as appropriate: reports were considered to be informative and persuasive. For coding reliability, a second trained coder recoded 15% (45) of the units. She coded for subjective traits like the rhetorical function of the units, but not for less subjective traits like unit's location on the Web site. She coded nearly all of the units on the Toxic Two Step Web site in order to allow for a variety of units, and our inter-rater reliability rate was 94%.

I counted the units supporting various goals or demonstrating various rhetorical techniques; I also noted characteristics complicating a quantitative analysis. For example, in addition to knowing how many units support a particular goal, I also noted whether they were on the homepage or the fourth layer of a Web site or had visual elements to draw attention. Coding for many characteristics and analyzing the results in several ways provided a rich understanding of the function of the SEED Coalition's Web sites.

6.3 Results

I coded 287 discourse blocs on the two Web sites. While I completed a word count of each discourse bloc, unit size is not as important as function, design, and placement. A picture and ten-word caption can have more impact than a link to a thirty-page report. When I coded the blocs to see how they reflected the group's goals, I found the following percentages of units, in descending order:

- Inform/Build awareness of the environment (87%).
- Persuade/Build support (61%).
- Provide means for taking political or personal action (51%).
- Provide tools (30%).
- Ask for time (invite to a meeting or rally; 22%).

- Inform/Build awareness of the SEED Coalition (9%).
- Provide a fax-action opportunity (8%).
- Ask for donations (3%).

The SEED Coalition's Web sites have characteristics pertaining to all the group's goals, and the number of units reflecting each goal basically reflects the group's priorities. For example, only about 61% of the units are persuasive, reflecting that the SEED Coalition does not intend its Web sites to persuade people who lack environmental leanings [18]. My analysis also allows me to comment on how accessible the units relating to each goal are. For example, most of the information about the group itself is on or linked directly to the Web sites' homepages. Therefore, although information about the group appears in only 9% of the units, its prominent placement (41% of these units are on a homepage) allows visitors to find and be aware of it. The following sections discuss how the results of this analysis provide insight into how the Web sites reflect the desire of the group to inform people and to promote action.

6.3.1 Why Might the Web Sites Heavily Inform?

The SEED Coalition has a primary goal of encouraging people to take action and a secondary goal of educating people about the environment: "Our Web sites are really geared towards action-oriented, not a comprehensive information source" [18]. The SEED Coalition informs people by linking to its own and others' reports, summarizing issues, and providing press releases and other linked documents. In contrast, it promotes action by inviting people to attend SEED Coalition events and providing fax actions. Although the group states it wishes to encourage action over informing, the SEED Coalition's Web sites devote 87% of their units to informing and only 51% to enabling action (totaling more than 100% because some units do both). Not only did more units provide information than encourage action, but these units also tended to be longer (like multi-page reports) than those (like one-page petitions) that encouraged action.

The SEED Coalition's Web sites may, by sheer quantity, emphasize educating people about environmental issues over promoting action because (1) it is convenient – information is readily available for online publishing, (2) the group hopes educating people will make them more confident about taking action [220], (3) available space allows the group to store information online, although Web site visitors may not access much of it, (4) the group hopes a sheer volume of information will help build its ethos, or (5) the group mimics other Web sites, the rationale for many of its Web site decisions. A combination of these factors is the most likely explanation.

Based on online trends and SEED Coalition self-reports, one reason for the SEED Coalition's information-laden Web sites is to build ethos. All the advocacy Web sites McAndrews studied included large quantities of information, including outside information intended to increase their credibility [360]. Political Web site planners believe people who visit candidate Web sites demand detailed content as they are doing research [72]. One SEED Coalition staff member explains how the SEED Coalition placed a *20/20* transcript about mercury online, even though sh- doubts anyone will read it: she hopes letting people know that the issue is imp tant enough for *20/20* to address will help [197]. Like many advocacy Web sites, the SEED Coalitions' choose to persuade indirectly, by building a Web site persona of rational expert, then letting that rational expert ask people to take action. The desire to portray the SEED Coalition as a source of rational expertise may explain why the SEED Coalition's Web sites present as much information as they do: reports, links to newspaper articles, pictures, graphs, and maps are all intended to convince visitors that the group has done its research, its research is credible, and, therefore, we should do as the group asks.

Advocacy groups cannot always count on the press to inform people of their activities and issues, so they often use Web sites to communicate. Davis argues, "The Web is a significant new tool in group communication with the public. It offers the opportunity to communicate with an interested general public, and to do so with greater depth than the groups could through news-releases-turned-stories" [126, p69]. Eighty-three percent of the coded units do provide information in more detail than the local media are likely to do. In addition, the SEED Coalition's Web sites give media representatives information for newscasts and articles. Again, this is something other groups do: 22% of interest group Web sites included a press section [126]. The SEED Coalition has a section directly targeting the media, with a "Media" link from its main homepage. In addition, about 84% of the units either directly provide information for the media (such as press releases) or less directly provide information that might be valuable to the media (such as copies of relevant government documents). The group's Director knows that media representatives visit its Web sites for information [17]. Research also indicates that media representatives often use Web sites for more information about issues and organizations [299]. Therefore, it is not surprising that the SEED Coalition uses its Web sites to provide information for and to work around its under-representation in the media.

The SEED Coalition also believes that informing people makes them more confident about acting. While this seems logical, Cantrill [85] cites multiple studies demonstrating that being informed about the environment does not necessarily lead people to take action for it. One explanation is Wiebe's concept of well-informed futility, in which the more people are educated about the environment the less action for the environment they take [85]. Cantrill explains that informing people about the environment may "redefine [their] civic obligation to assist in solving environmental problems as being, rather, the duty to simply become informed about some issues" [85, p37]. In addition, Cantrill cites Novic and Sanders' explanation that well-informed futility explains the troubling finding of no or even an inverse relationship between knowing about an environmental problem and being committed to help solve it [85]. The complexity of environmental issues may overwhelm people, who may hope that more information will keep them from choosing incorrectly or contradicting their earlier positions.

If people do reject action as they learn more because of overwhelmingly complex issues, then the SEED Coalition may be able to encourage action through education. Visitors already interested in taking action are unlikely to fear contradicting their earlier positions. Also, because the SEED Coalition's Web sites provide just one perspective on environmental issues, they may not contribute to people's cognitive paralysis. Indeed, this may be a reason not to link to Web sites advocating different positions.

Finally, while informing may not always be effective, advocacy groups need to be able to inform people. Giving people freedom to speak is insufficient; real communicative freedom includes "access to an audience", something denied if "unacceptable" worldviews are rarely printed or aired [554, p14]. Our society usually dictates that only experts can talk [554], and so our communication is too often one-way (expert to non-expert). Because media ownership has concentrated and the media is devoted to "fairness" it tends not to use non-official sources; 50% of mainstream US press articles quoted the government, corporations were the second largest group cited, and less than 4% of sources cited represented environmental groups [410]. This is one reason groups find the Internet attractive for disseminating information [299; 410]. Therefore, while placing large amounts of information online might not be the best strategy for the SEED Coalition in an ideal world, because Web sites can inform the public when the media chooses not to, online information makes sense.

6.3.2 How Might the Web Sites Inform and Promote Action?

Although the SEED Coalition's Web sites have more units that primarily serve to inform than that primarily serve to promote action, how much the Web sites emphasize informing more than taking action is more complicated than simply looking at the number of units pertaining to each purpose. Coding for the presence of appeals is not the same as determining their relative strength. While there are many more informative than action units on the SEED Coalition Web sites, the action units often have privileged placement; for example, 29% of action versus 19% of informing units appear on the homepage. The sites may functionally stress action while more text primarily informs or builds the group's ethos.

Furthermore, not only are the action units often prominently placed, but a call to action often also consists of visual and textual units. We should rely upon the visual along with the verbal online [314]. For example, the Seed Coalition's Web sites use visual units to draw more attention to action-related text than to informational text: 36% of the units encouraging action contain strong visual elements, like pictures, boxes, or icons, compared with 24% of the units providing information. One member of the staff enlarged a picture of Big Bend on the main SEED Coalition homepage to highlight the call for action related to it [219], a kairotic decision based on its being of temporarily greater political urgency. Similarly, the SEED Coalition's homepage has a bright red button that says, "Donate".

The ways in which the SEED Coalition enables people to take action easily also help it encourage action. For example, the group provides fax actions so people can fill out electronic forms and quickly send messages to policy-makers, using them as it uses postcards that people sign at SEED Coalition events for the group to mail, except that fax actions may be modified and can be accessed by people the group's staff would not see in person. Modifying a fax action takes less time than writing a letter from scratch, printing, stamping, and mailing it. More groups now use these tactics: Davis [126] reported that 14% of groups used online forms to send comments to policy-makers, although in his study few groups provided modifiable messages, and more groups almost certainly use these now. The US Forest Service's recent (April 2003) decision to ignore all group e-mail may lessen this use of Web sites, but that remains to be seen.

In addition to fax actions, the Web sites contain flyers, logos, and petitions that people can download and use locally. The group's director indicated that he would like to increase use of such tools, contacting members and asking them to use the materials [16]. This function of the SEED Coalition's Web sites seems newer (there are more examples on the Toxic Two Step Web site and newer sections of the main Web site than on older portions) and likely to help the organization work strategically across the state.

Finally, while more units on the SEED Coalition's Web sites inform, and although the group's main Web site goal is to induce action, when we consider (1) the need for information on the SEED Coalition's Web sites to communicate with the media and work around the SEED Coalition's lack of assured access to other media venues, and (2) the strategic placement of calls for action on the Web sites and their use in coordination with visual elements, the SEED Coalition's Web sites may indeed reflect the SEED Coalition's goals, asking first for action, then providing information.

6.3.3 Limitations

This study has two primary limitations: it is a case study and it was completed in 2002. We cannot know from an in-depth study of one Web site whether what we learn about it pertains to other Web sites. However, because little work has been done in environmental advocacy Web site analysis, a case study was a good first step to developing larger comparative studies, such as my current project, a survey of groups asking about their Web site goals and the constraints keeping them from meeting their goals. While this research has limited application, it provides an important starting place and methodology.

While the age of the study may appear problematic, I do not believe it is. Spencer [522] indicates that not-for-profit groups often adopt and adapt technology more slowly than groups in other sectors. The SEED Coalition is still using a Web design from before 2000. Not only does Web design in general seem to be changing less rapidly than it once did, but also not-for-profit groups, with financial and staffing constraints, are even less likely to quickly adopt new Web designs. Instead, they are more apt to, as SEED has, simply add to a current Web site. Indeed, because most employees in small organizations are not Web design experts, they are not likely to be on the advance edge of Web design. Nor are they likely to have the resources to redesign their Web sites completely very often even if they are aware of newer, more effective designs (and many new, cutting edge designs may actually be less effective for their use – if, for example, they use technology that some users cannot appreciate with their older machines or dial-up connections).

Also, users may not need new designs; an audience survey I conducted for the SEED Coalition in spring 2002 indicates its users are generally happy with its Web sites. Finally, the study's being a case study seems a greater limitation than its age, although, groups wanting to change their Web site designs based on the advice given here should consider whether it is still or ever was applicable to their situations and audiences.

6.4 Conclusions

The preceding sections explored how the SEED Coalition Web sites reflect two of its goals: to inform people about environmental issues and to promote personal and political action. While the number of units on the Web sites that primarily inform outweighs the number that primarily promote action, those promoting action often have privileged placement and are combined with strong visual elements. Looking at the quality rather than the quantity of its Web sites' rhetoric, we see that the SEED Coalition does, in fact, subordinate informing to promoting action and non-interactive Web elements to interactive Web elements by strategically placing information on its homepage and using visual elements. To draw an analogy, assuming that because more space on the SEED Coalition's Web sites informs than promotes action, the SEED Coalition's Web sites will encourage visitors to passively learn rather than take action might be like assuming that because there are more food vendors in an amusement park than there are roller coasters, people will eat (and identify their experience in the park with food) more than they will ride roller coasters.

Many of the group's online strategies, like including information to build ethos, may work because its targeted audiences are primarily people interested in its issues. As these people probably have previous environmental knowledge and already want to support the group's efforts, the group can provide information without worrying about paralyzing people by presenting new complex situations. Also, because people can easily leave a Web site (and probably won't stay at one they don't initially find interesting), the Internet is generally not effective for pushing views on people who do not already have some inclination towards them.

Still, though, the SEED Coalition and other advocacy groups (including non-environmental groups with similar goals) should consider how much of the information they provide meets their goals and how to consistently draw more attention to parts of Web sites that work toward their primary goals. While I think the SEED Coalition's Web sites reflect its goals fairly well, a glance at the SEED Coalition's homepage in October 2003, for example, shows a larger number of visuals on the homepage that draw attention to units that primarily inform than existed in 2002. This may indicate that the SEED Coalition does not consistently privilege promoting action over informing when updating its Web sites, although doing so may be critical to meeting its primary goal. I recommend that all Web designers and advocacy groups research Web rhetoric and consider how their Web site content meets their goals. In particular, they will want to consider how much information to include on their Web sites and why, and which Web site content they prioritize through use of visuals or their placement on the Web site.

Chapter 7
Usability Evaluation of an Environmental Web Site and its Support of the Organization's Mission and Goals

Ingrid Pohl

Abstract. *The usability of the Web site of the Agricultural Resources Center and Pesticide Education Project, a non-profit environmental advocacy organization, was investigated to determine if the site effectively communicates the mission and goals of the organization. As successful Internet communication is strongly dependent on the presentation, navigation, and quality of information, these three constructs were used to guide the evaluation. Through surveys, interviews, and the observation of eight participants interacting with the site, this study found that participants perceived the content on the site to be valuable, trustworthy, and of high quality. However, the usability of the site interfered with participants' ability to successfully experience or learn from the content. Participants requested more opportunities to take action and a clearer, stronger message about the objectives and services of the organization. Participants' perceptions guided a number of recommendations to improve the site's ability to effectively support its mission and goals.*

7.1 Introduction

Effective communication is essential to a non-profit environmental advocacy organization that endeavors to effect social change. By raising awareness of an issue in both the general public and interested parties - which then facilitates education, action, support, and the building of a community of activists - advocacy groups can successfully achieve their mission and goals. In fact, these activities constitute the very reasons advocacy groups exist. A Web site can provide an environmental organization with a highly effective device to reach and impact a wide and diverse audience. Any organization that uses the Internet, however, must pay close and ongoing attention to the presentation, navigation, and quality of its site in order to stand out among the wealth of competing information accessible on the Web and to capture, retain, and encourage the return of users.

This chapter presents an evaluation of the Web site of the Agricultural Resources Center and Pesticide Education Project (ARC), located at www.pested.org [645]. ARC is a North Carolina-based, private, non-profit, public-interest organization dedicated to preventing human and environmental exposure to the use of toxic pesticides in farming and food production, in the home and in urban and rural environments, through research, public outreach, by monitoring and lobbying state farming and environmental agencies, and by empowering people to make sound decisions about their health and environment. The ARC Web site provides information on agricultural issues, local environmental advocacy events, state regulations, and the organization's efforts to effect social change. A usability study of the site - comprising surveys, interviews, and the observation of eight participants - was conducted in January 2003. The primary research question sought to determine if the site effectively communicates ARC's mission and goals. The data collected was used to redesign the ARC site and to provide guidance and heuristics for other environmental organizations in the development of effective, mission-supportive Web sites.

7.1.1 Non-profit Organizations and the Internet

The National Environmental Directory [747], which lists 12,704 organizations in the United States concerned with environmental issues and education, shows that only 3,187 members, or one-quarter, provide Web sites. A number of potential advantages of using the Internet, unique to non-profit organizations, have been identified: network with peers, raise funds, recruit volunteers, advocate change, build a community of supporters, induce action, and promote participatory decision-making [322; 495; 522]. As more people go online for their information, environmental groups must have an online presence that speaks to and meets the needs of the individuals they want to educate and influence, that clearly communicates the services and mission of the organization, and that expresses why the site exists.

A study by Sehmel [495], one of the few that investigates Web use by non-profit environmental organizations, explores the use of Internet tools, strategies, and design features that can be especially useful for environmental advocacy groups seeking to educate, influence behaviors and values, encourage cooperation, and induce action. Sehmel explains that an environmental advocacy group can use its Web site to encourage visitors' involvement by providing a number of online opportunities to take political action, by suggesting changes individuals can make in their personal lives, and by providing online opportunities for community dialogue. Web sites can provide numerous ways for visitors to stay informed - through e-mail updates and newsletters. Additionally, an organization can present information at varying levels of detail so that users have easy access to information at the specificity they require. The techniques Sehmel delineates can be effective guidelines for an organization hoping to promote its mission through its Web site.

7.1.2 Web Usability and Information Quality

To successfully capture the interests and support of its Web site visitors, organizations must pay attention to Web usability, a combination of factors that influence a visitor's experience with a site. Usable Web sites begin with a design that meets the needs of the audience, the organization, and the objectives for the site. Web sites should be developed so that they are useful and usable; the design, content, structure, and navigation should facilitate findability and support user tasks and goals.

What constitutes a usable Web site is largely dependent on the purpose of the site and its intended audience [379]. However, professionals agree that successful sites are: accessible, appealing, consistent, unified in design, clear, simple, navigable, safe to use, easy to learn, easy to remember how to use, and forgiving of user errors [379; 388; 444]. Usability testing, designed to identify, confirm, or reject potential problems and to compare design alternatives [408], is one method to measure the effectiveness of a Web site. Though there are a number of techniques, a popular approach is to collect empirical data through the observation of users interacting with the site in typical ways [461]. Typically, five to twelve users can sufficiently discover the majority of usability problems on a site [141].

Another critical component of successful Web site development – information quality – must also be considered. The integrity of information presented on the Internet is essential to effective communication because Web sites do not have the implied authority of a publisher to endorse a resource. Five specific criteria – accuracy, authority, objectivity, currency, and coverage – play a fundamental role in the user's evaluation of information quality [11]. A number of additional criteria can also influence a user's perception: ease of access to specific information, navigability, and attractiveness [289; 327]. Although few methodologies exist for assessing information quality, there is a significant amount of overlap when measuring usability and information quality.

7.2 The ARC Web Site Usability Evaluation

The evaluation of the ARC Web site was guided by four primary evaluation questions:

- Does the presentation of ARC's Web site attract visitors, allow them to quickly understand the content, and entice them to remain on the site?
- Does the navigation of ARC's Web site facilitate information finding, ease of use, and ease of movement?
- Does ARC's Web site exhibit high quality in the areas of currency, accuracy, purposefulness, and understandability?
- Does ARC's Web site effectively communicate the mission and goals of the organization?

Eight adults, thought to be interested in or at risk from environmental hazards and pesticide use, participated in the study of the effectiveness of the ARC Web site. During the study, sessions were audio- and video-taped to capture verbal communication, facial expressions, hand movements, and the browser screen. After signing a consent form and completing a demographic questionnaire, participants independently browsed the ARC site for up to 20 minutes, during which time they were asked to think aloud. The think-aloud protocol, which asks users to voice what they are thinking, wondering, and attempting to do, is a popular usability technique that allows researchers to understand and get access to human cognitive processes [444; 461].

Five information-finding tasks, representing potential uses of the site, were then given to the participants to complete using only the ARC site. The information-gathering tasks included:

- You have a pesticide-related question and think that the Agricultural Resources Center and Pesticide Education Project would be able to help you. Find ARC's contact information.
- Find statistics about the effect of pesticides on children.
- You read a newspaper article in the *News & Observer* in October 2002 discussing ARC's report "Bitter Rains", which is about aerial pesticide spraying in North Carolina. Where on the site can you download a copy of "Bitter Rains"?
- Two weeks after you saw a power company employee spraying the power lines near your front yard, you notice that some of your plants have turned brown and died. Find information on the site that discusses what the power company may have been spraying and if it is legal for the power company to spray without notifying you first.
- You want to send a letter to the state to protest about pesticide use in public schools. What state organization should you contact?

These five tasks are typical of four distinct reasons individuals would visit the ARC site: to contact the organization, to get information, to solve a problem, and to take political action. By asking participants to find this information, the researcher could analyze how easily they were able to navigate the site and understand labels and link titles. Participants were asked to think aloud while completing the tasks.

Participants next responded to eight structured interview questions to ascertain users' perspectives on their experience with and understanding of the site. The researcher asked the participants what features of the site they perceived to be particularly valuable or detrimental and what suggestions they had to improve the site. The researcher also asked the participants to state, in their own words based on what they saw and read on the site, what they understood ARC's mission, services, and intended audience to be.

During the final study component, participants completed a 25-item post-test questionnaire to measure users' perceptions of the presentation, navigation, and quality of the site. The instrument, designed by the researcher, was based on three constructs – navigation, presentation, and quality – of an existing instrument developed by Zhang, Keeling, and Pavur [642]. Although many of the survey items from the Zhang et al. instrument were retained, some were rephrased to be more relevant to the features being evaluated. Additional survey items, developed from the literature of information quality [11; 327], Web usability [388], and the effectiveness of Web sites [509], were added to the instrument. The instrument was successfully validated with a pilot study involving three individuals from the target user group who did not participate in the user study – ensuring the questions were worded clearly and that users interpreted the questions as intended [440].

Data gathered from this study was analyzed both quantitatively and qualitatively. Qualitative data were assembled from the think-aloud reactions during the browsing and task activities, and from responses to interview questions. Quantitative data included demographic information, measurement of task performance, and responses to the post-test questionnaire.

7.3 Results and Discussion

7.3.1 Demographics

Participants were all English-speaking adults with some Internet experience and included farm workers, organic farmers, parents with school-age children, educators, and Environmental Studies students of the University of North Carolina. Five females and three males between the ages of 22 and 53 participated; the average age of the participant group was 31 years. Four participants were residents of suburban areas and four were residents of rural areas. Four participants rated their level of computer and Internet use comfort as "very comfortable"; three as "somewhat comfortable"; and one as "neither comfortable nor uncomfortable". Length of use of the Internet varied considerably, from less than six months to seven years or longer.

7.3.2 Evaluation Question 1: Presentation of the Site

Measuring Web presentation requires evaluating the graphics, colors, amount of information displayed, the organization of information, and the aesthetic appeal of the Web site. Data on users' perceptions of the site presentation were collected through responses to interview questions, which asked participants to explain what they initially liked best and least about the site, and through the results to seven post-test questionnaire items. Favored most by participants, as revealed during the interview, were the photographs that supplemented information, as well as the presentation, categorization, and annotation of the "Links" page. Favored least by participants were the excessive page lengths that required too much scrolling and the size and font of the text.

The questionnaire items asked users to evaluate elements of the site's presentation (Table 7.1). On a seven-point scale, a mean of 4.5 was set as the target score. The highest mean score in the presentation section indicates that participants found the text font and size to be easy to read. Other survey items failed to meet the target.

Table 7.1: Post-test questionnaire – presentation results

Survey item	Mean	StdDev	Min	Max
Overall presentation	3.2	1.8	1	7
The logo on the site's homepage is attractive.	2.4	2.0	1	7
I find the text font and size easy to read.	4.8	2.4	1	7
Colors used in the Web site are pleasant.	3.6	0.9	2	5
The amount of information displayed per page is just right.	2.8	1.8	1	7
Graphics, icons and photos contribute to my understanding of content.	3.1	2.0	1	7
The screen layout of this site is attractive.	2.9	1.1	2	5
The organization of the information is easy for me to understand.	2.9	1.4	1	5

7.3.3 Evaluation Question 2: Ease of Navigation

Ease of navigation was measured by looking at how effectively users could move through the site to find hyperlinks and information. The task activity measured the navigational efficacy and information-finding ability of the ARC Web site. Results were analyzed by looking at the success rate of each task and by comparing the mean number of hops taken for each task with the fewest number of hops actually required to complete each task (Table 7.2).

Overall, the task activity results suggest that site navigation and information organization need improvement. Participants stated that the site did not facilitate finding the answers; they thought the information was scattered illogically within the site, which required them to search more than they would have liked. All participants mentioned that they would have had more trouble completing the tasks had they not first browsed the site.

Table 7.2: Information-finding task results

Task	Target number of hops	Mean number of hops	Standard deviation	Number of hops (min)	Number of hops (max)	Success rate
1	0	0	0	0	0	100%
2	1	3	2.51	1	8	87.5%
3	2	4	1.41	2	6	87.5%
4	1	1.5	1.41	1	5	100%
5	1	2.8	2.48	1	6	62.5%

When asked to identify what elements of the site made completing the tasks particularly easy or difficult, participants identified the navigation system as their least favorite element of the site, corroborating qualitative results from the task activity. Some of the labels on the navigation bar were found to be incomprehensible and not reflective of the content of the page to which it linked them. One participant recommended adding a text-only site map for navigation.

The navigation section of the post-test questionnaire, shown in Table 7.3, received the lowest aggregate score of the three constructs. Significantly low scores indicate that the site design was ineffective in providing location cues and that navigation bar labels and page titles did not help users understand the page content.

Table 7.3: Post-task test questionnaire – navigation results

Survey Item	Mean	StdDev	Min	Max
Overall Navigation	3.0	1.6	1	7
The labels on menu items are understandable.	2.6	1.5	1	6
Menus and embedded links are helpful in navigating the site.	3.6	2.1	1	7
I find it easy to navigate this site.	3.3	1.5	2	5
I always know where I am within this site.	2.0	0.8	1	3
No matter where I am in this site, I can easily return to the homepage.	4.0	1.9	1	7
The internal directional links are consistent on each page.	2.9	1.9	1	6
Page titles help me understand the content of the page.	2.9	1.7	1	6
Information on this site is easy to find.	3.0	1.2	2	5

7.3.4 Evaluation Question 3: Information Quality

Information quality is measured by analyzing the credibility, reliability, accuracy, understandability, and currency of the site, its content, and the organization that provides it. The content on the ARC site was favorably received by most participants. They considered the content presented to be clear, in-depth, and at a level that is easy to understand. Only four suggestions for improvement involved the quality of the site. Participants asked for more information about ARC's activities, more frequent updates, the provision of Spanish and English content, and for content to have a more urgent, social advocacy tone.

The quality section of the questionnaire, Table 7.4, received the highest aggregate of the three constructs. The results indicate that, although participants felt more information about ARC and its mission should be provided, the information included was perceived as credible.

Table 7.4: Post-test questionnaire – quality results

Survey item	Mean	StdDev	Min	Max
Overall quality	3.9	1.9	1	7
This site provides enough information about ARC's mission.	4.6	1.9	1	7
This site provides enough information about ARC's main issues.	4.8	1.7	2	7
This site provides enough information about ARC's services.	3.3	2.1	1	6
Information on the Web site is appropriate for ARC's mission.	4.7	1.8	2	7
The site's homepage helps me understand what the site is about.	2.9	1.9	1	6
This site's menu helps me understand the content in this site.	3.0	1.5	1	5
I understand for whom this site is intended.	3.0	1.5	1	5
Information on this site is current and up-to-date.	2.8	1.7	1	6
The quantity of information provided in this site is sufficient.	4.8	1.9	2	7
I trust the information presented on this site.	5.1	1.2	4	7

7.3.5 Evaluation Question 4: Communicating Mission and Goals

Three interview questions were used to determine how effective the information on the ARC site is at communicating the organization's mission and goals. These questions asked users to state, in their own words, what they understood ARC's mission, main services, and served populations to be, based on the content they saw and read on the Web site. Participants' comments were then compared with information provided by the Executive Director of ARC.

Of the three questions, participants were most accurate in describing ARC's mission. As stated by Fawn Pattison, Executive Director, "ARC's mission is to prevent human and environmental exposure to toxic pesticides, which is done through two routes. The first is outreach and education to the public – teaching people about attainable alternatives and helping them work for change in their communities. The second is a policy arm where we watchdog state government and agencies. We make sure they are acting within the law and push them to enact a better policy. Each effort complements the other" [409]. Results showed that, overall, the content on the site gave participants a generally accurate – if somewhat incomplete – understanding of ARC's mission. Every participant perceived

ARC as an educator or information-provider on pesticide concerns and issues, and two participants understood that ARC's mission is to enact change in public policy and oversee state agencies. Participants were generally unable to identify specific target groups served by ARC. Pattison explained that ARC's main target groups are the public and North Carolina residents in general, and farm workers and people who are affected by toxics [409]. While all participants identified that ARC serves the public in general, few participants correctly identified North Carolina residents, farm workers, individuals who have been chemically injured, or individuals who are at risk to pesticide abuse as part of the target audience.

The ability of participants to identify the services provided by ARC was tested in the last interview question. As set forth by Pattison, "ARC's main tangible service is providing information about what is going on at the state level; we clarify issues and maintain an independent perspective. We work with individuals who have been chemically injured – we understand the causes and severity of victims' situations. ARC can also help victims report their problems to the state and try to help them deal with ongoing long-term problems. We also educate public agencies that can serve the chemically injured, so that they can be a more effective front line resource. ARC visits groups to talk about whatever pesticide related issue they are interested in" [409]. Participants accurately perceived ARC as an information-provider, but additional services were not clearly understood. Each of ARC's services forms an integral part of the success or failure of the nonprofit's mission, and as such should be clearly expressed on the site.

7.3.6 Summary of Results

As evident from the results of the study, the ARC site was failing to effectively communicate its mission and goals due to two main reasons. First, some of the organizational, navigational, and labeling elements detracted from the value of the content and kept users from finding mission-related content. Visitors will not be able to experience or learn from high-quality content if they are unable to find it, if design elements are distracting, or if individual pages do not highlight crucial information. Second, though the quality of the content was evaluated highest by participants, the site did not include sufficient content to facilitate a complete understanding of ARC's mission and goals.

7.4 Web Site Recommendations

The study revealed a number of recommendations that could be used by ARC and other environmental advocacy organizations to create a more usable, useful Web site that effectively communicates its mission and goals:

- Accurately provide and reinforce information about the organization's mission, goals, services, and target populations. Participants in this study expected a clear, strong message about the mission and goals of ARC. Participants perceived ARC's site to focus on issues of concern to the organization rather than on services and objectives that may be of greater interest to the user. ARC should explicitly and noticeably state and describe its mission, goals, and services on its site. Additionally, it would be effective to include a short missic statement on the site's top navigation bar.

- Design a homepage that is representative of the organization. ARC's homepage, which received a significantly low rating when tested, should engage visitors and better indicate what the site is about, its purpose, and why visitors should want to delve further into the site.
- Retain the introductory, non-technical content, but add a more social-activist tone to the language. Participants appreciated that the content was at an introductory level so that novices to the subjects could understand the issues. However, a few participants suggested that the content better reflect the organization's objective of taking action and working for change. While they asked for a more influential tone, participants were appreciative that the site was not overly emotional, forceful, or radical, which can be seen as intimidating to some visitors.
- Provide a number of ways for visitors to take action. Individuals visit advocacy Web sites, in part, for suggestions and opportunities to effect change in their lives, communities, and beyond. The ARC Web site should move beyond a purely informational site and reflect the organization's objective to induce grassroots action. ARC can provide opportunities to e-mail government and regulatory offices regarding pesticide issues, provide information on viable alternatives to pesticides, or instruct parents and teachers how to recommend low-toxicity programs to their school officials.
- Decrease the amount of content on each page and link to secondary pages that provide additional information. Participants commented that pages were too long and provided more information than they wished to read. More headlines, larger spacing, and smaller chunks of text would highlight important information, facilitate scanning and reading, and help visitors to feel comfortable with the amount of data provided on each page. It would be appropriate for the main page of each section within the site to contain only introductory or summary information, with secondary pages that provide more comprehensive and detailed information for interested visitors. Information organized in varying levels of detail is effective for audiences with different levels of expertise on the topic.
- Develop a section containing ARC's publications and reports. Decision-makers, members of the media, and interested individuals should be able to quickly locate a non-profit's publications and materials.
- Develop a section containing news articles about ARC. As environmental organizations must prove themselves to be reputable and worthy of the public's involvement, providing published news articles about the organization and its activities can show visitors that they are an active organization that is making efforts to effect change.
- Provide content in Spanish as well as English.
- Provide a number of ways for visitors to contact the organization and/or receive newsletters or e-mail updates.
- Provide staff biographies and pictures, if appropriate, so visitors can get a sense of who is running the organization.

7.5 Conclusions

The goal of this study was to determine if ARC effectively communicates its mission and goals through its Web site. This user study achieved its research goals in that it identified a number of strengths and weaknesses in the design of the site, and participant comments and actions provided insight into how visitors perceive the site. Other environmental organizations can learn from these research goals, methods, and results to develop and conduct their own usability studies, and use the recommendations resulting from the study to design an effective, mission-supportive Web site. The recommendations conceived as a result of this study guided a redesign of the ARC site. The new site with revised content, design, and information architecture is located at www.pested.org [645]. The site on which the usability study was conducted can be viewed at www.pested.org/archive. A usability study, using the methodology outlined here, on the new ARC site would be an appropriate next step to determine if the revisions are indeed successful in promoting ARC's mission and goals. Likewise, usability evaluations conducted by other environmental organizations on their Web sites will bring us closer to understanding effective Internet strategies and design features for non-profit environmental advocacy organizations.

Acknowledgements

The author thanks Fawn Pattison, Michal Percival, and Barbara Wildemuth for their support, advice, and assistance. The author also thanks the Interaction Design Lab at the University of North Carolina for use of their facilities.

Chapter 8
EFS Global Media Network: Exploring the Boundaries of Sustainable Education

George R. Smith

Abstract. *Information and education on the current status and future viability of human life on Earth is of vital importance. There is an urgent need to provide reliable information on this planet's life support. The Education for Sustainability Global Media Network (EFS) project investigated three primary research questions, including development of management strategies for online digital learning networks, investigating environmental impact of use of technology in learning environments, and determining critical advantages and disadvantages of employing digital media in online learning networks in a global context. One of the major results of this research is the assembly and synthesis of a sizable body of information on pedagogy and curriculum development for online education focused specifically on sustainable education.*

8.1 Introduction

In 2001 an ambitious research program was conceptualized, involving a partnership of organizations and individuals with similar interests, all focused on sustainable education. This collaborative effort was envisioned as being global in scope involving a diverse group of educators as a method of creating an international education for sustainability network. This year and a half long program commenced in January 2002 with a focus on exploring issues and opportunities related to computer-based sustainable education. The following discussion addresses this research process and critically examines related findings and conclusions.

Dr. Rita Colwell, Director of the National Science Foundation, recently stated that environmental education is of utmost importance to the world and will determine whether our future is sustainable [116]. The basic foundation for the *Education for Sustainability (EFS)* vision is the belief that the global learning context of the future is achievable in the near future provided that the will and the means exist to explore its boundaries and opportunities. The underlying premise of this research is the concept of sustainability. In a world of decreasing resources, technology increasingly affords us the means to investigate and provide unrestricted access to global learning opportunities. Accessible and usable environmental information of high quality is essential for effective decision-making at a grass roots level in society [569]. Priority must be given to providing learning opportunities that are intellectually compelling, timely and feasible, and important to the progress of knowledge and of society. Contributions and collaborations must be facilitated, involving many fields [433].

The EFS Network addresses these goals, focusing on the critical gap between complex environmental information and its accessibility and usability for the non-scientific community. An examination is provided of the context of education for sustainability and use of digital media for education in a global environment characterized by both technical innovations and technical inequities. Case study examples are presented and analyzed. Arguments are presented for the need for a rigorous pedagogical approach, highlighting key issues, advantages, and disadvantages of utilizing digital technology for learning, with specific reference to the problems and potential for sustainable education. Conclusions are provided related to a proposed framework for computer-based education for sustainability that is grounded in science and built on a solid pedagogical foundation. The EFS Network is intended to be a central clearing house that provides a portal, a range of learning options, and a global connection linking resources and providing opportunity for communication on sustainable education. The development of a unique pedagogical basis is a key component of this endeavor alongside its potential for success in the dynamic emerging context of both education for sustainability and global digital learning environments.

8.1.1 The EFS Concept

In the context of this discussion, the term "sustainable" is simply defined as balancing the environmental health with economic and social needs of the Earth's current population in order to ensure that future generations can similarly achieve this balance. Providing educational opportunities in an accessible and timely manner is an important tool in achieving sustainability. The EFS Global Media Network proposed a global education network that employs technology to enable a rich learning environment for sustainable education. Given this context it is interesting to note that some researchers use ecological metaphors to describe appropriate methods for developing and delivering the information technology curriculum. For example, Tatnall and Davey [555] state that concepts of ecology address complexity and interaction that provide an appropriate framework for research with the goal of testing and developing the information technology curriculum. The EFS has proven to be a complex and interrelated concept, necessitating that a clearly defined foundation for concept implementation be developed. The EFS vision of a technology-based "global university" offering almost unlimited possibilities for teaching, research, and learning at a local-to-global level was substantiated. Powerful learning / teaching opportunities can be facilitated by technology. On the other hand, these findings also identify a paradox between the advantages of computer-based education for sustainability and negative impacts on the global environment resulting from use of the technology, illustrating clearly that some gaps exist between EFS concept development and implementation.

The current research need is to take advantage of emerging and dynamic opportunities in education for sustainability in a timely manner. The National Science Foundation (NSF), a major driving force in scientific research in the United States, has recently received a significant increase in government funding. In January 2003, NSF published "Complex Environmental Systems" [433], a ten year outlook expressing a commitment to research and education in the numero complex areas where human systems and natural systems overlap. This publication establishes a National Science Foundation agenda for a science-based approach to education for sustainability. Closely related, the United States National Council for Science and the Environment (NCSE) has recently published its "Education for Sustainability and Security Conference Report", which contains recommendations that "collectively form an educational agenda for the United States and for the UN Decade of Education for Sustainable Development, slated to begin in 2005" [384]. These events are evidence of the emerging awareness of the importance of achieving a major goal established a decade ago with the publication of Agenda 21 [569], to facilitate the availability of accessible, usable, and credible information at a "grass roots" community level as a basis for effective sustainable decision-making.

8.1.2 Primary Research Questions and How They Were Addressed

8.1.2.1 Stage #1 Research Questions

The EFS Network research program was funded by the Ball State University iCommunication Global Media Network initiative. During the process of refining the original iCommunication project, three key research questions were identified for purposes of focusing the research process. These questions have remained substantially the same throughout the research program, providing a broad framework for structuring the research program during Stage #1. They are listed as follows:

- How can the EFS Global Media Network be managed?
- What can be learned from the research and development process about advantages and disadvantages of digital learning, especially in a global context?
- How "green" is the technology?

These questions were addressed using several methods. A review of literature and Web-based sources was initiated and became an ongoing process throughout both research stages. Exploration of precedence and models was undertaken. Consultations were initiated with internal and external experts in areas of technology and education. A tool that was regularly used was "group brainstorming", including undertaking the International Think Tank discussed below. Data assembly and synthesis was also ongoing throughout the research process, via the literature review process and related methods. Network conceptualizing and preliminary design development was fortified with in-field testing of concepts and technology.

8.1.2.2 Stage #2 Research Questions

Based on Stage #1 findings it was necessary to expand and refocus the original key research questions to provide specific answers in response to issues identified. Consequently the original research questions were expanded in Stage #2 to include the list of questions in Table 8.1.

Table 8.1: Summary of Stage #2 research questions and related testing methods

Research question	Testing methods
What is currently achievable and what is not?	Determined substantially by using the prototype EFS Network, including field-based Web casts. Delivery and effectiveness of digital knowledge was specifically tested by user surveys.
How successfully can the current constraints be overcome?	Explored by various methods that tested and compared effectiveness of in-class versus field study versus digitally-based learning experiences.
How can current opportunities be maximized?	Substantially addressed through the range of testing proposed.
How far can the limits of the existing condition be stretched to accomplish the long-range EFS Network vision?	Future testing and comparing a wide a range of curriculum delivery to identify resolvable versus non-resolvable limitations, to be used as design and development parameters.
How can this type of Network be managed and coordinated?	The next stage will examine management techniques for educational online delivery systems. Testing results were analyzed and recommendations made for system management & coordination.
How can the extent of "green" of the proposed Network be measured?	Energy and materials consumption data was researched. Traditional classroom delivery scenarios versus virtual classroom situations were compared. Environmental impact of computers was cross-compared with learning opportunities and benefits as tested in Stage #2 as a basis for determining how "green" EFS is.
What is learned about the learning advantages and disadvantages of digital media, especially when communicating with a global audience?	Effectiveness of virtual field study versus actual field study was tested and student experiences / learning results compared. Survey and interview techniques were and will be tested and compared related to attitudes towards digital media, combined with traditional aptitude and comprehension testing of curriculum delivery. Future phases will test and compare EFS curriculum delivery in a range of global delivery contexts and scenarios.

8.2 Discussion of Issues and Opportunities

The ecological metaphor referred to earlier has also applied to the process that the EFS Network research has followed. The EFS concept evolved significantly from the concept of a clearinghouse of global information on sustainable education over the year and a half duration of the research program. During this time education for sustainability "officially" entered the mainstream of national research and education policy in the United States. Technology has continued to evolve rapidly, fueling the "burgeoning visual culture" and its radical alteration of the dissemination and production of information and knowledge [207]. The conclusion is that given the issues and opportunities revealed in this study there exist extraordinary opportunities to contribute to this dynamic evolutionary process.

8.2.1 Issues Identified and Addressed

Towards the end of Stage #1 several primary issues began to emerge related to the lack of a pedagogical basis for the Education for Sustainability concept and related curriculum development, and second, the apparent paradox between the underlying principles of education for sustainability and the environmental implications of use of computer technology. On completion of Stage #2 and the ploration of these issues, major findings and conclusions were reached. These were grouped into four topic areas that address the three original research questions, including conclusions on managing the global media network, advantages and disadvantages of using digital media for learning, environmental impact of technology, and discussion of a pedagogical foundation for technology-based sustainable education.

8.2.2 Managing the EFS Network

The long-term viability of the Education for Sustainability Network was a primary concern of this research since its commencement. In Stage #1 this issue was addressed substantially during a week-long brainstorming session with the research investigators and consultants David Wilde from the Earth Centre, Doncaster UK, and Glenn Strachan from GESEarth Sustainable Education consultants, Wales UK. The overarching conclusion was that development of a sound business plan is the best way of ensuring the EFS Network has long-term success. The need for a sound business plan became obvious in the early phases of the project. It is of primary importance that future research on the EFS Network development includes some form of market testing undertaken with partners who contribute to decision-making on the implementation of the EFS, identifying costs and income sources.

To date it has been identified that the potential sources of income will conceivably provide a diverse economy of revenue, grants and sponsorships, including revenue from introducing a subscription service for members of the EFS Network. There could be a range of membership options depending on the types of services the member wishes to access and whether the member is an individual, a private company or an academic institution. Other feasible income sources include revenue from selling products and services such as teaching / learning modules. Revenue from Web advertising is also a possibility, although not without its inherent limitations, contradictions and pitfalls. Engaging a digital media partner capable of supplying technological assistance and funding additional research and development is also a clear possibility, as is engaging an "appropriate" commercial organization to sponsor the EFS Network. As a starting point, applications for funding to relevant trusts, foundations and grant-awarding bodies in the countries of the various partners are currently being prepared.

8.2.3 Advantages and Disadvantages of Digital Learning

On May 15th 2003, the final Stage #2 testing of the technical concepts and capabilities of the EFS Network was undertaken. Three Ball State University faculty traveled to the UK to pursue individual research efforts, with the agreement that each would also conduct field research and prepare seminars on the historic urban and architectural characteristics of the City of Bath, England. The facilities to stage and Web cast the seminar were located in Poundbury UK. The faculty prepared and delivered half-hour lectures utilizing PowerPoint presentation techniques. The seminar was Web cast live back to an audience in Indiana at Ball State University. Staff and faculty, and a group of 39 first year college undergraduate students participated. A survey of learning experience was conducted immediately following the event. Each of the 39 participating students answered questions related to use of videoconference technology, live Web cast interaction, and quality and value of the overall learning experiences. The Poundbury Web cast seminar was revealing and informative from several perspectives, including what the survey results revealed about the learning advantages and disadvantages of using digital media on a global basis, and environmental implications of computers used in education.

8.2.3.1 Learning Advantages

Research results substantiate the following conclusions. Burbules and Callister [79] argue that technical systems for organizing and presenting educational information challenge traditional notions of learning experience, for example, what "text" is and what it means to read different media or information sources. Further, the quantity of presented information can have a direct effect on the quality of learning. These can be positive factors. In an article on Web-based distance learning for education for sustainable development, Eastwood [145] states that virtual courses and E-Learning offer several advantages including "space flexibility" (the ability to participate from any place that has access to the technology), and "time flexibility" (the ability to access information and study any time). Cohen [113] elaborates on these points, adding that there are multiple advantages to anywhere / anytime E-Learning, including the option to participate real time in lectures, chat rooms and videoconferencing (synchronous learning) and alternatively, the option for students to access video materials and records of the lecture online at their convenience (asynchronous learning). Even the more critical voices in the E-Learning debate admit that the use of computer technology for education offers valuable opportunities to enhance all types of teaching, and current issues and limitations will be overcome in time [405].

Several key learning advantages and disadvantages of use of digital media were illustrated in the Poundbury experiment, which was intentionally a process of delivering a clearly defined package of information in a straightforward, simplified manner employing commonly used lecture delivery tools such as PowerPoint. Given this context, a total of 67% of students surveyed felt that their learning experience was enhanced by the use of technology. Of those surveyed, 70% felt that the ability to interact with the "long distance" presenters was a positive contribution to their learning experience. Not surprisingly, a total of 86% of survey participants felt that the use of graphic images was useful in illustrating key points of the presentations. A total of 63% believed that video and audio contact with presenters in another country was the most interesting feature of the research experiment. Related to this, 45% of those surveyed felt that live real time video of the location being discussed (Bath) would add value to the learning experience. Conversely, a total of 44% felt that the question and answer session following the presentations was the least interesting component of the seminar.

It was concluded that primary learning advantages of use of digital media include flexibility, accessibility and convenience. The ability to access and participate "anytime / anywhere" in the educational experience is a primary advantage. The digital learning experience has a distinct and dynamic nature because of th- ability to employ and utilize a range of options including multi-media, discuss-group format, lecture presentations, group or individual participation. The student has choices not usually found in traditional course delivery, which enable customization of the learning experience to suit individual lifestyles. In addition, technology-based education offers the potential for sharing educational resources from many sources accessed from one location, for example, taking related courses from multiple universities towards one degree.

8.2.3.2 Learning Disadvantages

Burbles and Callister [79] argue that technologies powerful enough to do good on a broad scale are also dangerous and add that information technology is particularly susceptible to over-promise and hyperbole. The findings of this research indicate that issues of information credibility, accessibility and usability are commonly cited as disadvantages of the use of digital media for educational purposes [79; 113; 207]. The question of accessibility brings to the forefront the related issue of inequity due in part to the rapidly evolving capabilities of the technology and conversely, uneven distribution of current technology resources. Further to an issue raised previously, as the commercialization of the Internet advances online education may face financial constraints that will necessitate seeking commercial funding to stay viable, a situation which has inherent negative aspects that are in conflict with educational goals. The issue of cyberspace privacy is also increasingly problematic [79].

On a final note, the Poundbury experiment highlighted several potential disadvantages to technology-based learning, including the danger that technological innovation could substitute for quality in the learning experience, and the potential for the instructor to become disconnected from the students. For example, only 25% of respondents felt that the opportunity to discuss the presentation with the on-screen presenters was the most interesting feature of the seminar, whereas 63% were of the opinion that visual and audio contact were the most important features of the seminar. In addition, 44% of participants felt that the question and answer session at the conclusion of the Web cast was the least interesting feature of the Web cast seminar.

In summary, primary learning disadvantages of the use of digital media include the potential for exaggeration and over-promise (for example, as a method of selling online courses) negatively affecting the learning experience. Also there are issues of credibility, accessibility, and usability of information sources for online education. There also exist issues of equity of distribution of technology resources among potential users, which can be a limiting factor in distribution of the educational materials. Disadvantages related to issues of information privacy and potential pitfalls of commercializing online education are identified as problems associated with computer-based learning. Physical isolation of each student from other students and instructors is a potential problem, as is the danger of technological innovation substituting for quality of learning experience. On a cautionary note, a study of student learning situations comparing interaction with computers versus observation concluded that the level of participatory interaction with computers significantly depends on individual student characteristics, as opposed to group dynamics [510].

8.2.4 The Environmental Implications of Technology

There is little doubt that the manufacturing, use, and after-use of computers have negative impacts on our environment. McDonough and Braungart [361] point out that the average computer contains more than a thousand different kinds of materials including toxic gases and metals (such as cadmium, lead, and mercury), acids, plastics, chlorinated and brominated substances. Regardless, there is every indication that computers in one form or another will continue to be widely used. In addition, there are clear advantages to computer-based education, which appear to balance the disadvantages.

The issue is not whether we should or should not use computers for educational purposes. Rather, the issue is how computers are manufactured, used, and ultimately disposed of. Given their widespread use and eventual disposal there is strong indication that environmental and associated human health issues linked to computers are a real and growing concern.

8.2.4.1 Environmental Issues Associated with Computer Technology

As early as 1996 credible research results indicated that the extent of resource consumption related to computer life cycles is clearly illustrated by examining computer manufacturing. Producing a 2-gram microchip consumes approximately three and a half pounds of fossil fuel and chemical inputs, plus approximately ten gallons of water and one and a half pounds of elemental gases, mainly nitrogen. The average computer components factory in the United States consumes 20 million gallons of water per month that becomes a "waste product" [232]. These findings were collaborated as recently as 2002 by the United Nations University in Tokyo Japan. Dr. Eric Williams, the research team leader, reported that computer chip manufacturing requires 700 times the weight of the original chip, totaling 1,400 grams of materials and fossil fuels to produce one 2-gram 32MB microchip [175].

On the positive side, the University of Arizona reports advances in research on chip fabrication processes, which hold promise of greatly reducing fluid waste during manufacturing [637].

Once produced, computers continue to have an impact on the environment, consuming resources such as energy, paper, chemicals, plastic, metals and water. Toner cartridges are a significant concern. Ongoing research of university facilities by the author reveals that many ink toner cartridges are not recycled. They contain among other things, health hazardous nickel, cobalt, and mercury. Regular inhalation of these poisonous "ingredients" has potentially negative effects on user health [361].

Computers are also a consumer of energy. In early 2002 Dr. John Vann of Ball State University performed several tests on the computer use of energy in faculty offices. The BSU test results show that the cost of energy consumed to keep 1,000 machines *and monitors* on continuously for one month would be approximately US$ 1,800.

Some of the most interesting claims come to light by examining the disposal of used computers. Fisher [182] claims that a 15-inch computer monitor contains approximately five to eight pounds of lead per unit. It is estimated that dispos computers and related electronics contribute 40% of lead found in landfill sit the United States. The positive news is that companies specializing in computer recycling are slowly growing in number because of recent innovations in recovery of metals such as lead and copper, as well as reclaiming more profitable metals including gold and platinum from central processing units.

8.2.4.2 Footprinting a Computer

In order to evaluate the environmental impact of the use of computer technology in sustainable education and draw conclusions, it is advantageous to have a tool for measuring that impact. "Ecological footprint analysis is an accounting tool that enables us to estimate the resource consumption and waste assimilation requirements of a defined human population or economy in terms of a corresponding productive land area" [595, p9]. Ecological footprinting is not a crystal ball but can clearly indicate the relationship between resource consumption and environmental impact caused by that consumption, and as such is an appropriate tool for illustrating the relative environmental impact of using computers. Following is an example of the application of the ecological footprinting concept to illustration of the environmental impact of computer use in sustainable education.

Judging by the Poundbury seminar post-Web cast evaluation survey, the participants found their learning experience to be rewarding and interesting. But what was the environmental impact of this educational experience relative to a more conventional seminar delivery? To answer this, a representative ecological footprint was calculated for the Poundbury / Bath test and compared with estimated footprints for seminars using more traditional educational methods. Two factors have been used to estimate the ecological footprint of the test seminar and compare this to traditionally delivered seminar, these being energy consumption for computer use, and energy consumed for travel combined with degraded lands impacted by travel.

The assumption is that both the Poundbury and Ball State University computers were approximately equivalent, and were on with monitors for 24 hours per day each, with total energy consumption 5.6 kWhs. Based on footprint calculations, computer use for the Poundbury-to-Ball State test would have a negligible footprint of approximately 0.00062 hectares. For the Poundbury experiment, travel by the three faculty members included air and ground transportation. For comparison purposes the following scenarios are offered. In "traditional" seminar scenario #1 all students would accompany the three professors on a field study trip to England involving greatly increased travel. Computers would not be a significant factor in this scenario. In scenario #2 all students and the three professors would participate in an online seminar delivered by an appropriate individual from the Poundbury UK facility. This scenario would greatly reduce travel. The underlying assumptions for scenario #2 are that the participants at Ball State would each travel by automobile to and from the seminar, and that computer usage at Ball State would be equivalent to the Poundbury Web cast test location. The individual in the UK who hypothetically researched Bath and delivered the seminar from Poundbury would travel by train. Table 8.2 offers a summary of ecological footprint assessment calculation results for the Poundbury / Bath experiment and the two hypothetical scenarios.

Table 8.2: Ecological footprint assessment for computer use in sustainable education (in hectares)

Scenarios	Computer use	Travel by car	Travel by train	Travel by air	Seminar session
Poundbury Web cast test (3 faculty travel to UK)	0.00062	0.08	0.02	2.3	2.4
Scenario #1 (42 students & faculty travel to UK)	n/a	1.05	0.30	32.26	33.61
Scenario #2 (42 students & faculty stay at Ball State)	0.00062	0.02	0.01	n/a	0.03

8.2.5 Towards a Pedagogy-based EFS Curriculum

The EFS research process has uncovered evidence of significant global interest in information technology-based education, and also documentation of extensive research efforts in information technology and information systems education focused on advancing the pedagogical basis of those disciplines [89; 113; 564]. Similarly, this research has gathered significant evidence of the early stages of emergence of an educational framework and related curriculum for sustainable education [494; 559; 596]. Literature is also available on technology-based sustainable education [145], although this is a less-published area.

The conclusion is that in order for the EFS Network to be successfully implemented and remain sustainable in the long term it is necessary to define a pedagogical basis for technology-based sustainable education. However, to date few suitable models have been found to build on. The EFS research process has relied on existing successful models and related precedence. The conclusion is that this research must address this gap by proposing a pedagogical basis for computer-based sustainable education. Consequently, extensive research has been undertaken on existing pedagogy and its relationship to technology-based sustainable education. Goldfarb [207] asserts that pedagogy "draws attention to the process through which knowledge is produced. Pedagogy addresses the 'how' questions not only in transmission or reproduction of knowledge but also in its production". With this as a starting point the question would be how technology-based sustainable education knowledge can be produced and disseminated.

8.2.5.1 Literature Review

During the last year and a half an extensive literature review has been undertaken. A major focus of that effort has been investigation of models and precedence for creating an online education for sustainability network. Where possible, meetings have been held with universities and institutes in the United States and Europe that currently offer successful online educational opportunities in sustainable education and related educational areas. Based on this work, selected examples of existing sustainable education courses and related courses on education were analyzed with emphasis on curriculum delivery characteristics, learning emphasis, and learning strategies employed by each program, as a basis for proposing a foundation for an EFS pedagogy / curriculum. This analysis was further strengthened through cross-reference with several recent publications on Education for Sustainability [385; 433].

8.2.5.2 The Principles and Values Underlying Online Education for Sustainability

There are common principles and values to be drawn from examining examples of online course offerings as explored by this research. Collectively these examples indicate patterns and characteristics that a well-designed online Education for Sustainability Network will incorporate. For example, flexibility and choice is a desirable characteristic as is the ability of a student to participate anywhere and anytime. The ability to customize a course of learning to one's lifestyle and learning style is very attractive in an online course offering. In summary, considering the unique context within which an EFS curriculum would be offered online, there are several fundamental conclusions to be drawn from this review, providing an informed starting point for proposing pedagogy for online sustainable education. The values underlying the concept of sustainability should be inherent in education for sustainability. These values include:

- The interdisciplinary nature of sustainability
- The rich complexity of sustainable issues
- The scope of sustainability from local to global
- The interconnected nature of the context of sustainability
- The exploratory nature of seeking the sustainable balance.

Based on this research a learning environment characterized by seven principles and the underlying values that support these principles will be provided by the EFS Network:

- Accessibility to information and learning tools
- Usability of the learning environment and the information provided
- Reliability and credibility of information provided as a basis for dissemination of knowledge and educational services
- Flexibility of choice
- Accommodation of different learning styles
- Facilitation and promotion of communication of information, ideas, and knowledge through motivated interaction and participation
- Personalization of the learning experience to the individual student through customizing that experience, and where possible, grounding the learning experience in the local context.

8.3 Conclusions

The journey towards the global university of sustainable education of the future will, without doubt, have bumps in the road. The EFS research program had its beginnings in many related areas of academic interest. The research process generated and investigated three questions. The journey is not over. Phirman [433] states emphatically "anecdotal evidence indicates that teams that sequentially develop new theories, integrate knowledge and practice, and conduct experiments generally require eight to ten year financial commitments to produce the best results". There is more work to do. However, the EFS Network framework has been firmly established to address these opportunities.

The next important research question may be related to the issue of quality of learning experience. The findings of this research indicate that there are learning advantages as well as learning disadvantages to be derived from computer-based education. Based on the initial testing conducted there is good indication that the quality of learning experience is enhanced by use of computer technology. But which of the three scenarios discussed above will produce the best quality of learning experience? Evidence to support an answer to this question is a missing component in any discussion of the use of computers for sustainable education. Sustainability is a balance between social, economic, and environmental concerns [633]. Further research is necessary to explore several primary issues related to technology-based education for sustainability, including research on the quality of learning related to use of computers in the sustainable education process. In summary, it can be argued that sustainable education offers rich opportunities for us to explore the complex links between human systems and natural systems, and in this context a further more in-depth exploration of issues related to use of computers in sustainable education would be a very rewarding journey.

Acknowledgements

The author would like to thank the following individuals and institutes for providing major funding support for this research: Prof. David Ferguson, Director of the Center for Media Design, Ball State University; Prof. Robert Koester, Director of the Center for Energy, Research, Education, Service, Ball State University, and the Center for International Programs at Ball State University.

Chapter 9
An Online Community of Learners in Rural, Regional, and Remote Australia

Fiona M. Haslam McKenzie

Abstract. *Many rural, regional and remote areas of Australia have struggled with environmental degradation for some time. At the same time, it has been a challenge to maintain regional development and environmental management professionals in the field because of a concern for lack of professional support and a dearth of opportunities to formally upgrade professional skills. The Institute for Regional Development at the University of Western Australia has developed an online post-graduate program that offers opportunities for professionals to remain in their communities while also upgrading their professional qualifications. The unexpected benefit of the program has been a nationally based network of students working in the area who are openly discussing issues and working together to find solutions to environmental problems irrespective of location.*

9.1 Introduction

This chapter will outline how an online postgraduate program is meeting the needs of people living in non-metropolitan locations who do not normally have access to university education or training facilities but who wish to keep abreast of environmental trends and education programs while remaining in their rural, regional or remote locations. Through the development of an online community of learners, the students are creating a network of professionals across all regions and states of Australia. Using communication technology, they are discussing mutual areas of interest, developing problem solutions, and offering ideas about each other's regional environmental problems. The collegiality developed through the program is overflowing into their work lives where they are using the networks developed to integrate environmental solutions across regional and state boundaries. This is highly cost effective, developing better integrated sustainable development programs across government and non-government agencies.

9.2 Background

Australia is a highly urbanized society with 64% of the population living in one of the seven capital cities [29], however, non-metropolitan Australia contributes more than half of the national export income [292]. There is broadly documented evidence that people living in non-metropolitan Australia experience distinctive economic and social problems [30; 115; 228; 294] including restricted access to ertiary education opportunities. Rationalization of services and infrastructure tinues to be a key strategy of governments to achieve cost restraint, economies of scale, "efficiency" and effectiveness of service delivery. A survey of services in the Australian non-metropolitan areas shows that there has been a decline in the level of service provision and general infrastructure in the region over the past three decades [539]. The provision of government infrastructure such as transport networks, postal, health and education services, as well as corporate services such as banking have been referred to as crucial for the survival of a viable rural Australia [277; 556; 563; 582]. It is recognized that improved educational outcomes can increase employment opportunities, but it is also accepted that rural education facilities in non-metropolitan Australia are worse than those available in metropolitan areas [115; 198].

9.3 Online Environmental and Sustainable Agricultural Post-graduate Program

The economics of sustainability and therefore of regional development, go beyond the efficient allocation of scarce environmental resources and should include equity and social values as well as acknowledging the role of uncertainty and irreversibility [251]. Access to environmental education is therefore an integral part of the sustainability debate.

In 2000, the Institute for Regional Development at the University of Western Australia developed a post-graduate educational program dedicated to rural, regional and remote communities utilizing communication technology. It is providing up-to-date environmental education and training for practitioners in the field through a wholly online program, although educational facilities and the teaching of theories and trends is only a small part of the educational process. Practical depth to the educative program has been fostered through interactivity between students and with key environmental policy and decision-makers from government agencies, who are actively involved in mentoring the program. The overall aim is to develop a community of online learners [466] whose expertise, interests and concerns will contribute to the learning of others in the program and the ongoing sustainable development of rural, regional and remote Australia.

The program was developed in partnership with the Western Australian government as a response to a recognized need to maintain skilled people in regional and remote areas, while also providing the opportunity for people in non-metropolitan locations to upgrade or broaden their academic qualifications without having to leave their community. By providing an online learning facility, the Institute for Regional Development is responding to unique Australian regional conditions; typically, small communities separated by large distances with limited infrastructure and tertiary education facilities. Currently, 42 students are enrolled in the programs and this is growing annually.

WebCT is the online platform utilized in the teaching program with few modifications. The lecture material is designed to give a basic outline of the current issues associated with the topic and poses a number of research activities or tasks, directing the student to do some independent work, exploring issues in their own communities and regions. Lectures are prepared by Institute for Regional Development staff and recognized experts in the particular lecture topic area. In this way, students are provided with a variety of perspectives and views from government practitioners, academic experts, expert consultants working in the area

researchers. Their research and the experiences of applying the theory are posted to the asynchronous bulletin board for general discussion on a weekly basis. It is this bulletin board that excites the most academic and practical activities and exploration of solutions to regional development problems. The Masters of Regional Development program is constituted of four course-work units, followed by a research component whereby students investigate a regional development issue or develop strategies for the more percipient implementation of regional development that will satisfy the economic, socio-cultural and environmental development of rural, regional and remote Australia. This research is developed into a thesis.

In order to encourage a community of learners, the asynchronous bulletin board has been a vital element of the program's delivery platform. Problems, issues and debate topics based on the lecture material and topical environmental events are regularly posted in the course work and students are expected to respond to these through the asynchronous bulletin board. Generally, there is an expectation that the topics discussed will require a practical solution based on the theory being presented in the lecture material. It is in this forum that students discuss their professional responsibilities, environmental issues and the challenges they face on a daily basis to achieve ongoing sustainable development. Networks are developed and strengthened through this facility and innovative ideas are discussed and debated, many of which are later trialed in practice. Being asynchronous, the students can access the bulletin board whenever convenient, irrespective of location, time differences and other commitments. The collegiality developed through the asynchronous forum has been exciting, particularly when it is appreciated how distant students are from each other.

The success of this program spawned another. In 2001, the Department of Agriculture, a Western Australian government agency identified several areas of need for departmental staff in regional communities. In partnership with the State government agricultural agency, the Institute for Regional Development developed a Graduate Certificate in Regional Agricultural Development Planning, an accredited award for practitioners in a wide range of agricultural domains. The award was based on the need to develop an approach to land-use planning that is consistent with the approach outlined in the Food and Agriculture Organization's guidelines. Again, students are separated by great distances but face many of the same challenges, working to stall environmental degradation from poor agricultural practices and developing strategies that will maintain the viability of the agricultural industry while also improving the ecosystem health of the environment that supports that industry.

9.4 Sustainable Environmental Issues

Environmental issues across Australia incorporate a wide diversity of problems and "what may once have been dismissed as 'green' issues are now central to economic development" [618, p24]. The areas of high priority are the overall health of the ecosystem, land and water salinization, the maintenance of biodiversity, erosion, land contamination, loss of fringing vegetation and stratospheric ozone depletion [154], none of which can be maintained or are cheaply ameliorated. Land degradation is estimated to cost the nation approximately $3.5 billion annually.

There is a broad diversity of approaches to dealing with environmental issues which increasingly acknowledge the inter-related nature and multiple cross-linkages of the issues. As a consequence, environmental, social and economic policies are constantly shifting and evolving and within Australia there is a "huge diversity of sustainable regional development initiatives in terms of their approach, organizational structure, powers and resources. Improving on them by sharing experiences, suggestions, challenges and analyses makes economic, social and environmental sense" [137, p161] has been an overall goal for the Institute for Regional Development.

One of the key environmental issues identified by Institute for Regional Development students is the need to involve local people in the resolution of local environmental problems. One of the unexpected outcomes of the teaching program has been the ongoing discussion through the asynchronous bulletin board of strategies to actively engage with local stakeholders in environmental rehabilitation and awareness programs. Students from diverse communities have shared their ideas, what has worked for them and what has not and why different initiatives have or have not worked. Other students have taken up the suggestions, adapted them to their own communities and issues and then regularly reported progress on the bulletin board. A particular success has been the knowledge shared by students who regularly work with indigenous groups, sharing insights regarding protocols and cultural understandings in working with indigenous people and incorporating traditional environmental practices with scientific knowledge. Strategies for building trust and developing two-way communication with indigenous people have been picked up by other students in the learning group and and suggest useful community outcomes in the future.

9.5 Conclusions

Australia's biodiversity is recognized as highly significant globally, and its conservation is a prerequisite for sustainability. Furthermore, much of Australia's wealth and sense of identity comes from the use of natural resources. Concerns about sustainability and long-term ecosystem health are now prominent policy issues at every level of national and even international government. In rural, regional and remote Australia, how we use natural resources now and in the future will influence the economic and social health of the regions. There are, however, significant challenges for environmental agencies and professionals to provide integrated, cost-effective solutions over very large and sparsely populated areas.

The commitment by the Institute as a resources centre for regional communities and the determination to develop enduring relationships has led to the development of strong relationships throughout Australia. This has come about because students in the Masters of Regional Development program, all of whom come from diverse backgrounds and are currently employed in all three tiers of government and the private sector, are sharing information for their academic needs which inevitably overflow into their work environments. Consequently, this has facilitated improved dialogue between government agencies, across regions and even across state boundaries, greater resource and information cooperation and generally, a cost-effective and better integrated approach to environmental solutions for what is a very costly national problem.

Chapter 10
Self-assessment of Consumptive Behavior Based on Material Intensity

Andreas Niederl • Roman Mesicek

Abstract. *This chapter describes the development of a questionnaire for the self-assessment of consumptive behavior. After pointing out the importance of the material flows created by society, we argue that one way to reverse the current trend is a change in consumption patterns. An important precondition for such a change is problem awareness. Offering online information can contribute to such awareness. Therefore an online questionnaire has been developed identifying direct and indirect resource consumption. The questionnaire allows to spot the material flows caused by their consumptive behavior in selected areas and inspires to rethink current practices. Material flow analysis (MFA) serves as the methodological basis and material intensity per unit of service (MIPS) as the respective indicator. The consumption clusters with the highest relevance are those which are environmentally significant and can also be influenced by the individual. After identifying these clusters, the process of creating the questionnaire is presented.*

10.1 Introduction

Economic production has been growing enormously since the beginning of the industrial revolution. This growth not only led to a high standard of living for most people in the Western world but also to a very high use of natural materials. These materials are provided by the natural environment, which makes it the ultimate source for all production processes as a source of raw materials and a sink for them after their use in human society. The natural environment is not only the source of production – it is also a necessity for human life. And while the scale of material flows continues to grow, the global ecosystem is limited [239]. As human intervention in the natural environment has consequences that cannot be foreseen, it is important to lower the absolute amount of environmental consumption, thereby decreasing material use and the impact on nature [239]. Strategies to achieve dematerialization, mainly of the industrialized Western society, should aim at increasing the resource productivity and changing the current consumption patterns to provide the current wealth with less resource input.

The current amount of environmental consumption at a level of roughly 76 tons per person and year in Austria [6] is a consequence of the way the environment is used by both consumers and producers. Consumers play a crucial role: the use of a mobile phone for 30 minutes a day on average, for example, creates annual material flows of 1.13 tons. Considering consumption as the ultimate rea- for all production processes shows the importance and high influence in a national economy. Even if consumers cannot influence all production processes their consumptive decisions are nevertheless relevant. At least they can influence the range of products that are produced and to another, in fact much lower, degree they can promote the production of products and services with low material intensity by consuming them.

It is important that people become increasingly aware of the fact that the material flows of their consumptive decisions are relevant and that they get an idea of how intensive their use of the environment really is. To raise the awareness of individuals regarding their consumptive behavior, a questionnaire was created. The aim of the questionnaire is on the one hand to visualize the material rucksack the people are creating by their actions and on the other hand to present various possibilities for reducing their resource consumption.

10.2 Method

Various methods exist on how to grasp the biophysical basis of societies [123]. Material flow analysis (MFA) turned out to be one of the most important methods [169; 204]. Within the framework of this project other methods were also assessed but to simplify the creation of the tool and make the results easier to understand, MFA was chosen as the single method.

MFA allows to measure the total amount of resources which are required to execute and maintain specific human activities. These activities can be the production of a certain good or the resource use of a household or individual in a year. On a higher aggregated level MFA allows the measurement of the whole metabolism of a region or a country. According to the question under consideration it is possible to focus on the resources used (input side) in a country or region and on their origin, which can be either domestic or imported. The focus can also be on waste and emissions (output side). MFA includes the collection of all material inputs from nature and all material outputs to nature.

Data collection of resource use usually occurs in the same unit (so energy and matter are measured in the same unit, usually tons) and flows include not only products that are actually used in the economy but also unused material (such as overburden from mining). These flows of unused material are important because they are also responsible for environmental impacts (for example, acidification of ground water).

MFA data is collected in separate categories. On the input side, typical categories are fossil fuels, abiotic and biotic materials, soil, water and air. Water and air are usually illustrated separately because they are much more important in terms of weight. On the output side, distinctions are made typically between emissions into water, air or land.

The most important material flow-based indicator on the product level is material intensity per unit of service (MIPS). It was proposed as an indicator that is able to provide a solid estimation of the ecological stress potential of goods and services [480]. The underlying assumption is that the environmental stress potential can be roughly estimated by the specific resource use.

MIPS measures the aggregated inputs caused by the creation of one unit of a specific service. It is a life cycle-wide indicator as it incorporates the material input over the entire lifespan from manufacturing, processing, packaging, opeing, re-using, recycling, re-manufacturing to disposal [482]. The material input of five categories is computed: mass of water, air, soil, biotic and abiotic materials. Energy is included in the form of mass required to provide the necessary energy. The material input considered consists of the product itself and the amount of resources that were used for its production but are not part of the product itself – the so-called ecological rucksack. MIPS allows to operationalize important aspects of sustainability at the micro- and the meso-level in a way that is easy to understand and interpret. In particular it is very useful to compare the material associated with the provision of a service unit by different means.

10.3 Consumption Clusters

Private consumptive behavior is jointly responsible for many environmental problems. But the actual impact of different consumptive decisions on the environment varies significantly. Bundling of consumption goods is a favorable way to allow the analysis of consumption on the environment.

Lorek and Spangenberg [343] developed a matrix with consumption clusters and their importance for the analysis of the effects of private consumption on the environment. The consumption clusters in the matrix are those found most frequently in the relevant literature: clothing, education/training, food, health care, construction/housing, hygiene, cleaning, recreation, social life and transport. These clusters are responsible for more than 95% of the resource use by private households. From them, three clusters were identified which are highly environmentally relevant as well as accessible to consumer's choices: food, construction/housing, transport. Therefore they are the priority fields of action towards sustainable private consumption patterns.

Lorek and Spangenberg concede that there are consumer goods that are not notably resource intensive and hence particularly environmentally relevant but which have a symbolic function as they represent the overall consumption pattern [343]. This fact is accounted for by an additional category: lifestyle. Despite the symbolic function of the lifestyle category it aims at the creation of some awareness of the dimension of effects of different behavioral patterns and especially behavioral changes. The material flow associated with a particular consumptive decision measured in tons might be quite abstract for people not dealing with physical accountings. The introduction of the lifestyle category allows to illustrate the different scope for effects by behavioral changes in different consumptive clusters. It is for example much more important to rethink transportation patterns than clothes shopping. This is not to say that so-called lifestyle decisions do not matter. Some materials used for typical lifestyle goods like jewelry generate enormous material flows (e.g., gold) and many are produced in a socially unsustainable way.

Thus the consumption clusters considered in the analysis are: *Food*, *Construction/Housing*, and *Transport* because of their key positions within private resource consumption, and *Lifestyle* because of its symbolic and interpretational value.

10.4 The Questionnaire

As consumptive behavior is a key to dematerialization, the awareness of this fact by consumers is crucial, as it is necessary to have an idea of the environmental stress potential of specific consumption goods to be capable to personally evaluate the impacts of decisions. An online questionnaire that is easy to use and understand allows that.

The questionnaire was especially created for use in schools and universities. The main group of users will therefore be people aged between 10 and 30 years although in principle it is applicable for all age groups.

The questions it contains can only give examples of the represented consumption clusters. These examples have to be representative in order to state something about a bundle. To give the questionnaire some weight, the examples should be considered interesting and important by consumers, so that participants do not have the impression that major things are left out. The above is necessary to increase the positive effect on actual decision-making. Furthermore the examples given should also be able to be influenced by the target group.

Ideally the creation of the questionnaire for self-assessment would mean looking for the most representative, interesting, relevant and also influenceable consumption goods in the four clusters identified above. A lack of data does not allow that, so compromises were necessary. Nevertheless, for almost all interesting areas it was possible to find some data – and with the work done on the area in future the questionnaire could easily be improved.

All calculations undertaken in the wake of creating the questionnaire are based on existing, published data referring to material intensities in Central Europe. Thus the set of questions was influenced very strongly by the availability of data. As usual only biotic and abiotic material and soil were taken into account. This is partly not to distort the results (as water and air dominate the other resources by far) and because data on these categories is hardly available. Additional assumptions were made when necessary. This emphasizes that all values of material use calculated represent estimates – because of the way the questions are raised and because of the assumptions inherent in the data used. The material use for each consumption good/service is calculated on a yearly basis as this allows the comparison of different types of service provided. So the unit of service is always the average yearly use. More detailed information about the questions in the four clusters is provided below.

10.4.1 Transport

The questionnaire asks for the average distances traveled in a certain time period by car, tramway/underground, train, bus/coach, plane, bike, and on foot. According to Mertz [368], the material flows in kg resulting per person and km are given in Table 10.1.

Table 10.1: Material flows in kg resulting per person and km

Traffic mode	kg/person km
train (local traffic)	0.759
train (long-distance traffic)	0.722
car (urban)	0.847
car (inter-urban)	0.460
Bus	0.192
Coach	0.103
tramway / underground	0.783
short-distance flight	0.278
Bike	0.162

Walking is also included in the questionnaire as a transport option because it is important in daily life and also represents a possibility to reduce resource consumption due to transportation. Due to a lack of data the material intensity of walking was assumed to be zero.

Although a minimum estimate of short-distance flights is presented by Mertz this estimation is used for all flights in calculations for the questionnaire because it is the only estimate available. The calculations for the questionnaire required some further assumptions: for distances traveled by train between 101 and 200 km, it is assumed that 30% are long-distance traffic and for larger distances this share is assumed to be 70%. For distances traveled by car between 51 and 100 km, 50% are assumed to be inter-urban and for larger distances this number is 30%.

Options presented to the consumer on how the material flow resulting from transport can be decreased are the increase of the passengers using a vehicle and of course to question whether the journey is necessary at all and if yes – whether it is possible to use a less material-intensive mode.

10.4.2 Food

In the food section the user is asked about three different things: the frequency of meat consumption, the number of beverage cans and bottles consumed and the bags used for the shopping. The material intensity of food is given in Table 10.2 [45].

Table 10.2: Material intensity of food

Product group	kg/kg
grain, cereals	4.7
rice, legumes, potatoes	3.0
sugar, glucose, cacao	14.1
vegetables and fruit	2.4
meat and meat products	17.7
fish and fish products	2.3
milk and milk products	7.6
vegetable oils and fats	13.1
eggs and egg products	5.2

These material intensities together with the average share of the product groups in overall food consumption [45] and the information that the average person in Austria consumes meat almost daily [151] served as a basis for someone who consumes meat "regularly (several times a week)". The shares of the product groups for vegetarians are estimates based on Leitzmann [329]. The other profiles are assumptions as is the fact that the amount of food consumed per person in each group is equal (Table 10.3).

Table 10.3: Assumed share of product group (in kg) per category of meat consumption

How often do you eat meat?	Never (vegetarian)	Never (vegan)	Seldom (several times a month)	Daily
grain, cereals	12%	29%	13%	9%
rice, legumes, potatoes	18%	31%	18%	13%
sugar, glucose, cacao	6%	7%	5%	7%
vegetables and fruit	22%	28%	18%	11%
meat and meat products	0%	0%	5%	20%
fish and fish products	0%	0%	2%	2%
milk and milk products	39%	0%	35%	30%
vegetable oils and fats	2%	5%	2%	4%
eggs and egg products	2%	0%	2%	4%

Food is very important in terms of its environmental relevance. Therefore it is considered indispensable for the purpose of the questionnaire, which justifies the required assumptions.

The questionnaire does not represent the water required for food production. Nevertheless the user is made aware of the fact that food production is actually responsible for most water use world wide.

The questions about the habits of beverage shopping are based on the analysis by Wurbs et al. [638], whereas the question about shopping bags is based on the estimates of Baedecker et al. [33].

10.4.3 Construction and Housing

While construction and housing might not be areas of action that the target group of the questionnaire can influence in the short run, these are very material-intensive areas of private consumption. Because younger users still have to decide where to live at some point in time, awareness of that fact is particularly important. The main message is that the size of their home and the number of people living in it are the main determinants of the resource requirement per person.

The construction of a multi-occupancy house is less material-intensive per user *ceteris paribus* than the construction of a single-occupancy house. The material intensity for a single-occupancy house is 3.3 t/m² of living area and year and only 1.6 t/m² of living area and year for a multi-occupancy house according to Schmidt-Bleek et al. [483]. For the calculations the life span of a house is assumed to be 40 years. The material intensity of energy sources is given in Table 10.4 [481].

Table 10.4: Material intensity of energy sources

Energy source	kg/kg
brown coal	9.68
hard coal	2.36
hard coal coke	4.22
steam (16bar)	0.39
steam (4bar)	0.39
natural gas	1.22
light fuel oil	1.36
heavy fuel oil	1.50

The material intensity for electricity is 0.80 kg/kg, according to the Material Intensity Calculator [351], and 6 kg/kg in the case of wood [33]. The energy requirement for a single-occupancy house is assumed to be 170 kWh/m²/year and 10% less for multi-occupancy houses.

Furniture is usually not considered to play a very prominent role although the example of a kitchen demonstrates that plywood furniture is three times more material-intensive than dimension lumber [336].

10.4.4 Lifestyle

In the lifestyle section the user is asked about the consumptive behavior of various consumption goods from gold rings to clothes and mobile phones. The material intensity of the included products is given in Table 10.5.

Table 10.5: Material intensity of lifestyle products and services

Product or service	Measure	Source
gold products (kg/kg)	539254.0	[71]
silver products (kg/kg)	7504.6	[71]
shoes (kg/kg)	3.5	[481]
t-shirt (kg/kg)	1.2	[481]
shirt (kg/kg)	2.2	[481]
jeans (kg/kg)	6.7	[481]
watch (kg/kg)	19.5	[481]
book, online (kg/book)	2.3	[538]
book, conventional (kg/book)	3.1	[538]
mobile ownership (kg/year)	134.0	[177]
mobile calls (kg/min)	0.2	[177]
text message (kg/text message)	0.6	[177]

Most products or services in this cluster are not very material-intensive. This is important to know for environmentally conscious consumers. But the material flow created by lifestyle consumption per year is highly dependent on the average time of usage of a product. As this is a key point, the usual time of usage is asked for most products and the user is made aware of the importance of usage time of oducts and the interrelation of product quality and the time of use. A complete of the questions asked and their sources can be found in the next section.

10.5 Realization of the Online Questionnaire

The questionnaire is available online in English and German. The English version can be found at www.umweltoekonomie.at/mfa/en [741]. As a starting point the questionnaire gives a rough idea about the environmental relevance of private consumption and on physical accounting as a method of gathering the biophysical basis of the societal metabolism. As a further step it is possible for users to get more detailed information about the theoretical background and the method applied. The questionnaire itself is divided into four parts according to the clusters identified above. In each part a number of questions are presented and users are asked to choose a consumption level that corresponds most closely to their actual behavior.

Each chosen consumption level is then added to the overall material use in tons per year, which is displayed at the top of the screen. In addition, background information about some specific characteristics of each cluster is provided in an optional window.

After answering all the questions, an overview is provided. This shows the estimated material flow associated with each consumptive decision and compares the results with the average user of the questionnaire. For registered users it is also possible to compare the result with the one of a previous session, allowing the user to get an estimate of the change in material flows caused by a change in behavior. Results for a hypothetical user are presented in Table 10.6 where the last column shows the material flow caused by the average user in the areas questioned.

10.6 Conclusions

The questionnaire is available for public use, for both anonymous and registered users, at www.umweltoekonomie.at/mfa/en [741]. Registered users are able to fill out the online form and save their submitted data. If they change their consumptive behavior in areas that are covered by the questionnaire it is possible for them to return to it, fill out the form again and view the effects of their behavioral changes. This repeated use of the questionnaire also gives some impression of the impact of the questionnaire on actual behavior that can, in principle, serve as an (important) source of data. Because of the short time that the questionnaire has been available online for public use, an evaluation of its acceptance and influence has not been possible so far.

Another benefit of the creation of the questionnaire is the production of an overview and database containing a substantial number of material intensities of the products and services in the selected sectors. Future research in the area can easily be incorporated in the questionnaire.

The average material flow caused by the users in the areas questioned is approximately 27 tons (corresponding to information available in October 2003 on the Web site of the online questionnaire), which is one third of the total of around 76 tons in Austria [6]. This share is quite substantial, considering that only some examples in each cluster are measured.

Table 10.6: Presentation of a hypothetical user's consumption behavior, annual material flow created, and comparison with average user

No	Question	Answer	Tons (t)	Av. t/user
Mobility [368]				
1a	How many km a week do you go by car?	20–50 km	1.48	2.45
1b	How many km a week do you go by tramway, municipal railway or subway?	51–100 km	2.94	1.37
1c	How many km a week do you go by train?	51–100 km	2.85	5.61
1d	How many km a week do you go by bus?	51–100 km	0.72	1.44
1e	How many km a year do you go by plane?	500–1000 km	0.21	0.42
1f	How many km a week do you go by bicycle?	20–50 km	0.28	0.28
1g	How many km a week do you walk?	less than 5 km	0.00	0.00
Food consumption				
1a	How often do you eat meat? [45; 329]	seldom	5.54	5.54
2a	How many beverage cans do you buy a week? [638]	1–2	0.09	0.28
2b	How many glass bottles do you buy a week? [638]	1–2	0.02	0.15
3a	Do you use a plastic, paper or cotton bag for your shopping? [33; 481]	own bag	0.00	0.01
Housing				
1a	How many persons live in your household?	1 person		
1b	How big is your living space?	40–60 m^2		
1c	Do you live in a single-occupancy or in a multi-occupancy house? [483]	single occupancy	4.12	0.00
2a	What is the heating system in use? [33; 481]	gas	0.86	0.00
3a	Does your kitchen consist of dimension lumber? [336]	no	0.23	0.00
Lifestyle				
1a	How often do you get or buy a gold ring on average? [71]	rarely	0.02	0.04
2a	How often do you get or buy a silver ring on average? [71]	never	0.00	0.02
3a	How many pairs of shoes do you buy a year?	1–2 pairs		
3b	How long do you wear your shoes? [481]	1–2 years	0.03	0.00
4a	How many t-shirts do you buy a year?	3–4		
4b	How long do you wear a t-shirt on average? [481]	6–12 months	0.01	0.00
5a	How many shirts do you buy a year?	3–4		
5b	How long do you wear a shirt on average? [481]	1–2 years	0.01	0.00
6a	How many pairs of jeans do you buy a year on average?	1–2		
6b	How long do you wear a pair of jeans on average? [481]	less than a year	0.04	0.00
7a	How many wrist watches do you buy a year?	1–2		
7b	How long do you use your wrist watch? [481]	3–4 years	0.01	0.00
8a	How many books do you buy online a year? [538]	5–8	1.50	0.01
8b	How many books do you buy in a bookstore a year? [538]	3–4	1.08	0.02
9a	How many mobile phones do you own?	2		
9b	How long do you use your mobile phone on average? [177]	1–2 years	0.18	0.08
9c	How many hours a day do you spend talking on your mobile phone on average? [177]	less than one hour	1.13	6.80
9d	How many SMS do you send a day on average? [177]	more than 15	4.61	2.31
	Total		**27.96**	**26.82**

Part II
Environmental Science

Data Sharing and Distributed Internet Computing
Geographically Referenced Data

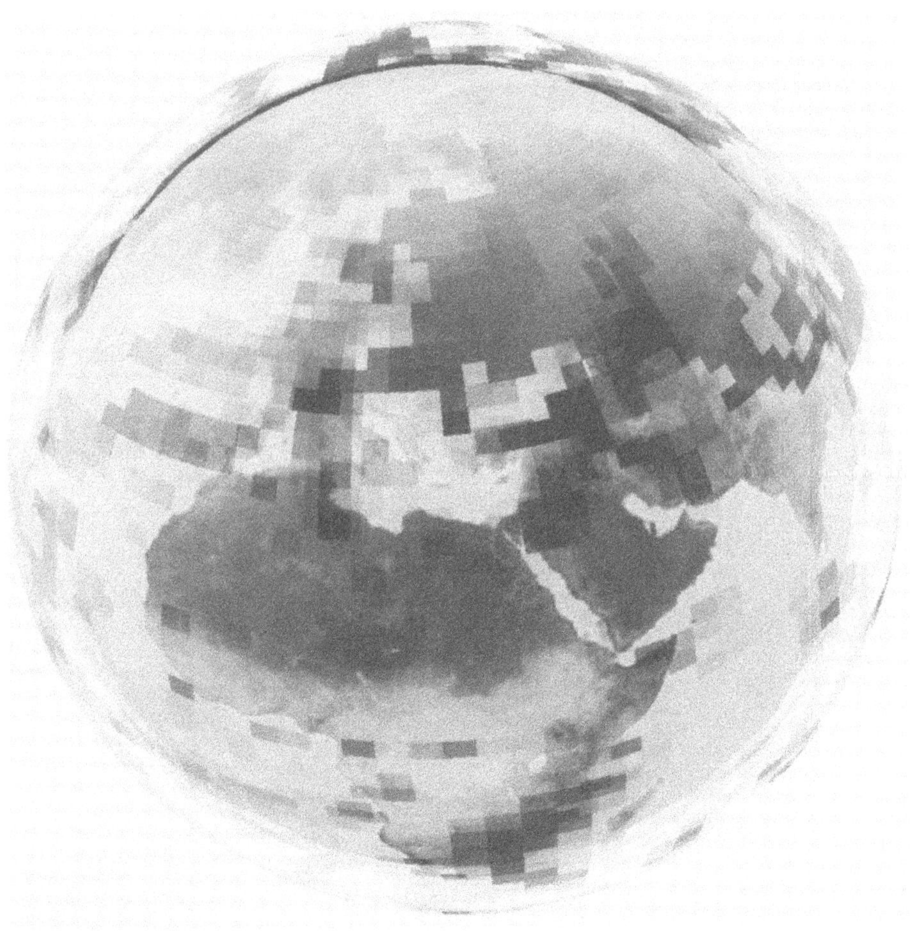

Chapter 11
Biodiversity and the Internet: Building and Using the Virtual World Museum

Edward O. Wiley • A. Townsend Peterson

Abstract. *Biodiversity information has been collected by natural history museums for almost 300 years and is estimated to exceed three billion records. Electronic capture of data and the use of distributed information technology allow scientists to access this information on a global scale. We trace the development of this technology in one community from paper to the Internet and show how biologists are using biodiversity information to address fundamental questions ranging from predicting species distribution to predicting the effects of global climate change.*

11.1 Introduction

Biodiversity information, in the form of specimens and their associated data records, has been collected for over 300 years and carefully (or not so carefully) stored away in the world's museums and other repositories. Krishtalka and Humphrey [315] estimated that the number of such specimens exceeds three billion globally. Most of these specimens are what might be termed "high quality" records because they consist of specimens as well as associated information, allowing subsequent generations of scientists to examine the specimens, interpret them against changes in taxonomy, and compare them with discoveries of new species. Together, these museum resources constitute the most complete and authoritative record of global biodiversity [411]. Prior to the 1980s, much of this information existed only on paper, in the form of catalogues, indices, and publications. Computing changed the environment. From modest beginnings in the mid-1970s, museums began the task of capturing collections data electronically. This step was critical because it allowed all subsequent progress towards sharing data and making data available through the Internet.

A microcosm of the progression from paper to distributed information systems is the ichthyological community, a community that has been particularly progressive and that played a leadership role among natural history collections in capturing and sharing data. The push to computerize records in the fish community began with a program named MUSE [744] and two young ichthyologists, Berry Chernoff (then of the Academy of Natural Sciences, Philadelphia, ANSP) and Julian Humphries (then of the University of Michigan). Chernoff obtained an NSF grant to create a relational database for the ANSP fish collection. Humphries wrote MUSE with input from Chernoff and William L. Fink (University of Michigan). The ANSP fish records were not the first to be captured electronically, nor was MUSE the first relational database platform for biodiversity collections, but it did have three key features: (1) it was relatively quick at retrieving records, (2) it was easy to enter data, and (3) everyone's fish data had a common format and data structure. This third feature (not to the liking of all) provided both an advantage and a limitation. The major limitation was a lack of flexibility; the program was great for freshwater fishes, but had limitations for marine collections. The advantage was that everyone's data looked the same in terms of retrieving the data. This advantage was exploited by the creation of FishGopher, what might now be described as the first Internet-based biodiversity data-sharing system linking fish databases at different institutions. The existence of MUSE and FishGopher provided a powerful incentive for other fish collections to computerize simply because it allowed a vision of sharing data among collections. It was also seen as progressive and even fashionable. Fish curators who had resources but resisted computerization ran the risk of seeming "old fashioned"; funding resources were an additional attraction. Government agencies in several countries encouraged computerization through generous grants to museums and the platform of MUSE provided curators at fish collections with a sturdy and proven product on which to base their computerization grant applications.

MUSE and FishGopher gave rise to NeoDat [748], a more sophisticated network linking data on neotropical fishes and creating the impetus for fish collections in Latin American countries to computerize their collections. By the mid-1990s, over 50 major and minor fish collections were computerizing, and the list was growing every year. Further, the fish community developed a willingness to form cooperative partnerships whose basis for cooperation was the sharing of information.

MUSE, FishGopher, and NeoDat created the atmosphere of cooperation, but they had (and have) their limitations – principally among them that not everyone used MUSE. Indeed, the diverse set of database management platforms used in the systematics community has standards that may be identical or similar, but structures that may be idiosyncratic relative to one another. Many of these collections (e.g., Division of Fishes, US National Museum) have computerized databases and make their computerized records available online, but searching each museum and picking out data for species individually is a time-consuming task. What was needed was a system that could search many databases with relatively idiosyncratic structures simultaneously in a reasonable amount of time and return those data in a common format for use by the researcher.

11.2 Distributing the Information: The Species Analyst

Newer technologies provide a way in which idiosyncratic databases can be queried and the results returned in a common format; the first of these possibilities to be explored was the ANSI/NISO Z39.50 information transfer protocol [793]. This protocol allows for creation of distributed information systems, using the combined search capability of individual servers based at local institutions to return data from queries to an investigator in real time on the order of seconds. David Vieglais (University of Kansas) began exploring the application of this protocol in 1998 in collaboration with Peterson, who wanted to improve access to avian databases. The result was The Species Analyst (TSA), an Internet-based Z39.50 interface that fulfilled much of the promise of linking natural history databases into fully implemented distributed information network [777].

ANSI/NISO Z39.50 is the same technology used for bibliographic searches. It defines a standard way for two computers to communicate. One computer is the client. It requests information from another computer, the server. The server searches the database, generates a set of records that meets criteria specified by the search request, and returns this set of records to the client for use by the investigator who generated the request.

In TSA, the request may take many forms built around a common core of information fields, the Darwin Core [680], a standard suite of data fields deemed common to all natural history specimens. Fields such as taxonomic categories (family, genus, species), locality (country, state, county, specific locality), date of collection, latitude, and longitude are examples. Not all museum records have all of the fields, but they must have at least one of the fields in order to be queried. A typical TSA query might take the form of asking for all records for a particular species from all relevant servers (e.g., fish museum databases) connected to the network.

The system is fast compared with a centralized database because the tasks of searching each database and compiling the results are distributed among the servers. This arrangement not only cuts computing time, but also allows for almost infinite expansion of the network – the workload at any single server is only a function of the size of its own database. The trick relative to databases using different platforms is that each server has two pieces of Z39.50 software. One is configured to read the resident database and one to put the data from the search into a format specified by the Darwin Core. Thus, even if each server uses a different platform and has a different organization of fields, the results of a search will appear in a common format on the client. TSA receives these packets of information, collates them, and makes the results of the search available to the investigator as Excel, XML, or ESRI shape files.

11.3 Sharing the Information: FishNet

FishNet [705] is a political consortium of fish collections at natural history museums that agree to share data through The Species Analyst [588]. The partnership was formed in response to a call for the creation of a distributed biogeographic information network known as OBIS [752], the Ocean Biogeographic Information System, an initiative begun by Fred Grassle (Rutgers University) and Jesse Ausebel (Sloan Foundation) and sponsored by the Sloan Foundation, the National Oceanographic Partnership Program, the Census of Marine Life, and the NSF. A similar initiative has been developed for mammal collections (MaNIS [739]), one is under development for herpetological collections (HerpNet [716]), and one is currently proposed for avian collections (ORNIS).

FishNet differs from many OBIS projects in that it has a very simple function; providing specimen data to investigators. It does not provide bibliographic information, taxonomic histories, descriptions, species illustrations, or other narrative material. It also differs in that it forms a distributed, rather than centralized, network. Distributed networks have both advantages and limitations. Among the advantages are the following:

- The network is scalable and expandable without appreciable increase in query time.
- Institutions with electronic databases can be added to the network quickly, without having to reformat or otherwise manipulate the data, provided that basic data standards are observed.
- Each partner institution retains full stewardship of the information contained in their database. Thus the institution has the capability to mask information it does not wish accessed (e.g., exact locality information for an endangered species).
- The information is readily available to anyone who has access to the Internet, globally. This access is particularly important to investigators in countries outside developed countries because it allows access to specimen data that are housed in museums but which originated in other countries. This concept could be termed data repatriation; access to specimen data taken from one country and housed in another.
- Data are returned in a common format for processing, as specified by the investigator.

Among the potential disadvantages are the following:

- Data are made available to any interested user, which some may feel reduces their value [212].
- Not all data may be displayed if these data are not incorporated into the standard adopted.
- Not all collections may be accessible at any one time due to network problems at the local level.

11.4 Using the Information

Once distributed, biodiversity information can be used for a variety of purposes. Following the theme of data repatriation, workers in one country can have immediate access to specimen records for native species collected and housed in museums outside the country. For example, some of the more comprehensive collections of fishes from Belize are housed in the United States. A major fish collection made in Western Australia is housed in New York at the American Museum of Natural History. Such collections may be of vital importance to biologists in the host country. Below we briefly review four of many potential research programs using biodiversity information.

11.4.1 Basic Biodiversity Research: Where and When

Single institutions rarely have the specimen data needed to assess the current biodiversity information available for a particular species, fauna, or flora. For example, Peterson et al. [382; 383; 422] assembled some 64 avian collections databases to gain an adequate understanding of Mexican bird biodiversity. Peterson et al. [431] reported that no single bird collection contained more than 16% of th total biodiversity information for Mexico. The effort was immense, because e database had to be concatenated manually into a central database before research could commence. Such immense time sinks can be avoided if databases can be accessed in a distributed manner.

Ichthyological databases are similar. Many museums specialize in specific geographic regions and the distributional limits of many species exceed these limits. Distributional records of the largemouth bass (*Micropterus salmoides*) of North American freshwaters are an example (Figure 11.1). This species is native to eastern North America, but has been introduced in more than 60 countries, as well as to many other parts of the United States. By querying many different collections and visualizing the data in a geographic information system environment, we can see that no one fish collection portrays the current distribution of this species in North America but that combining the information from many museums builds a more complete description of its actual distribution.

Once we can access multiple databases and build a more complete picture of the known occurrences of species, we are in a position to move beyond simple visualization to detailed analyses, predictions, and projections. Such synthetic analyses have taken many distinct paths – extrapolation of biodiversity estimates, predictions of species invasions, projection of potential geographic species distributions in future climate scenarios, predicting the spatial dynamics of disease transmission cycles, etc. These applications are but a few of the possibilities that result from the development of a predictive understanding of distributional phenomena of species. Below, we present three specific examples to illustrate the benefits of the shift to a full-access, integrated biodiversity information system such as Species Analyst and FishNet.

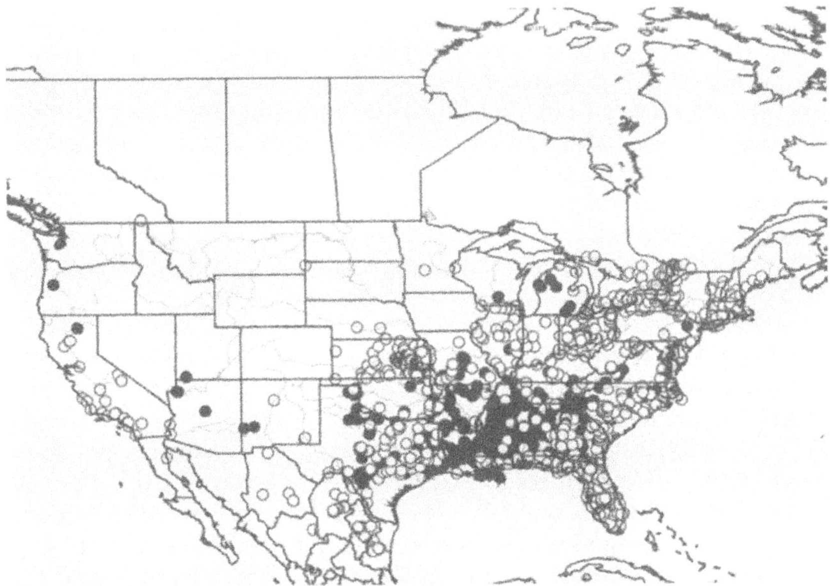

Figure 11.1: Georeferenced data points for the largemouth bass (*Micropterus salmoides*) in North America downloaded from FishNet partner institutions. Black dots are point occurrences ~m the largest collection, Tulane University Natural History Museum. Open circles are additional records from ten additional museum collections (see Acknowledgements).

11.4.2 Predicting Species Distributions

Although geographic distributions of species may seem to be a feature that is already well understood, and not in need of complex applications, the simple fact is that even the basic details of geographic distribution of most species known to science remain poorly understood. Many species are characterized by records from sets of points across regions, and yet these points tend to be few in number – in the Mexican bird example cited above, the modal number of unique locality records per species was less than 25. Considering all of the complexities of a geographic distribution of a species, including both range limits and internal complexities, this number is vanishingly small. Twenty-five points, nevertheless, are plenty for development of highly predictive ecological niche models, which yield hypotheses regarding potential geographic distributions of species [428].

As an illustration, Wiley et al. [622] demonstrate that statistically robust geographic range predictions could be developed for marine fishes. They used occurrence data harvested from FishNet partner institutions and environmental data from the World Ocean Atlas [390] and Smith and Sandwell [513] data sets for a region that included the Gulf of Mexico, Caribbean Sea, and Central Western Atlantic Ocean. Niche models were generated using DesktopGARP [682], a program that uses a number of algorithms to develop a niche model for a species using the information provided by the occurrence points and the environmental coverages. The niche model is projected onto the selected geographic landscape and those geographic areas where the niche model is realized become the predicted geographic range of the species within the landscape. Such research programs are termed "ecological forecasting".

The example shown in Figure 11.2 illustrates part of the predicted range of the benthic shark *(Etmopterus schultzi)*. This shark inhabits the continental slope from -384 meters to -732 meters in depth. The prediction was based on the environmental conditions at depth, achieved by projecting the conditions at each depth range onto the bathymetric surface of the ocean bottom. The models were tested by reserving know points from model building and using these points to test the results (Figure 11.2).

Ecological forecasting in the terrestrial environment is equally successful and has, without exception, revealed excellent predictivity of geographic distributions of terrestrial species [19; 20; 21; 419; 420; 428; 540; 541; 542].

A particularly interesting initiative in predicting species distributions is Lifemapper [735]. Lifemapper uses TSA technology to access databases and produce predictions of distributions automatically, with results stored for future exploration and analysis. It is a distributed computing system that works in a manner similar to the SETI distributed computing system [309; 768], using the distributed power of individual computers to map actual and potential distributions. It has the potential to develop into an important resource of species' distributional information for many applications in education and in science, such as an early warning system for documenting the presence of or predicting the threat of invasive species.

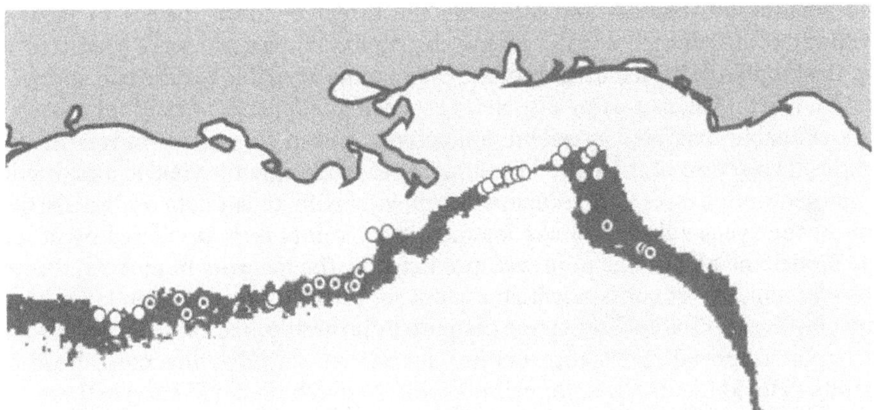

Figure 11.2: Geographic range prediction of the benthic shark (Etmopterus schultzi) in the northern Gulf of Mexico off Louisiana, Alabama and Florida. Gray is the continental landmass. Black is the convergence of ten independent models predicting "presence" of the species. Open circles are training occurrence points used in modeling. Circles with dots are independent test data points used to test the model predictions. Redrawn from Wiley et al. [622].

11.4.3 Predicting Species Invasions

The technical definition of an invasive species is one that is introduced and established outside its home range and that causes significant impacts on human or natural systems. A conservative estimate of economic costs of invasive species was set at $137 billion per year [417]. One problem with invasive species is that, once introduced, they are difficult or impossible to eradicate. This suggests that research in this field should focus on methods for anticipating or forecasting invasive threats *before they occur*, so that governmental regulatory agencies can take steps to avoid creation of new problems.

Our research group is addressing this problem by taking advantage of the predictive nature of ecological niche models to understand distributional phenomena of species, regardless of *where* the particular population of the species is located. We use occurrence information gathered through Species Analyst and other sources to build niche models of potential invasive species on their native geographic distributional areas. After testing the ability of these models to predict independent data from the native range (e.g., specimen occurrence data withheld from modeling), models can then be projected onto any selected landscape to see if the potential niche is manifested in a new area. Peterson and Vieglais [430] used this technique to predict the invasion of the Asian longhorn beetle into the United States; a team of Mexican researchers has used similar techniques to anticipate the invasion of a moth *Cactoblastis cactorum* that causes enormous damage to cactus (*Opuntia* spp.) populations [514]. This technique has now seen broad application to a great diversity of invasive species, with the universal result that models predicted invasions with excellent success [244; 406; 407; 425; 457; 502; 506; 641].

To provide an illustration of these analyses, an international team composed of merican and Japanese researchers has recently extended this technique to tic organisms by modeling two North American bass species (largemouth and smallmouth basses) and assessing the threat of these species to Japanese freshwater environments [256]. Native distributional patterns were predicted using DesktopGARP based on over 1,600 data points available for the two species in eastern North America using FishNet. Tests using independent random subsets of data available indicated excellent predictivity within the native ranges of both species. Projection of these models to Japanese environments yielded a prediction for largemouth basses that coincided excellently with actual data on the distribution of the species there (all 634 Japanese data points were predicted by at least one model and all but one point was predicted by the majority of models). Projections for smallmouth basses (which are not yet broadly distributed in Japan) indicate ample potential for invasion of Japanese freshwater habitats. This study is of particular importance because neither species has invaded the more northern island of Hokkaido, but the projections indicate that both could survive there.

11.4.4 Climate Change and Biodiversity

With the observation that ecological niche models summarize what is effectively a long-term stable constraint on the distributional potential of species [250; 427; 429], numerous possibilities for further extrapolation open up. One of the most interesting of these possibilities is that of predicting changes in the potential distributional areas of species in the face of changing climates – here, ecological niches are modeled for the present distribution of species, and are projected onto modeled future climates to outline a potential distributional area for the species. This general approach has now been applied to a diversity of taxa and regions [36; 52; 87; 155; 250; 370; 412; 418; 423; 424; 426; 429; 445; 504].

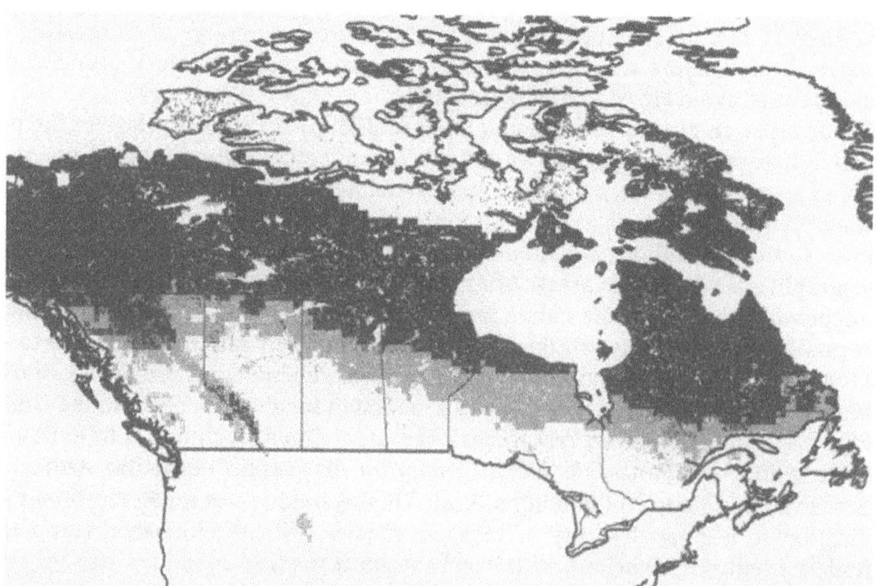

Figure 11.3: Predicted poleward shifts in species distributions as a consequence of global climat change: *Oeneis polixenes.* Present predicted distribution in gray, future predicted distributio black (pictured on top of present predicted distribution). Redrawn from Peterson et al. [

To provide an example of this sort of exploration, we modeled the distribution of a Canadian butterfly *Oeneis polixenes* under present-day climates. Testing the validity and predictivity of this model using random independent subsets of the available present-day distributional data for the species indicated excellent predictivity. Projection of this model onto modeled future climate surfaces indicated a generalized northward shift of appropriate conditions for this species, as well as a nontrivial reduction in the habitable area for the species as shown in Figure 11.3 [421].

11.5 Conclusions and Outlook

Investigators can now access biodiversity information on a global scale thanks to recent advances in the integration of biodiversity information. Distributed networks such as the Species Analyst and others will enable new information to be added to the knowledge base without a high cost in access time. Research applications based on these distributed biodiversity information resources demonstrate the immense value to science and society that these data hold. Nevertheless, much still needs to be accomplished. The single biggest obstacle to these advances, nevertheless, is the creation of the electronic data – far too many natural history museum specimens remain undocumented in electronic format, limiting the information base available to users. Shifting the Species Analyst and other distributed information networks to the new DiGIR [683] protocols will greatly improve access and efficiency for data. Increasing access to such data will pay huge dividends, both scientifically and economically.

Acknowledgements

Funding for this study was provided by the US National Science Foundation (DEB-9985737), the Office of Naval Research (N0014-00-1-0887) through the Census of Marine Life/Ocean Biogeographic Information System Program and the US Geological Survey Caribbean Science Center. The Biodiversity Research Center, University of Kansas provided the biodiversity informatics infrastructure that made this study possible. We thank David Vieglais, C. Richard Robins, Joel Kolatch, Aimee Stewart, and the curators at the 29 participating FishNet institutions. The following museums contributed specimen data via FishNet: Canadian Museum of Nature, Cornell University Ichthyology Collection, Florida Museum of Natural History, Harvard University Museum of Comparative Zoology, Illinois Natural History Survey, Natural History Museum of Los Angeles County, Ohio State University Museum of Biological Diversity, Texas A&M University Cooperative Wildlife Collection, Tulane University Natural History Museum, University of Kansas Natural History Museum, University of Michigan Museum of Zoology. Our thanks to all institutions.

Chapter 12
Climateprediction.net: A Global Community for Research in Climate Physics

David A. Stainforth • Myles R. Allen • David Frame • Jamie Kettleborough
Carl Christensen • Tolu Aina • Matthew Collins

Abstract. *The climateprediction.net project is undertaking climate research by harnessing spare computer cycles volunteered by businesses and members of the public. It is running a massive ensemble of climate simulations using a state-of-the-art climate model with the aim of quantifying the uncertainties in predictions of future climate change. Such a probabilistic forecast will support and improve decision-making in government and industry. The computational design builds and expands on the concepts of distributed computing, as demonstrated by projects such as SETI@home, but along with higher computational demands comes a much greater requirement and desire to involve participants in the details of the research. Issues of collaborative learning, educational tools, and participant involvement therefore form some of the core activities.*

12.1 Introduction

On 12th September 2003 the climate*prediction*.net project [525] was launched to the global public. For the first time anyone with a standard personal computer (PC) could run a full-scale, three-dimensional Global Circulation Model (GCM) of the climate system. And they responded in large numbers. Within 48 hours 20,000 individuals were using a research model which until recently could only be installed on supercomputers costing millions of dollars. The project's opportunities for research in physics and computer science and for community building and education could begin to be realized. Future academic and commercial collaboration will now determine the degree to which this project is used as a mechanism to increase our (society's and researchers') understanding of the climate system and man's impact upon it.

The project was conceived in 1999 [15], stimulated by the launch earlier that year of the first large-scale distributed computing project, SETI@home [309; 768]. A one-page Web site was created to quantify the scale of interest. This received more than 15,000 registrants in the first weekend and laid the foundation for climate*prediction*.net's creation. There followed four years of fundraising, computer programming, experimental design, climate modeling, and "lobbying" of the atmospheric physics and computer science communities to gain acceptance that the project was both feasible and valuable. The effort came from a small team of dedicated individuals, backed by a much larger group of supporters from the various academic communities involved.

Climate*prediction*.net works by providing a software package which individuals and businesses can download from the Internet and install on their personal computers. The software includes a state-of-the-art model of the atmosphere / cryosphere / land surface / ocean system – commonly referred to as an Atmosphere Ocean Global Circulation Model (AOGCM). Once installed it connects to the climate*prediction*.net servers, registers the individual and his/her machine, and is given a "simulation definition file" which perturbs the way the model represents the physics of the climate system. The individual PC then runs the model as a background process and simulates a period of 45 to 50 years, taking between three weeks and three months depending on the machine's speed. During this period there is no requirement for the user to be connected to the Internet, although if they are status information is returned to the servers and is available for display on the Web site. As a background process it does not interfere with the regular use of the PC for most other applications and in this early stage of the project users are requested to leave their machines on as much as possible to make quick progress towards the research aims. However, if the machine is closed down and later re-booted the model simulation picks up from within three model-days of where it left off. When the simulation is complete the software again connects to the servers, returns a subset of the data created by the model, and downloads a new simulation definition file to allow it to begin again.

12.1.1 Project Goals

The climate*prediction*.net project involves three separate disciplines: climate physics, computer science, and online education/collaboration. Each aspect is equally important because failure in any one would jeopardize the success of the project as a whole. Within these disciplines there are four fundamental aims:

- To quantify uncertainty in forecasts of 21st century climate change.
- To improve our understanding of the climate system and the ways it can be represented by computer models.
- To design and develop software and systems which facilitate distributed modeling; including distributed and massively-distributed data analysis.
- To raise awareness and understanding of uncertainty in climate forecasts.

This chapter represents a case study of the project and discusses the approach taken to achieve these aims. Section 12.2 describes the project in the context of previous distributed computing projects. The background and project methodology for probabilistic climate forecasting are presented in Section 12.3, with Sections 12.4 and 12.5 outlining the computational design and public involvement aspects. This need for public involvement is fundamental to the project's success since without it many people are unlikely to continue participating for the relatively long period required (three weeks plus). One of the mechanisms by which this involvement is achieved is by inclusion of a visualization module through which the users can see the temperature, rainfall, clouds, etc. in *their* model-version while it is running (Figure 12.1). Finally, Sections 12.6 and 12.7 outline some of the future activities by which the project hopes to expand the potential of distributed modeling for climate research.

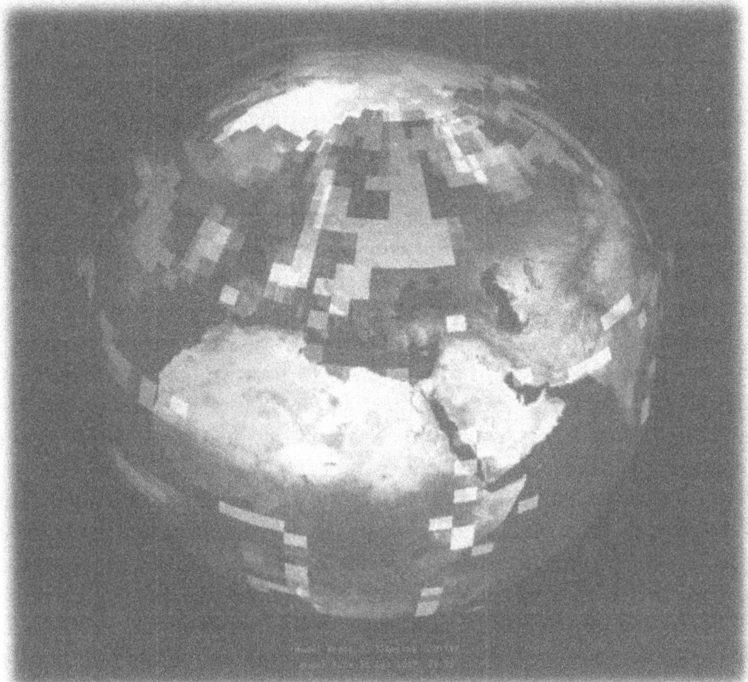

Figure 12.1: Integrated visualization package showing the clouds within the model on two layers in the vertical and at each horizontal grid point

12.1.2 The Climate Model

To fulfill the first two aims requires the use of a cutting edge and internationally respected AOGCM. Initially the project is using versions of the UK Hadley Centre's climate model: HadSM3 [623], HadCM3 [119; 210], and HadCM3L [275]; all global set-ups of the UK Met Office's Unified Model [120]. These all include a full three-dimensional representation of the atmosphere and a separate representation of sea ice. They differ in the level of detail included in the ocean component; ranging from a simplified two-dimensional representation at $3.75^{\circ} \times 2.5^{\circ}$ (longitude x latitude) resolution, to a full, three-dimensional, dynamical representation with a resolution of $1^{\circ} \times 1^{\circ}$ and 20 vertical levels. The atmosphere has a resolution of $3.75^{\circ} \times 2.5^{\circ}$ (longitude × latitude) and 19 vertical levels, giving more than 130,000 grid-points. At each point the basic dynamical equations are solved at every model timestep (30 model-minutes). Furthermore, at every timestep a series of parameterizations are run to calculate the effects of processes which take place on scales too small to be included in, or emerge from, the basic equations in the model. These include land surface processes, clouds, certain wave motions, etc. Finally, the effects of radiation (short wave radiation from the Sun and long wave radiation from terrestrial sources) are calculated at each point every eight mesteps (three hours). These models represent state-of-the-art tools for study- climate change and were widely used in the 2001 report of the IPCC, the Inter- nmental Panel on Climate Change [246].

12.2 Climateprediction.net: A Distributed Computing Project

The first major distributed computing (DC) project was SETI@home [309; 768], which since May 1999 has been using computing capacity on volunteered machines to analyze radio-telescope data for signs of extra-terrestrial intelligence. Since then a large number of other similar projects have been developed in fields ranging from the physical sciences [774] to life sciences projects such as FightAIDS@home and "Screensaver Lifesaver" [125; 455] to mathematics and cryptography – e.g., GIMPS, the Great Internet Mersenne Prime Search [711] and Distributed.net [684]. As part of this proliferation a number of platforms have been constructed to facilitate the simplified development of such projects – e.g., BOINC, the Berkeley Open Infrastructure for Network Computing [662], Fida Distributed Computing [704], and the Cosm Project [678]. Climate*prediction*.net builds on what has gone before but it has not yet been able to use any of these generic platforms because it is different in a number of crucial ways.

Firstly, it involves running a highly complex and computationally intensive simulation model, thus placing much greater demands on the participating machines (in terms of minimum processor speed, memory and disk space), the participants (in terms of long-term commitment), and the project developers (in terms of designing and preparing experiments, as well as implementing complex software designed for supercomputers on household machines). Secondly, it creates large volumes of data, raising challenges for data collection and analysis not present in other DC projects. And finally the requirement for long-term participant commitment creates a need for tools to encourage active involvement in the project. Fortunately, this matches well with one of the project's original objectives of raising awareness of the issues surrounding climate research and in particular climate modeling and uncertainty analysis. Nevertheless, it also presents significant additional barriers such as the need for software to visualize the model while it is running; something which did not previously exist within the climate modeling research community.

The project has been fortunate in receiving encouragement and advice from the SETI@home team and it is hoped that ultimately a generalized distributed modeling platform will be developed which can handle these additional requirements.

12.3 The Climate Forecasting Problem

Like weather forecasting, projections of future climate use complex, three-dimensional computer models of the Earth system. Substantial effort has been invested in the creation of such models, and a handful now exist worldwide. Some of these have been shown to be able to simulate the large-scale changes in climate over the 20th century with remarkable accuracy [548]. They include details of external factors which change the way climate behaves, known as climate "forcings", and to achieve good simulations of the past they must include both natural forcings (e.g. volcanic emissions and changes in solar activity) and anthropogenic forcings (i.e. greenhouse gas emissions). However, even a good simulation of th 20th century does not necessarily demonstrate that they will make an accu projection of the 21st century under a given scenario for future anthropogenic greenhouse gas emissions and natural forcings. What is necessary is a method to quantify the uncertainty in a forecast so instead of giving a single value for, say, global or regional mean temperature in 2050, we present a probability distribution function (Figure 12.2) representing an objective range of probability for each point in time. This would not only present the range of likely behavior but, perhaps more importantly, would allow certain responses to be ruled out given our current understanding of the system. Such probabilistic forecasts, including ones of regionally based quantities, would be of significant value in social, governmental and commercial planning processes.

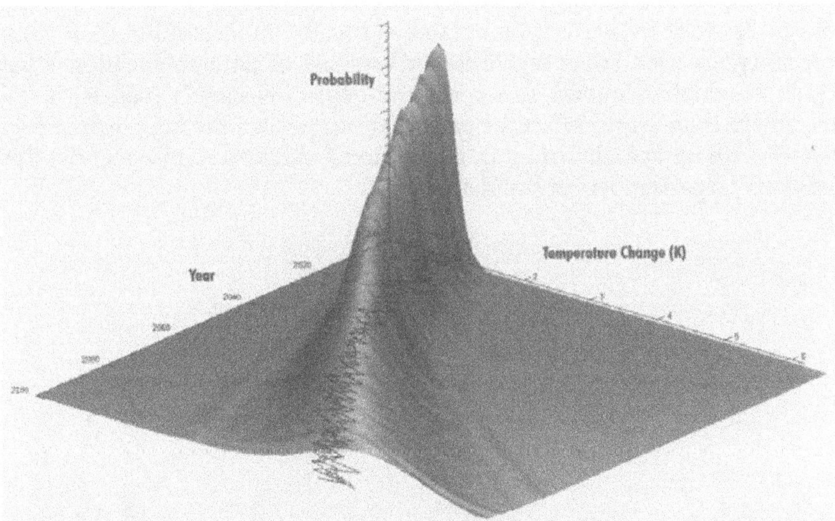

Figure 12.2: A forecast probability distribution of global mean warming above pre-industrial levels accounting for uncertainty in system response. The black lines show the impact of initial condition uncertainty. Data courtesy of P. Stott (Met Office) and J. Kettleborough (Rutherford Appleton Laboratory), using IPCC SRES A2 scenario; figure based on Allen and Stainforth [14].

There are three major sources of uncertainty in climate forecasts: (1) the chaotic nature of the atmosphere/ocean system as demonstrated by its "natural variability", (2) the external forcings, and (3) the way the system responds, i.e. model uncertainty. Forcing uncertainty is generally handled using a scenario for greenhouse gas emissions, usually following one of the scenarios outlined by the IPCC [380]. In the latest IPCC assessment report [246] climate change forecasts allowed for forcing uncertainty through the use of a range of emission scenarios, and system response uncertainty through the range of projections made by a simple model tuned to represent the global mean results of seven complex models (i.e. GCMs). Although an excellent, pragmatic solution at the time, such a method does not provide any means for quantifying the probability that the real world response to a given scenario will be within or beyond the range given by these models – is it the 5–95% range or the 30–70% range? Nor does it enable one to make predictions of the regional responses to climate change. The IPCC therefore identified as a high priority the need to "improve methods to quantify uncertainties of climate projections and scenarios, including long-term ensemble simulations using complex models" [246, p17]. This is the core of climate*prediction*.net.

Since the last IPCC report there have been a number of pieces of work in this area. Stott and Kettleborough [547] used uncertainty in the observed warming trend in the latter half of the 20th century to scale the projections from an AOGCM (Figure 12.2). Forest et al. [186] and Knutti et al. [302] both altered the parameters within an intermediate complexity model to explore system response uncertainty. Climate*prediction*.net is using a related methodology but by using a full AOGCM it includes a much wider range of feedback mechanisms as well as producing results at a regional level. The procedure involves an ensemble (a number of slightly different model simulations) with each simulation using a different model version in which the physics of the system is represented differently. This can be done by perturbing parameters in the model within their range of uncertainty, or even swapping different versions of parameterization schemes. Such an ensemble is known as a perturbed-physics ensemble [14; 525]. Unfortunately, there is no way to select the perturbations to span the full range of possible behavior without actually running many model versions. Consequently, this requires very large numbers of simulations.

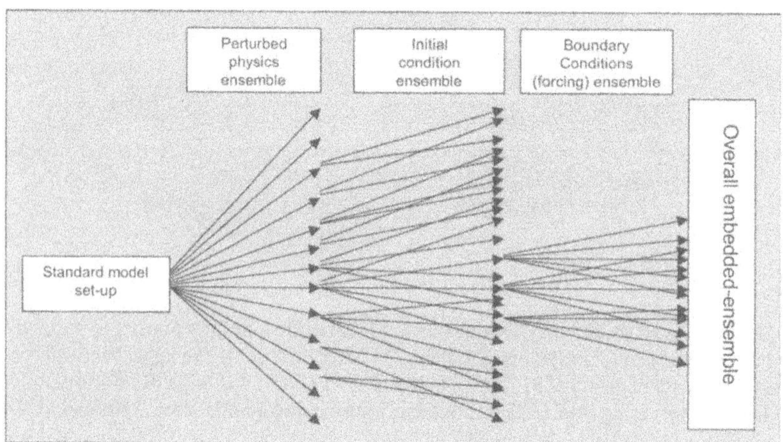

Figure 12.3: Schematic of an embedded ensemble exploring system response uncertainty via a perturbed physics ensemble, natural variability within the model via an initial condition ensemble and forcing or boundary condition uncertainty using a scenario-based ensemble. The embedded ensemble approach enables quantification of interdependencies in the sources of uncertainty.

Furthermore, to properly quantify uncertainty it is also necessary to carry out ensembles of simulations to explore natural variability (using "initial condition" ensembles which involve slight perturbations to the starting conditions of the model) and response uncertainty (using forcing scenario ensembles). Since these sources of uncertainty are potentially dependent on each other, it is necessary to explore them in one, combined or "embedded-ensemble" involving parameter perturbations, initial condition perturbations and forcing scenarios (Figure 12.3). This further increases the requirement for an extremely large AOGCM ensemble and it is hoped that the climate*prediction*.net experiment will involve simulatio of at least five million model years. The distributed computing approach prov a mechanism to access this level of computing power and potentially much more since greater ensemble sizes will enable better quantification of uncertainty. The use of supercomputers could not provide this scalability and such machines are in any case much better placed to carry out different types of experiment, such as ones involving very high resolution models.

The climate*prediction*.net experiment is an embedded ensemble experiment. Most participants receive a unique version of the model thanks to the selection of a combination of model parameters, the initial conditions, and a forcing scenario. As the experiment progresses the parameter values for each new simulation distributed will be based on the results returned to date in order to explore parameter space as efficiently as possible. The resulting data will be weighted appropriately [13], using observations as a constraint, to provide an objective and stable forecast which is principally dependent on the observations.

The project will proceed through a series of experiments beginning with an exploration of the climate's response to doubling levels of CO_2 using HadSM3 [623], and proceeding to a transient simulation of the second half of the 20th century, followed by a forecast of the first half of the 21st century with a range of emission scenarios, using HadCM3 [119; 210] and HadCM3L [275].

12.4 Computational Complexities

The software design posed a series of software engineering challenges, the first of which was to port the chosen AOGCM from a massively parallel processor supercomputing set-up to a Windows PC installation. The decision was taken early on to concentrate on a Windows implementation because the statistics from SETI@home indicated that Windows machines contributed an order of magnitude more than any other type of operating system. Fortunately there had already been substantial effort invested at the UK Met Office and the Universities of Reading and Oxford to develop a Linux-based installation of the model. Transferring a single set-up of the model to a Windows command line version was therefore achieved in only a few weeks. The issue was then to provide a friendly user interface, suitable model management and control, and a client/server design by which the simulations could be handed out and data collected [526]. Some of the design requirements are summarized in Table 12.1.

Table 12.1: Project requirements and their implications for software design (italicized points are not yet implemented)

	Design implications
	Digitally signed distribution package, secure data transfer, integrated model integrity checks, and analysis of initial condition sub-ensembles for signs of tampering.
Suitable for use with infrequent, low speed (modem-based) Internet connections.	Suitable for CD-ROM distribution. Requires minimal download. Very small minimum level of data collection. Efficient post-processing of data. Prioritization of data to be collected.

Design requirements	Design implications
Ease of participation.	Simple installation and registration procedure.
Large volumes of data to be collected.	Design for distributed data collection servers. Design for "trickle back" of data during the simulation.
Distributed analysis.	*Dynamic updating of post-processing code.* *Grid-based analysis software for servers [187].*
Flexibility to include other models.	Separation of user and server interaction processes from model control.
Maintain participant interest.	The ability to visualize the model fields while it is running. The ability to visualize model data on disk. A dynamic participant-driven Web site with community-building software.

These requirements were achieved using a design involving a central server and multiple data collection or "upload" servers which can be geographically distributed around the world (Figure 12.4). The allocation of an upload server, to which any individual machine returns its data, is made by the central server at the time when data is ready for collection. This provides flexibility in the data collection process to cope with varying numbers of participants as well as with future developments whereby there may be increased data returned from those people with fast Internet connections. Data storage capacity can be increased to meet demand, without placing an excessive load on any individual server-hosting institution, and distributed data collection also facilitates better load balancing as well as optimization according to available bandwidth and Internet conditions. Initially upload servers are being installed in collaborating institutions in the UK, Canada, Switzerland, the USA and New Zealand.

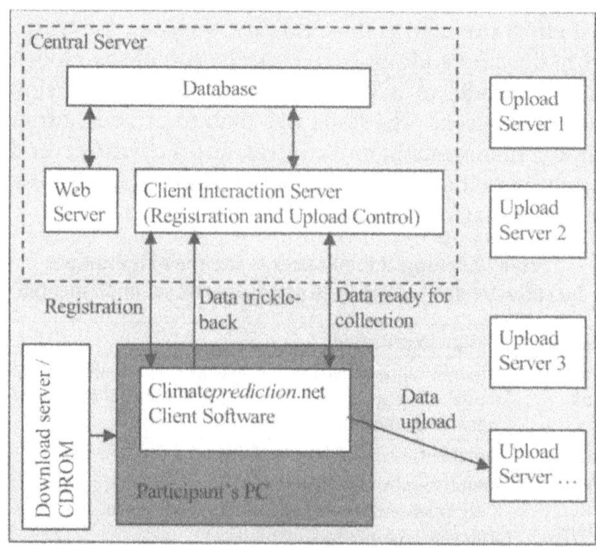

Figure 12.4: System architecture and software interactions in the climate*prediction*.net design, based on Stainforth et al. [526].

12.5 Public Involvement Mechanisms

The public involvement aspects of the project are both an independent goal and a core requirement for its success. For many participants, only their active involvement will maintain the necessary interest and commitment to the project to continue running the model. There are two fundamental mechanisms to stimulate their involvement: (1) PC-based visualization facilities to enable interaction with and study of their own model simulation, and (2) Web-based facilities to encourage community building and collaborative learning.

The requirement for a visualization module, whereby the progress of the model could be seen on a timestep by timestep basis, was identified early in the experimental design. However, this represented a new area of research as in climate physics there is rarely any need to view model fields at this level of time resolution; and never while the model is actually running. Furthermore, the visualization needs to be exciting to a non-scientific audience. It should therefore be three-dimensional and allow "zooming-in" to regions of interest, while being less than one megabyte in size to ensure that the package does not become too large to download over a modem. These restrictions required the use of a relatively low-level graphics library and consequently more involved and complex programming. The open source Coin [676] implementation of the Open Inventor API [617] was chosen to meet these requirements and although requiring a relatively high level of expertise for development, produced impressive results (Figure 12.1).

A secondary visualization package has also been built (Figure 12.5). This allows users to see the data fields contained in the model output files and to view how their simulation has developed over time. This will be used in a course on climate run by the Open University and it is hoped that with the related teaching material it will constitute the core of a range of educational links with schools and universities. It is also available for any participants to download but as it is not seen as critical to project involvement it is provided as a separate download and no restrictions were placed on the package size. It was developed using commercially available visualization software with an embedded distribution license provided as a form of sponsorship.

The climate*prediction*.net Web site forms the heart of the project and is the principle means of providing information, as well as being fundamental to community building. In addition to the usual Web pages of information, two aspects of the Web Site are worth highlighting. The community space discussion forum has been a popular discussion site for all aspects of the project since early in the software's beta testing phase. Discussions have ranged from detailed analysis of the software (which have proved invaluable in the development process) to wider discussions of the experimental design and climate modeling. The second aspect, and the source of ongoing development, is the user-area where participants can login and view details of the progress of their simulations. These provide details of the number of years of simulation completed as well as particulars of the ensemble members the individual is running and summary information on their results such as plots of the change in global mean temperature over time. Like other distributed computing projects, participants can also form teams and are ble to gather team statistics as well as, in the future, viewing team results in the text of the overall experimental ensemble.

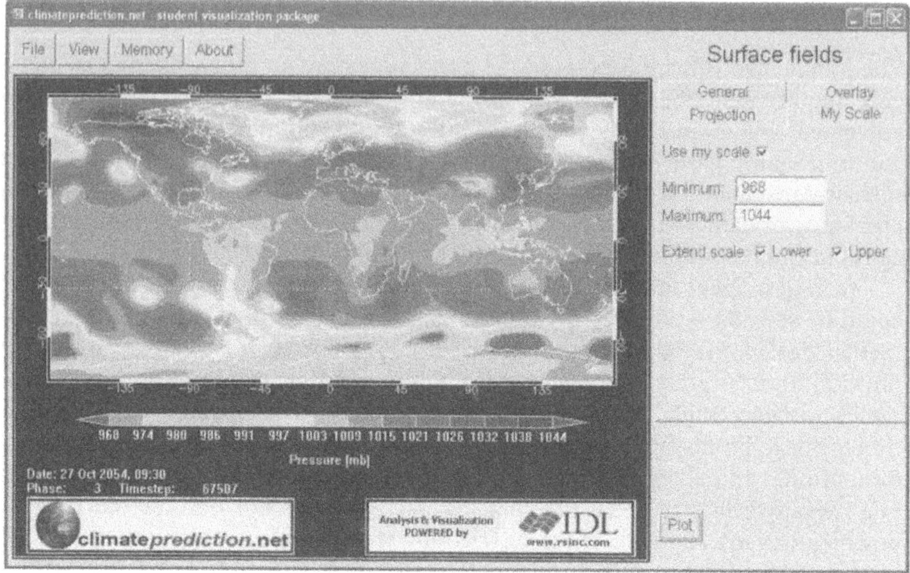

Figure 12.5: Optional secondary visualization package (developed by A. Heaps, Reading University)

12.6 Future Developments

At the end of October 2003, six weeks after the public launch of the project at the UK National Science Museum in London and the British Association for the Advancement of Science Conference in Salford, almost 40,000 individuals, from over 100 countries, had downloaded and installed the climate*prediction*.net software. More than 20,000 of these were feeding regular updates back to project servers. Almost 450,000 model years of simulation had been completed – a Pentium IV 2GHz processor can simulate approximately 1.6 model years in a real day – and nearly 1,800 people had participated in the discussion forum with over 6,000 postings. Furthermore, even during beta tests, which ran from May to September 2003, over 1,500 people supported the project by testing the software and providing feedback through the discussion forum. During this period over 500, 45-year simulations were completed, making this the largest GCM-based climate ensemble ever completed, even before the project was officially launched.

Ongoing work is required in a number of areas. There will be a continuing need to respond to questions from participants and to review and expand the Web site to reflect the queries received. More critical for long-term involvement and education is the need to provide software upgrades which enable more regular and detailed feedback of how each model simulation is progressing so that this can be presented on the Web site and promote discussion between individuals and teams. Anecdotal evidence from the discussion forum suggests that the visualization software (Figure 12.1) is a major factor in maintaining participant interest and the secondary package (Figure 12.5) will be even more so. Nevertheless, for the project's long-term success there is a need to involve appropriate researchers to study participant attitudes, factors affecting participation, attrition rat changes in comprehension, etc.

On the climate research side there is a need to prepare the next phase of experiments involving simulations of the 20th and 21st century using model versions with a full, 3D ocean (HadCM3L and HadCM3). These experiments raise a series of additional software complications. The volume of data produced by these models will be greater so there is a need for additional post-processing to analyze the data and reduce its volume on the client machines. Within this post-processing the simulation results will be compared with observations, raising a need to include the appropriate observational data sets in the distributed package without it becoming too large. If this comparison suggests that any particular model version cannot simulate the past sufficiently well, then that simulation may be stopped and the participant asked to continue, using a different model-version which would require the transferring of model state dumps (the internal model checkpointing procedure) possibly in a peer-to-peer fashion. There are also a number of additional complexities relating to the preparation of AOGCMs with full, dynamical oceans in which perturbations have been made to the way the physics of the system is represented. These will require using diagnostic data files from the first release as part of the package for the second. Matching up simulations with the same parameter perturbations and distributing the appropriate files adds complexity to the server-side database structure. Furthermore, preparation of a perturbed version of the ocean will require substantial preparatory simulations to "spin-up" the model versions to a stable state before they can be used to simulate the 20th century [275]. This will require significant levels of computing which may be undertaken "in-house" using PC clusters or may be built into a special Linux package of climate*prediction*.net.

Of course, the project has received many complaints from people running operating systems other than Windows regarding the lack of a package suitable for them. Certainly providing packages for Macs and UNIX-based machines is an additional aim, though highly non-trivial. Compiling the model, which consists of more than 500,000 lines of FORTRAN, on different architectures is time-consuming in itself, but combined with the system calls required for an integrated visualization and for network connectivity, this requires major investment of effort. Nevertheless, it is hoped to achieve a multiple platform release in the future by collaboration with colleagues working on the generic distributed computing platform BOINC [662].

The project is likely to produce TeraBytes of data on the upload servers (Figure 12.4) and analysis of this data will require Grid and DataGrid techniques [104; 188] since gathering all the data in one place may become infeasible. There is already substantial effort by a number of teams in this area and it is hoped that suitable collaborations will provide Web Services [788] tools which will be easily used by climate researchers.

Finally, there has been substantial interest from schools regarding the use of the climate*prediction*.net software as a teaching resource. It is very much hoped that support can be obtained to develop educational packs to facilitate this. Furthermore, the project is hoping to implement new Web technologies such as the semantic Web [144] for better education and participant involvement opportunities, and BuddySpace [75; 149] for collaborative learning, along with expanded isualization packages including peer-to-peer facilities.

12.7 Conclusions

Climate*prediction*.net was successfully launched in September 2003 and represents a major step forward in the technology of distributed computing and perhaps the development of a new theme of "distributed modeling". The step increase in computational complexity requires an inherently multi-disciplinary approach involving software engineering, a research discipline and collaborative learning and community building. This raises many issues including the management and funding of multi-disciplinary projects which are often treated cautiously by all research funding bodies because of their possible fit with another. Furthermore, commercial sponsorship is often hard to obtain because the risks are seen as high until a prototype has been fully tested. However, such projects have massive potential to provide computational resources to a certain class of research problems, and at the same time actively involve the wider community in the process of scientific research. In an era when science is often regarded with a certain level of distrust, this must be considered highly valuable.

In climate*prediction*.net, beyond the activities described above, it is hoped to expand the research to include a selection of alternative AOGCMs by collaborations with research teams around the world, and in the longer term to develop experiments involving regional climate modeling and climate change impacts and mitigation studies. With community support it could become a community tool for climate researchers worldwide and by involving the wider public the number of professional, and amateur, climate scientists may increase.

Acknowledgements

We would like to gratefully acknowledge the support of the UK Natural Environment Research Council and the Department of Trade and Industry for supporting this work within the e-Science and Coupled Ocean-Atmosphere Processes and European Climate (COAPEC) programs. We would also like to thank the many individuals who have contributed to the project development and the following organizations for their support: the UK Met Office, Research Systems Inc., Numerical Algorithms Group, Ltd., Tessella Support Systems, plc., Risk Management Solutions, plc.

Chapter 13
Web Portal and Markup Language for Collaborative Environmental Research

Xiaorong Xiang • Gregory Madey • Yingping Huang • Steve Cabaniss

Abstract. *The main purpose of existing Web portals is to promote information sharing over the Internet. In this chapter we present the design and implementation of a science portal, the Natural Organic Matter research portal, which is not only for information sharing but also provides tools to allow scientists to share remote computational resources and do science on the Web. An XML-based Markup Language, NOML, is provided to build the XML-based Web components and facilitate Web services development in the future. This portal also consists of a set of online simulators with an intelligent interface for guiding simulation configuration and various electronic communication tools. The development of this portal takes advantage of the J2EE and Oracle database technologies.*

13.1 Introduction

Scientific portals are intended to promote information sharing and discussion among scientists over the Internet. They allow scientists to share expensive research instruments as well as data and information stored at distributed sites, to exchange personal experiences, and to accelerate the development and dissemination of knowledge. Warr says that the term "portal" suggests a doorway or gateway to other places [601]. The author also states, according to a recent survey, that Internet users see portals as destinations rather than as gateways to the Web. Several scientific portals have been developed for various scientific purposes, including the Eco-Portal [694] for environmental sustainability, the Microbiology Information Portal [742], the Earth Sciences Portal [688], and the eSeFDee Marine Sciences Portal [702]. All of these portals include search engines, links to other Web sites, directories, and a set of communication tools. The main purposes of these portals are to disseminate information on regulations, policies, knowledge and news.

In this chapter we present the design and implementation of a science portal, the Natural Organic Matter (NOM) research portal, which is not only for information sharing but also provides tools to access remote computational resources and do science on the Web. The purpose of this portal is to build a seamless environment where scientists can work together virtually and to support the collaboration between scientists. Electronic mail, discussion boards, and electronic notebooks are typical asynchronous communication tools, while audio/video conferencing, chat boxes, and white boards are synchronous communication tools [310]. Determining which tools are suitable for building into a portal depends on whether the tools add value to the scientific interaction process. It also heavily depends on the requirements of end-users, the collaborators.

Besides these generic communication tools that can be integrated into the portal, the software packages that scientists use to do science are developed separately and integrated in such a way that they correspond to different scientific topics. The design and implementation of portals are important to create a flexible portal.

The NOM research portal provides capabilities for scientists to share computational resources (including large-scale databases and a high-performance simulation cluster), analysis tools, simulation results, and data and information. It also allows scientists to communicate with each other through a wide range of tools. The NOM portal also provides an XML-based Markup Language, NOML, which is used to manage molecule information and simulation configurations. Additionally, NOML is used to build the XML-based Web components to support collaboration. Using the universal XML-based data representations enhances component reuse, enables data sharing, and facilitates Web services development. The NOM portal provides several online simulation models to allow scientists to do various experiments via the Web. This portal has been built, evaluated, and used by several environment research groups [27; 82; 83; 347].

In this chapter, an overview of the NOM portal is introduced in Section 13.2. NOML, an XML-based format for describing the molecule structure and the simulation configuration, is presented in Section 13.3. The implementation of several portal components is described in Section 13.4. The evaluation of this portal is described in Section 13.5, and conclusions are drawn and future work is described in Section 13.6.

13.2 NOM Research Portal Overview

Natural organic matter (NOM) is a mixture of molecular compounds with different types of structures, compositions, functional group concentrations, molecular weights, and different degrees of reactivity. NOM comes from animal and plant material in the natural environment. It exists everywhere in the world, from terrestrial ecosystems to aquatic environments. NOM plays a crucial role in ecological and bio-geochemical processes such as the evolution of soils, the transport of pollutants, and the global biochemical and geochemical cycling of elements. The evolution of NOM over time from precursor molecules to mineralization is an important research area in a wide range of disciplines, including biology, geochemistry, ecology, soil science, and water resources. NOM, a prevalent constituent of natural waters, is highly reactive with mineral surfaces [83]. While NOM is transported through soil pores by water, it can be adsorbed onto or desorbed from mineral surfaces. Sorption of NOM is an important consideration in the treatment of drinking water.

The NOM project involves a group of NSF-sponsored researchers from different research areas, including chemists, ecologists, biologists, and computer scientists. As is often the case, these researchers are also geographically separated. A Web-based portal was built to enable scientists from different geographical areas and research areas to work together virtually. The NOM research portal is ru ning on a distributed cluster. This cluster has multiple dual processor PCs ning Redhat Linux 8.0 and Windows 2000 server operating system. The machines in the cluster include a HTTP server, application servers, database servers, a reports server, and a data mining server. The architecture is showed in Figure 13.1. Scientists can access this portal from any standard Web browser.

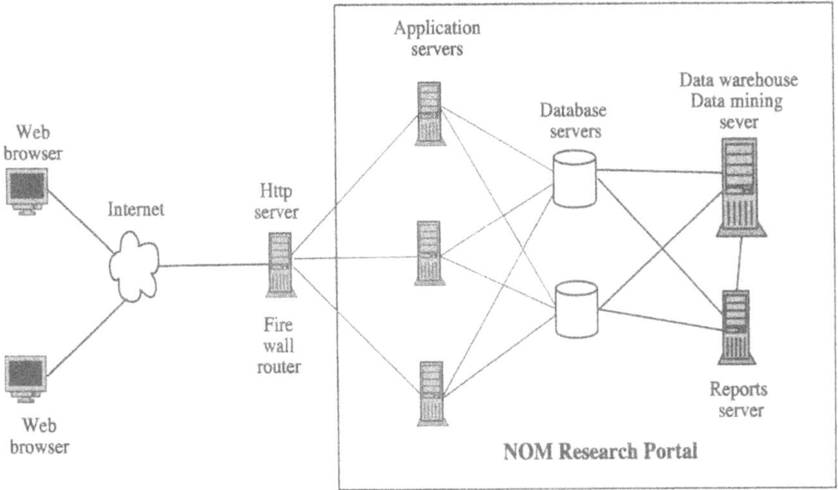

Figure 13.1: The architecture of the NOM research portal

13.2.1 Technologies and Platform

With the evolution of Internet technology, especially after Sun Microsystems introduced the J2EE technology [728], there is growing interest in using the Web as a new platform for scientific applications. J2EE provides developers with the development tools and runtime capabilities to build applications that meet security, scalability, and maintenance requirements. Employing the J2EE technology that includes Java Server Pages (JSPs), Servlets, JavaBeans, and Java Database Connection (JDBC) can overcome the disadvantages of traditional alternatives, such as Common Gateway Interface (CGI), for the dynamic creation of Web pages. The Oracle database management system was chosen for this purpose due to its popularity, reliability, and powerful functionalities. OC4J (Oracle9iAS container for J2EE) was selected for the application servers. It is a fast, lightweight, highly scalable, and easy-to-use J2EE environment. OC4J is written entirely in Java and executes on the standard Java Development Kit (JDK) virtual machine. It provides all the containers that J2EE specifies and can be downloaded from the Oracle Web site [756].

13.2.2 Design Pattern

In order to easily maintain the programs, add new components, and reuse existing components, the design of the portal follows the model-view-controller design attern [288]. All components of the NOM portal are implemented within a thin- ent, three-tier architecture by following the model-view-controller design pat- Servlets process the incoming requests and manage the session and then call the corresponding JavaBeans components to handle the application logic. Finally, JSPs generate the dynamic-presentations to users. Components can be easily added, removed, and modified without changing any other features.

A client tier interacts with end-users and displays information from the server to end-users. In general, HTML and Java applets in a client container implement this tier. For the design of our Web interface, we use standard XHTML, XML/XSL, and JSPs instead of Java applets. Although applets can make the Web page look more attractive, not all Web browsers support Java applets very well. Sometimes, users need to download the Java Runtime Environment (JRE) in order to run the Java applets correctly. Normally, the size of a Web page containing Java applets is bigger than that of a Web page written in XHTML/JSP and has longer response times, especially a problem when the Internet connection is slow.

A Web tier accepts users' responses from the client tier and generates the presentation logic. We used JSP pages for presentation logic and Servlets for session management.

The application tier handles the core scientific logic of the application. EJB components in an EJB container and JavaBeans are used for implementing the application tier. In the NOM system, we used JavaBeans. An Oracle database was designed as the backend to store all the data input and output.

13.2.3 Functionalities of the NOM Research Portal

The NOM portal provides the following functions:

- *Distributed computational resource utilization:* Users can configure and invoke their simulations through a Web interface, and a job manager allocates computational resources on remote sites transparently.
- *Data analysis:* Users can view their simulation data, generated by the NOM simulator, from a Web-based interface. These data and information are represented in various types of graphs (bar charts, pie charts, line charts) and statistical reports by employing data query and data mining technologies.
- *Information sharing:* Users can share the results of their simulation, the molecule definitions, and the simulation configurations through Web interfaces and a search engine.
- *Data repository:* Oracle databases are used to store the internal data that are generated from the NOM simulator. Additionally, external data, including publications, technical reports, and other forms of dissemination, which are uploaded by scientists, are also stored in the database.
- *Secure access:* Users do not have the same level of access privileges to all the tools in the NOM portal. Some particular tools can only be accessed by authorized persons. Users have access to their own simulations, and other users cannot access data that have not been authorized for public usage.
- *Communication tools integration:* A discussion board and a chat room are integrated into the portal in order to facilitate communications between users.

13.3 XML-based NOM Markup Language (NOML)

XML allows user-defined tags to be used to represent any kind of information. Users can develop their own XML document formatting, content, and specifications by defining a set of tags and attributes to represent their own data [363; 450].

The XML-based model ensures uniformity across components and helps to abstract the component structure and implementation from the component interface. The NOM portal provides a set of standardized XML DTD definitions that form an ontology for the simulation configuration and the molecular structure definition. This standard format can reduce efforts of information interchange between users. Users can create XML documents, conforming to the standardized NOML DTD definitions, to describe simulation configurations and molecular structures. Once users have stored the information in NOML format files, they can attach these files into e-mail or documents and share the data with others. Users can also upload files using the NOML uploader service that is embedded in the NOM portal. Without using the Web-based configuration interface, a simulation can also be invoked after a NOML-based configuration file is uploaded. Three DTDs are defined in NOML to describe various data information.

- *environment.dtd* describes the format of the environment information. Users can define a set of environment parameters using tags that are defined in the DTD and used to indicate the owner as well as the access privilege.
- *molecules.dtd* describes the format of the molecular information. Within an XML file, users can define more than one new molecule structure and specify the access privilege.
- *setup.dtd* describes the format of the simulation configuration. Users can define one set of parameters, including environment parameters, molecule types, and molecule distributions, to configure a simulation. The DTD is shown below.

```
<!--  simulationsetup  DTD    setup.dtd  -->

<!--  Root Element  -->
<!ELEMENT setup (environment, moleculetypes, owner)>

<!--  Main Element  -->
<!ELEMENT environment (environmentid)>
<!ELEMENT moleculetypes (moleculetype)*>
<!ELEMENT moleculetype (moleculeid, distribution)>

<!--  Basic Element  -->
<!ELEMENT environmentid     (#PCDATA)>
<!ELEMENT moleculeid       (#PCDATA)>
<!ELEMENT distribution     (#PCDATA)>
<!ELEMENT owner           (#PCDATA)>
```

Users can view the DTDs and download the NOML examples from the Web. The simple tree structures of the sample documents in NOML format are shown in Figure 13.2.

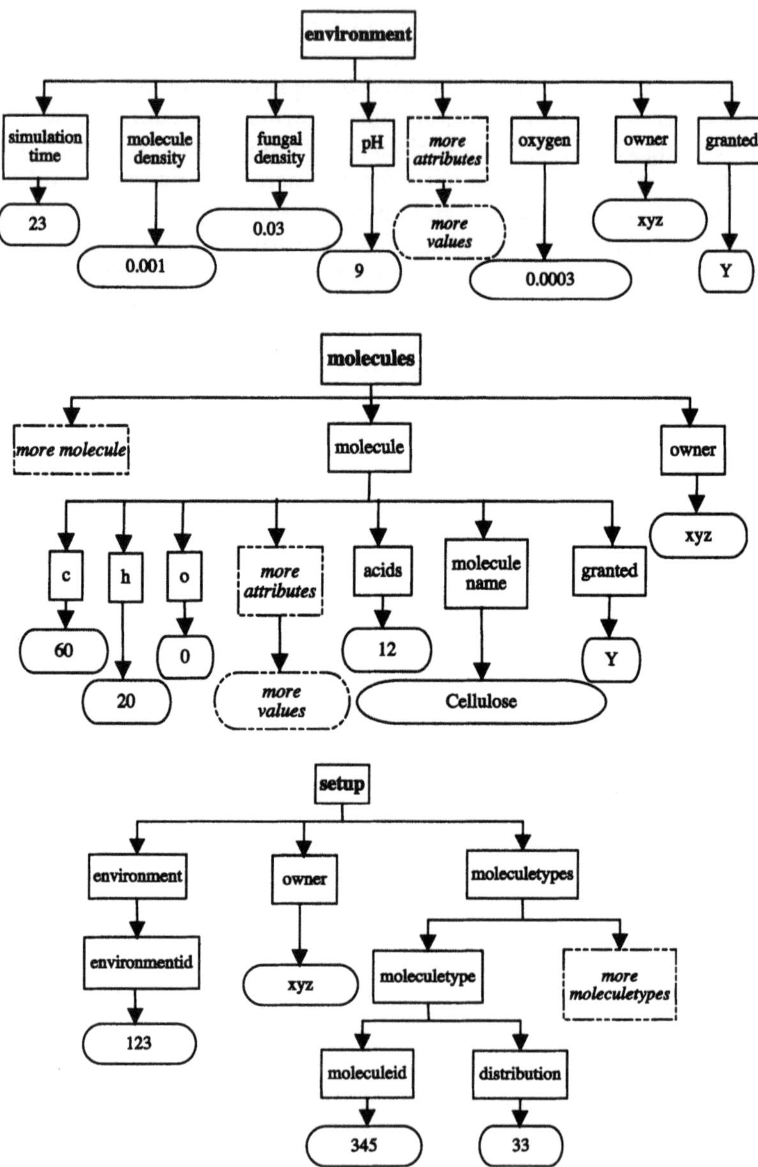

Figure 13.2: Tree-structured view of three types of documents in NOML format (not all the elements are shown in the graph, as indicated by the dash lines)

13.4 Portal Components

The NOM portal includes Web-based simulators, job manager, search engine, NOML uploader, molecule editor, molecule validator, chat room, discussion board, and document management tool. The details of these components are described in the following sections.

Figure 13.3 illustrates components in the NOM portal. Users can access these tools and services in the portal environment through multiple Web-based interfaces from a standard Web browser.

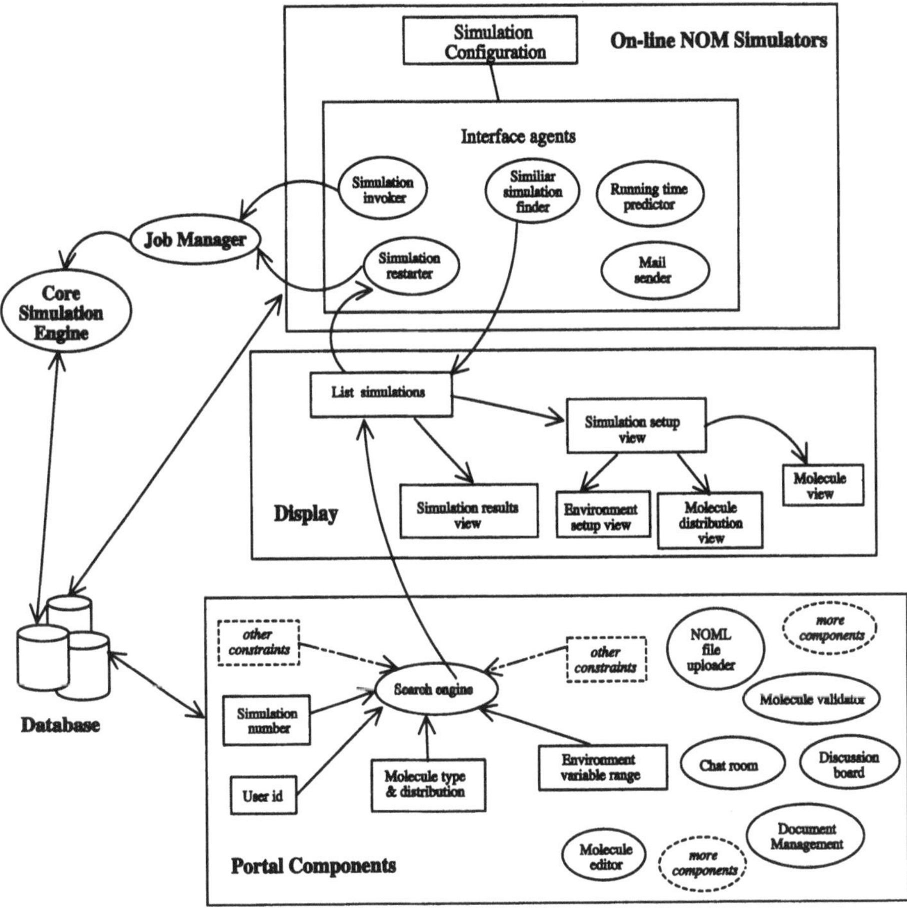

Figure 13.3: Components of the NOM research portal environment

13.4.1 Web-based NOM Simulators

NOM simulators are core parts of this portal. Users can access these online NOM simulators through an intelligent Web interface using a standard Web browser. Several simulation models can be chosen to do their experiments. They can configure their simulations with multiple XHTML and JSP pages and invoke their simulations on the remote computer cluster. The intelligent Web interface provides a set of intelligent agents to find similar simulations, stop and resume simulations, predict running time of simulations, and send e-mail to users after simulations are finished. These agents guide users to use the computational resources more efficiently. A user can submit one or more simulations and several users can ubmit their simulations simultaneously. A simple job manager assigns tasks on eral simulation servers to achieve load balancing.

Currently, in order to meet different requirements from different users, two different NOM simulators have been built into the portal. Each of these NOM simulators is an agent-based stochastic model with an intelligent Web interface that can model NOM, mineral surfaces, and microbial interactions near the surface of the soil. In the stochastic model of the evolution of NOM in discrete time and space, NOM is presented as a large number of discrete molecules with varying chemical and physical properties. Individual molecules can be transported through the soil medium via water flow, adsorb on the soil particle surfaces, and react with other molecules, micro-organisms, and the environment. The chemical and physical attributes, as well as behaviors of molecules are simulated using the Java programming language and Swarm library from Santa Fe Institute [765]. More details on the modeling and the implementation of core simulation engines are described in [640].

The purpose of these NOM simulators is to provide scientists with a test-bed for their theoretical analysis and experimental results for studying NOM. As the simulation system is refined, we expect that the system will help many scientists better understand the NOM complex system by providing them with information for predicting the properties of the NOM system over time.

13.4.1.1 NOM Simulation Model Overview

An NOM simulation model includes three parts: an intelligent Web interface that assists users in configuring simulation parameters, core simulation engines that do the computations and simulate the complex behavior of a large number of molecules, and a data analysis package that allows users to view their simulation results through a standard Web browser. Users can configure a simulation by inputting the simulation parameters either from the Web interface or from a NOML file. These inputs are stored in the backend Oracle database and a unique identification number is assigned to this particular simulation. The output data is stored into the database while the simulation is running. Figure 13.4 shows this process.

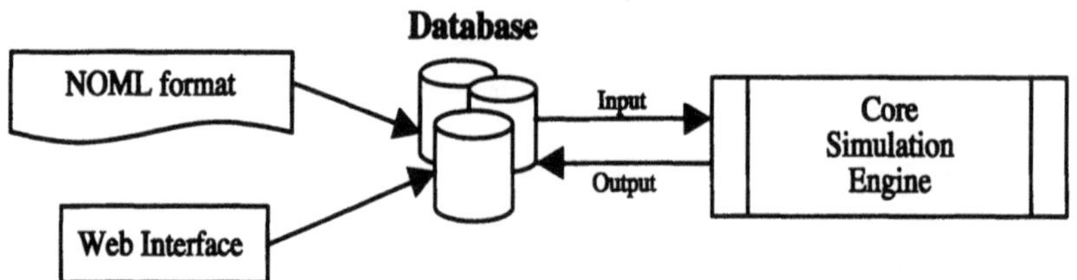

Figure 13.4: Simulation input and output

13.4.1.2 Intelligent Agents

Intelligent agents built in the Web interface can not only provide a novice user with guidance during the configuration process, but can also provide the experienced user with more information on how to better use the simulator. A summary of these intelligent agents is described below:

- The *sending e-mail agent.* The agent automatically sends an e-mail to the user after the simulation is done. This e-mail message gives the user information about the status of the simulation, either it was successfully completed or it was terminated due to an exception that was caused by a hardware malfunction or software problem.
- The *running time prediction agent.* A user can ask this agent to statically predict how long it will take to run the simulation according to the user's configuration inputs before the user invokes the simulation. According to the time returned by the agent, the user can decide if the simulation configuration should be adjusted. This agent also provides the capability for dynamically predicting the running time while the simulation is running.
- The *similar simulation finder.* Since executing a simulation is a time-consuming task, especially when the problem size is large and the running time is long, we provide users with the option to take advantage of simulation results that were generated earlier by the user or by other users. A similar-simulation-finder-agent helps users find finished simulations that have some similarities with the current setup in the database. Users can either retrieve the simulation results from the database without running their simulations or submit the simulation for execution according to their own preference. The agent can return an ordered list of similar simulations, with the most similar simulation listed first. Users can also view the parameter setup of similar simulations. The basic algorithm for finding the similar simulations is described as follows:

  After users issue a request to search for similar simulations, the agent retrieves the data sets in the database. These data sets can be viewed as a number of points $(a_1, a_2, a_3 \ldots a_i \ldots a_n)$ in a high-dimensional space with m attributes $(1, 2 \ldots j \ldots m)$. The task is transformed to compute the distance from the current parameter setup to the other parameter setups. Each attribute in these data sets is normalized to the [0,1] interval using the equation

  $$\frac{a_{ij} - \min(a_i)}{\max(a_i) - \min(a_i)}$$

  There are several efficient algorithms for finding the nearest neighbor on a high-dimensional plane. For simplicity, by employing the "Euclidean distance", the distance from the current input data set to other data sets retrieved from the database can be computed. The data sets that have the smallest distance have the highest degree of similarity with the current setup.
- The *automatic restarting agent.* The agent helps users extend their simulations for longer time steps by restarting their simulation from the point where it stopped. While one simulation is running, the data information about reactions is stored in the database every timestep for further data analysis. In addition to this, the state of the simulation system is stored in the database at different checkpoints. A table is created for storing the state. A series of SQL statements is executed through JDBC code. The insertion and updating of the state at each checkpoint are treated as a data transaction process to guarantee data integrity.

13.4.2 Data Analysis and Visualization

One of the primary goals of this portal is to provide an explanation of the simulation results, to help users better understand their data, and to predict the global properties of NOM over time. Large-scale scientific simulations produce massive amounts of data and take days to run. A user typically needs to run several variations and compare many sets of results with their experimental data. It is very difficult for users to do this with only raw data. Gray and Szalay show how important new data mining algorithms are in helping scientists to access their online data more quickly [213]. The Oracle9i tools are used to build the data warehousing and data mining and to incorporate them into the simulation model in order to organize, visualize, and analyze large data sets. Huang describes the detail of building the data warehousing and employing data mining technologies [248]. We enabled the display of simulation results independently from the processing. Users can examine the results and download the data in XML format from the Web site during the simulation and after the simulation completes running. Figure 13.5 shows examples of both input information and output from a simulation executed by a scientist.

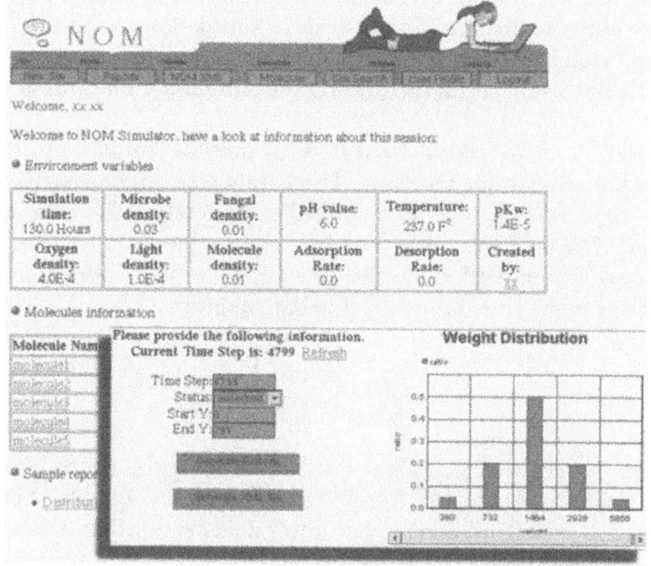

Figure 13.5: Configuration and sample output of a simulation

13.4.3 Job Manager

After users submit jobs, a simple job manager in this portal is invoked and assigns tasks on several simulation servers to achieve load balancing. The goal of load balancing is to distribute simulations evenly among application servers. There are many load-balancing schemes available in the literature. A simple round robin algorithm is used for job manager implementation. New jobs a assigned to the application servers sequentially. For example, suppose we hav jobs to be assigned to five application servers, then the first job is assigned to application server one, the second job is assigned to application server two, and so on. Job number six is assigned to application server one again.

13.4.4 Search Engine

Collaborative research demands sharing information and experiments to reduce the time it takes to produce new results. Executing a large-scale scientific simulation is a time-consuming task. The NOM portal allows users to share the configurations and results of their simulations by providing a search engine. Users can take advantage of the simulation's stored results from previous executions instead of doing the same simulation over and over again. By achieving this capability, the instruments and computational resources can be used more effectively.

Users can access the search engine through a Web interface by providing several search conditions. Users can either leave all the fields blank or make reasonable combinations and submit the request form to the server. The remote server processes the request and returns a list of simulations associated with the simulation configurations that meet users' requests. Users can view data information corresponding to each simulation in more depth by clicking on links. Users can also view the analysis data of a particular simulation. If the simulation results are valuable for users, they can extend the simulation by entering extra simulation time and resuming the simulation from the Web interface.

13.4.5 NOML File Uploader

The NOML file uploader provides an interface for users to upload a set of NOML format files. Users issue a Multipart Request by submitting a HTML form from the client side. On the server side, a Servlet was implemented to parse the request. Instead of saving files in the disk on the server, the Servlet reads incoming files and parameters and parses the incoming file's information to a XML FileInputStream. The FileInputStream is fed into a corresponding JavaBean according to the content of the incoming files. JavaBeans are implemented to validate and parse the NOML documents, establish a connection to the Oracle database, and write the data in the database. The Servlet returns the information to users for further reference. The design model is shown in Figure 13.6.

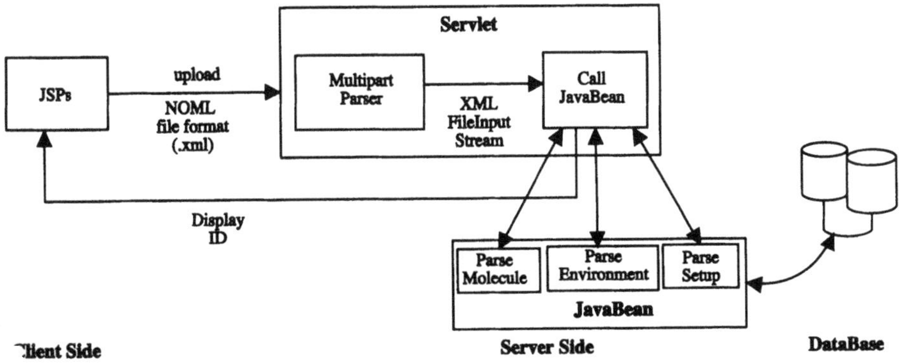

Figure 13.6: The design model for the NOML file uploader

Users can access the NOML file uploader by providing a file name. The server side JavaBeans does the data validation. No information is stored in the database if the data is invalid and corresponding error messages are returned to users. Users can create their NOML files using a text editor. Allowed formats are published on the Web. Users can download the example files and change the values. An example NOML file is shown as follows:

```
<?xml version='1.0'?>
<!DOCTYPE setup SYSTEM "http://tenor.cse.nd.edu/setup.dtd">

<setup>
    <environment>
        <environmentid>102 </environmentid>
    </environment>
    <moleculetypes>
        <moleculetype>
            <moleculeid>1 </moleculeid>
            <distribution>33 </distribution>
        </moleculetype>
        <moleculetype>
            <moleculeid>2 </moleculeid>
            <distribution>67 </distribution>
        </moleculetype>
    </moleculetypes>
    <owner>xyz </owner>
</setup>
```

13.4.6 Molecule Editor

The "molecule editor" tool provides a flexible way for users to define new molecules for their simulations. Several simple rules are encoded to automatically validate the newly defined molecule in order to prevent the specification of chemically impossible molecule structures. For example, a simple rule is defined as follows.

If the "benzene" ring structure is part of the molecule structure, then the total number of carbons (atom C) in this molecule structure definition must be equal to or larger than six times the number of "benzene" rings.

Besides providing a Web interface for the molecule definition, users can define new molecule types by uploading a NOML format file to the server. Users can also search particular molecules according to the molecule name and owner of the molecules.

13.4.7 Molecule Validator

Before one molecule type definition is ready to be shared by the NOM community, this molecule definition must be validated. However, the molecular validation is not only an objective task that can be accomplished by a piece of software, but also a subjective task that needs expert human involvement. General rules that can be encoded in software are not sufficient for validating a newly created molecule. Therefore, an administrator role is provided to an authorized person who can validate newly defined molecules manually.

The authorized user, a chemist in the general case, can access the validator by clicking a button. A set of molecules that is indicated for public usage by their owner is displayed. Users who participate in the portal can share only validated molecules. Non-validated molecules can only be used by their owners.

An authorized user can also define the molecule type from the separate molecule editor or upload a NOML format file to the server. This user can remove a particular molecule from the NOM system.

13.4.8 Chat Room and Discussion Board

In order to facilitate communications between researchers, a chat room (a synchronous collaboration tool) and a threaded discussion board (an asynchronous tool) are integrated into the portal. Users can enter the chat room through a Web interface by providing a user ID and password. The threaded discussion board helps users to easily trace discussion contents about one topic.

13.4.9 Document Management

Collaborative research often demands exchange of large data files and documentation for faster dissemination of the knowledge. Users can upload research papers and data sets through a Web interface into a shared place residing on the remote servers. These files are accessible to other researchers after they are uploaded.

13.5 Evaluation

We conducted a preliminary evaluation and testing of the NOM portal using several approaches: (1) a value analysis to the research scientist users, and (2) formal software testing. Each approach is briefly discussed below.

The potential value of the NOM portal to environmental research scientists can be broken into two categories: (1) support to the actual scientific research and (2) communication support. The NOM portal permits scientists to compare their field and laboratory data with predictions made by the NOM simulators within the simulator. Several simulations, each developed for a different scientific project, are implementations of models and theories developed by the scientists to explain their experimental data. For each simulation, the proposed explanatory model or theory could not be easily reduced to a system of equations and thus, a simulation was the practical way to compare the predictions of the model or theory against observed data. In all cases, the initial simulation did not fit the data and the underlying model was revised, resulting in a better model or theory of the NOM phenomenon being studied by the environmental scientists. Once the scientists were convinced that their revised explanatory model or theory satisfactorily fitted their experimental data, the simulation was made available to other environmental scientists who either compared the predictions of the simulations with their data, or "played around" with the simulation to gain an understanding of the proposed model of the original environmental scientist, then used the communication tools of the portal to exchange ideas and comments between the scientists.

A formal test plan was developed, implemented and used to assess the alignment between the design goals and the implementation of the NOM portal. Standard software engineering methods were employed [255; 472] resulting in minor revisions to the NOM portal.

13.6 Conclusions and Future Work

In this chapter, a description of a Web-based portal environment that supports new ways for conducting environmental research is presented. The NOM portal deploys collaborative technologies to NOM researchers geographically separated or in different disciplines. This system integrates a number of service components, including NOM simulators, a molecular validator, a molecule editor, search engine, document management, and communication tools. The architectural design of the NOM portal makes it easy to add, modify, and remove components in the portal according to the requirements of scientists.

In the future, some powerful third-party tools (such as video/audio conferencing) that rely on a fast and efficient high-bandwidth network technology [723] and advanced hardware may be integrated. Oracle Collaboration Suite [755] offers integrated e-mail, voice mail, phone, fax, scheduling, calendaring, meeting management, and file management. It could also be considered for integration into the NOM portal.

An XML-based NOM Markup Language, NOML, which provides a standard definition for the molecule and simulation configuration has been described. The NOML not only supports data communication between users but also give us flexibility in extending the online NOM simulators. It also supports building new XML-based Web services in the near future. Due to the limitations of DTD for describe the NOML, XML Schema (XMLS) will be used to describe the NOML. A NOML editor will be created and integrated into the portal.

By requests and suggestions from our users, a set of new simulation models for the NOM study will be built and integrated into the portal. The communication between these models can be achieved by the extension of the NOML. The NOM portal allows users from different places to collaborate by sharing their simulation data and configurations as well as by hosting discussions among the NOM community.

Acknowledgements

This research was supported in part by a NSF ITR Grant No. 0112820 and by the Center for Environmental Science & Technology at the University of Notre Dame. We acknowledge the contributions of Patricia Maurice and Leilani Arthurs of the Department of Civil Engineering & Geological Science at the University of Notre Dame for their respective discussions on the Web interface, input, and testing of the NOM portal.

Chapter 14
Web-based Exploration of Environmental Data and Corresponding Metadata, in Particular Lineage Information

Andri Baltensweiler • Martin Brändli

Abstract. *A Web-based platform for visualizing, querying, and analyzing environmental data focusing on metadata handling is presented. The chapter addresses the database design and Web communication of lineage metadata in order to comprehensively describe the origin of stored data sets. Departing from a detailed list of requirements for an Internet GIS solution, an architecture and implementation of an online communication platform is derived. The implementation is based on the Web components of the Java 2 platform and integrates ESRI's Internet mapping software ArcIMS. Experiences from a first prototype implementation show that providing metadata, in particular lineage information, enables environmental data exploration in a temporal, spatial and semantic context which is thoroughly accessible to the user.*

14.1 Introduction

Traditionally, geographic information has been distributed via paper maps and digital media such as disks or tapes which are costly and difficult to update. Therefore the demand for online geographic information services on the Internet has been stated by a considerable number of representatives of the GIS community. In particular, federal agencies are faced with expectations that information will be made available to the public and research needs will be satisfied by effective and efficient methods [414]. Eddy [146, p6] formulated this demand as follows: "We must put in place a global data and information system that makes environmental data, past and current, available to all who need them, in a form that they can use". In the last few years, major GIS vendors and several open source projects developed Web mapping servers which allow visualizing, querying and analyzing spatial data from multiple sources within a simple Web browser. However, accessing distributed geographic information via the Internet without any information about semantics, spatial references, data quality, formats, etc. makes it difficult to use data in an appropriate way. This information is provided by means of metadata which should be a fundamental component of current Internet GIS solutions [341].

In this chapter, a Web-based platform for visualizing and analyzing environmental data with the special feature of comprehensive metadata handling and communication is presented. The work described is part of an ongoing research project called "Data Center for Nature and Landscape" (DNL) conducted in collaboration with the Swiss Agency for the Environment, Forests and Landscape. The database of the DNL is conceived to store data and metadata of endangered biotopes in Switzerland and to provide access via Web-based geographic information services to Swiss federal and cantonal agencies and research institutes.

A particular feature of the Web-based platform and the database is that it addresses metadata for the description of the lineage or history of the stored data. Lineage is defined as "the recounting of the life cycle of a data set, from its collection or acquisition, through the many stages of compilations, corrections, conversions and transformation to the generation of new interpreted products" [111]. Comprehensive description of lineage is one of the key requirements for the operational use of geographic and environmental information within multi-user and multi-organization applications [381; 469]. The knowledge of the life cycle of the data is a crucial element of the DNL database as described below.

14.2 Data Collection and Storage

The Web-based communication platform described in this chapter is built on top of a database which provides environmental data from endangered biotopes consisting of both floristic protection areas such as mires, flood plains and semi-arid meadows, and fauna protection areas such as migratory bird habitats. Endangered biotopes are usually managed by inventories including biotope objects known by name, number, spatial extent and additional characteristics such as the type of vegetation. The realization of an inventory usually involves several steps:

- Process initialization by governmental decision.
- Determination of criteria for choosing endangered areas.
- Selection and production of base documents (images, maps) for digitizing.
- Verification and refinement of potential areas by field work.
- Political negotiations between land owners, administration and conservation organizations.
- Establishment of the legal base for the inventory by the government.
- Repeated monitoring of the effects of protection measures such as management restrictions.

Storing and communicating lineage shall deliver comprehensive knowledge of this realization procedure. This is essential to provide and ensure valid information necessary as a legal basis for lawsuits, subsidies or compensation payments, to be fully documented and to have all the data readily available. More generally, the implementation of the underlying database had the following requirements:

- Handling temporal and spatial information of a stored inventory object. This includes the complete storage of the facts that contributed to the final establishment of the object which allows the tracking of its history.
- Preserving the semantic context for data generation steps in order to handle data for long-term monitoring projects. This ensures that the environmental data are comprehensive and reproducible and existing procedures can be used for future tasks with equal parameters.

- Storage of several different data types (digital images, documents, additional geometric data and attribute data) which result from the individual steps of inventory realization.
- Storage of general spatial metadata according to the ISO International Standard for metadata concerning geographic information [266].

The data are stored in an Oracle relational database. Digital images and documents are handled as so-called Binary Large Objects (Blobs). Spatial data is managed and accessed using ESRI's Spatial Database Engine (SDE) [159]. General geographic information metadata are stored as regular relations according to the schema ratified by ISO. To represent the lineage a process-oriented database model was developed. Every step of delineating and determining protection areas is considered as a process which generates new information. For each process specific metadata such as the processor, event time, used software, hardware, etc. and the generated process data (e.g., images, documents, spatial, and attribute data) are stored. Subsequent processes are linked to form directed graphs which completely define the history of the origin of an inventory object. A detailed description of the design and implementation of the DNL database is presented in Baumberger and Haegeli [43].

14.3 Data Access and Retrieval

14.3.1 Requirements

The software for data access acts as an online communication platform which must be able to visualize, query and analyze spatial data and metadata, track lineage data and display base documents and images in one single application framework [38]. The following requirements must be fulfilled by this platform:

- *Browser-based Web application:* Since various Swiss federal and cantonal agencies as well as research institutions need access to the data, the software has to enable the users to explore the data via the Internet.
- *Easy access to the database:* Because most of the expected users are not familiar with database query languages, data must be searchable by an intuitive interactive and graphical user interface.
- *GIS functionality:* Common GIS tasks such as zooming, panning, spatial selections, etc. have to be implemented and provided in a way that users with no expert GIS knowledge can use them.
- *Graphic representation of lineage graphs:* Data history has to be represented as a graphical lineage network which shows the interrelations between processes.
- *Visualization of different types of data:* Since base documents and images may have any format (text files, PDF documents, TIFF, or JPEG images) the access software must be capable of rendering the various formats.
- *Security:* Since the inventories contain sensitive data about endangered species authentication mechanisms have to be implemented.
- *Performance:* A Web-based application will only find general acceptance if it is st and reliable.

14.3.2 System Implementation

14.3.2.1 Access to Spatial Data, Attribute Data, Base Documents, and General Metadata

The Web application developed for visualizing and querying environmental data and metadata is based on the Java 2 Platform using Java Server Pages (JSP), Java Servlets [23] and the Jakarta Struts Framework [95]. It takes advantage of ESRI's Internet mapping software ArcIMS [160] which is based on a typical three-tier architecture consisting of presentation, business logic and data tier (Figure 14.1).

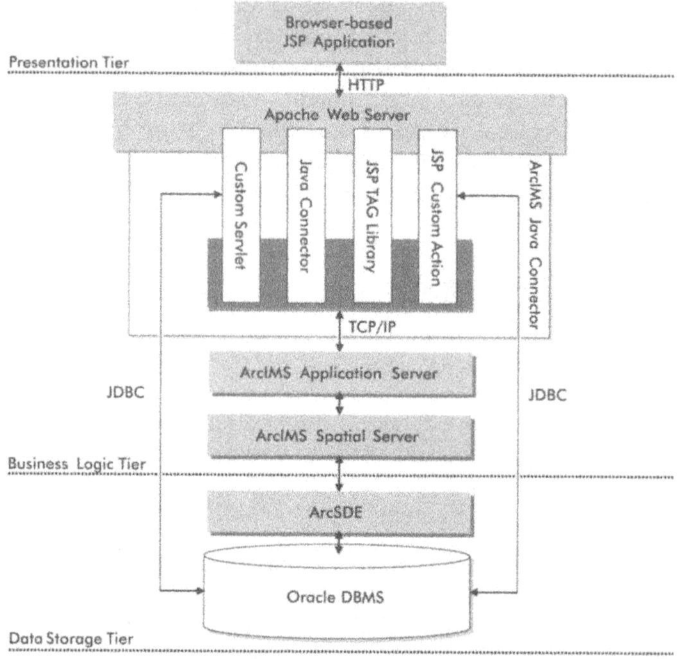

Figure 14.1: Scheme of the online communication platform of the DNL project

The visualization and querying of spatial data and corresponding attribute data for the Internet application is based on out-of-the-box components of the ArcIMS package. ESRI's Web mapping software provides an application and spatial server for map compilation and spatial query resolving. Requests to the server are handled by so-called Java Connector Beans and results are rendered choosing from a rich set of JSP tags provided by the ArcIMS JSP Tag library. For visualizing and querying base documents and metadata including lineage information, a custom servlet for request handling and custom JSP actions for information presentation are developed and incorporated in the application as shown in Figure 14.1.

The client interface (Figure 14.2) consists of a toolbar for interactive GIS functions, a map area and a table of content which organizes the inventories as map layers in a folder structure. Metadata about a specific inventory according to the ISO metadata standard can be retrieved by clicking on an icon which is associa to each layer. This act invokes a JSP custom action which executes the necessary SQL statements on the database and generates a resulting new Web page. To obtain information for individual inventory objects, the user can perform a spatial selection on the active layer. Again a JSP custom action is invoked which provides an attribute table containing attribute values and hyperlinks to related base documents or images. The hyperlinks point to subsequent custom actions which retrieve the associated documents from the database.

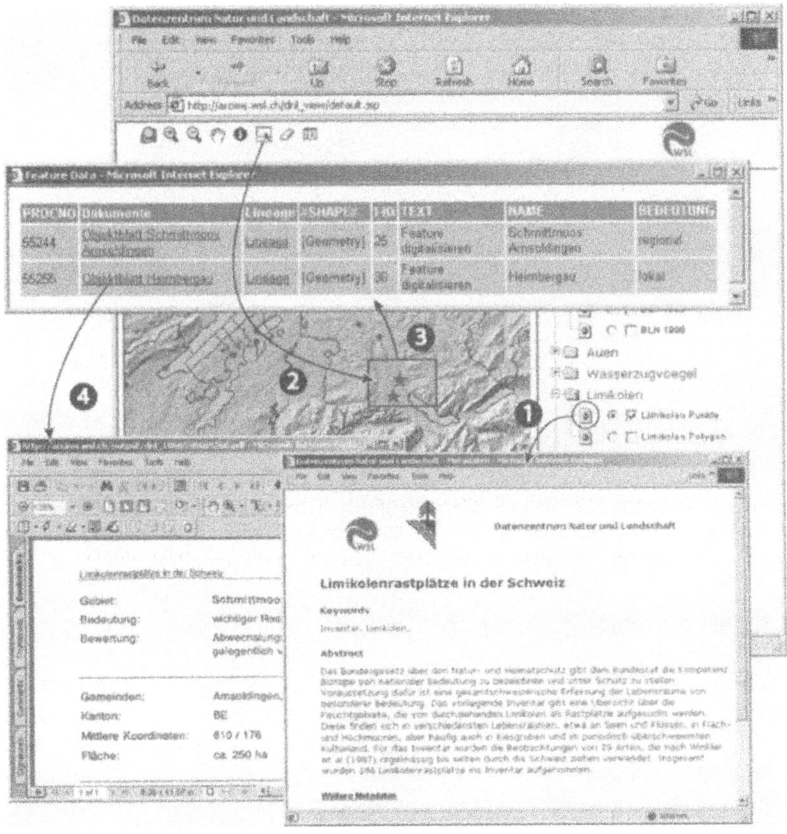

Figure 14.2: Web application interface for visualizing metadata: (1) retrieves metadata of a specific layer, (2) performs a spatial selection for retrieving (3) tabular data and metadata, (4) invokes a custom Servlet to retrieve base documents of a specific object

14.3.2.2 Access to Lineage Metadata

Of particular interest is the visual representation of lineage information. In contrast to offering a pure listing of processed steps, a graphical representation enabling hyperlink referencing the lineage or history of inventory objects is provided. This attractive way of examining lineage substantially promotes attention to this articular metadata element. Lineage information is stored as a directed graph of ocesses. Rendering the graph is achieved by applying a specific graph-drawing rithm [157] using straight and polyline edges and rectangular vertices for graph representation. As a specific requirement for the application, a graph vertex should contain clickable hyperlink information in order to allow retrieval of associated process metadata and data. Since the graphs may be of high complexity and a fine-grained access to graph components is necessary, the usual way of sending graphics as images is not appropriate in this case. Therefore, a solution based on the Scalable Vector Graphics (SVG) format was chosen [634]. SVG is an XML-based language for the description of two-dimensional graphics, in particular vector graphic shapes. It allows drawing of complex lineage graphs and interacting with every single vertex of the graph for our purpose (Figure 14.3), but comes with a drawback. Displaying SVG requires the installation of the Adobe SVG Viewer plug-in for Internet Explorer or Netscape Navigator.

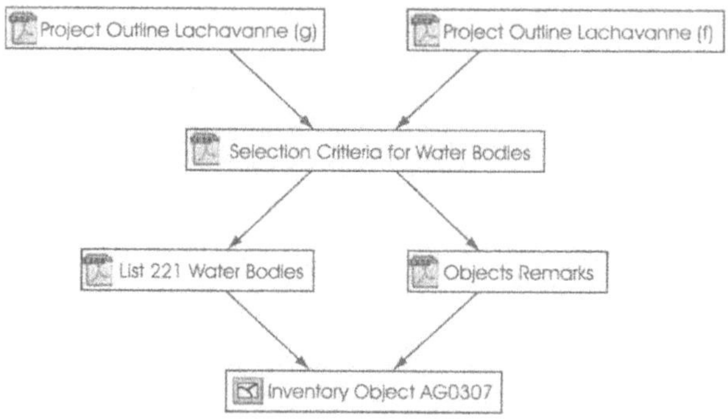

Figure 14.3: An example of a simple directed graph representing the history of creating the inventory object AG0307; several PDF documents such as selection criteria and further remarks for object delineation determine the extent of the inventory object

14.4 Conclusions

The implementation of the described Web application is in the state of a working prototype. First user experiences and feedback are positive and very promising in respect of requesting and serving heterogeneous data over the Internet. The user is able to dynamically explore data of different formats stored in a complex spatially-enabled database without the know-how of a database or GIS expert. In particular, including comprehensive lineage information puts the data into a temporal, spatial and semantic context, which allows an assessment of the data for further use. Moreover, providing lineage information also supports the work of data providers. By visually accessing entered data, lineage tracking allows to verify the content of the database for completeness since missing links between successive processes may be identified quickly.

The online platform satisfies Eddy's requirements stated in the introductory section for establishing an information system able to serve environmental data, past and current, to everybody in a form they can use. However, since the system is currently in the state of testing it is only available to key individuals within the institutions involved in the DNL project. After successful testing all non-sensitive data will be accessible to other interested parties and the general public in spring 2004.

Chapter 15
Sharing Environmental Maps on the Web: The Austrian EnviroMap System

Rudolf Orthofer • Wolfgang Loibl

Abstract. *This chapter describes a system for public dissemination of environmental maps through an interactive Web-based system. This "EnviroMap" system is a platform in which data holders can publish a raster version of their data and data seekers can get access to these data. For data holders, the system provides a tool for data documentation and upload. Data users can access the data with Web browsers and a Java Virtual Machine. The system is based on a raster file format with a predefined map projection and cell size. EnviroMap may serve as a marketplace for environmental maps, where data holders can make accessible an "EnviroMap raster version" of their original GIS data and advertise the availability of their original data. Usability analysis has shown that in order to fully use the functionality more emphasis must be given to a familiar and intuitive user interface for novice users.*

15.1 Introduction

"Environment is the complex of physical, chemical, and biotic factors that act upon an organism or an ecological community and ultimately determine its form and survival". This definition from *Encyclopaedia Britannica* Web edition [697] is straightforward, yet it does not take into account an important feature of environment, namely its spatial and temporal dimensions. These two dimensions are particularly important when environmental issues should be integrated with other aspects. This is important because decisions for improving future living conditions need to incorporate social, economic and environmental aspects. Those three "pillars of sustainability" can only be interrelated via their spatial dimension which underlines the necessity to provide environmental information with its spatial dimension – namely as maps. However, as maps are often complex representations of multidimensional contents, they are usually simplified by the data holders. Nevertheless such maps are a good and established format to communicate the spatial dimension of environmental data to the general public. Increasing public awareness and international agreements such as the EU Water Framework Directive [703] require the national authorities to make spatial environmental information available to the general public.

The complexity of spatial data makes usage difficult. Spatial data exist in a variety of geometric, thematic, and file formats which are often difficult to compare and combine. The rationale behind the development of the EnviroMap system was to provide a tool that gives the public an opportunity to retrieve, combine and display harmonized spatial data in a simple way. This should make users interested in the spatial dimension of environmental issues and their relations to socio-economic effects. EnviroMap was also developed with a view to allowing data holders a marketplace to share their maps with data seekers.

15.2 Spatial Data for Environmental Assessment

EnviroMap aims at providing comparable spatial environmental data to the users. Comparability of spatial data is a pre-requisite to assessing environmental quality and sustainability. There are several concepts to evaluate environmental quality such as the *pressure-state-response* framework [140], or the more complex *ecological footprint* concept [595] that considers multi-layered pressures resulting in different impact-ranges. Both approaches require data that can be compared and integrated in spatial and temporal dimensions.

The EnviroMap concept is based on the pressure-state-response approach. Human activities are driving forces for pressures on the environment. These pressures lead to a disturbed environmental state or disturbed natural conditions. The impacts of those pressures are observed or experienced by people, who then respond to those impacts. Responses are reactions on a personal or a political-societal level, aiming at decisions to overcome personally received impacts, or at mitigation strategies to improve the general situation, respectively [140].

A crucial aspect of environmental assessment is to examine the spatial and temporal coincidence of different pressures that lead to different states and impacts. EnviroMap may serve as a tool that allows observation and combination of data to examine pressures, states and impacts. Several requirements have to be met:

- Data about different environmental parameters should be prepared so that they can be related to each other with appropriate spatial and temporal relevance.
- Data should be visualized in a way that allows the users to identify areas of concern, or to monitor environmental parameters in areas of their interest.
- Spatial overlay of data should allow to combine pressure or state patterns to better understand pressure-impact loops.
- All data must be documented, so that data quality can be assessed (e.g. data sources, scales of data acquisition, units and uncertainties).

EnviroMap is expected to make a contribution to the general goal of Web-based environmental information, which is to provide a "kit of modular information units which can readily be combined to a while range of views on an environmental issue" [462]. The system follows the objectives of the European *Infrastructure for Spatial Information in Europe Initiative, INSPIRE [720]*, which aims to establish a framework that allows data users to identify and access spatial information from different data formats and spatial scales. EnviroMap Austria might be considered a national simplified component on the way towards a European spatial data repository.

15.3 Conceptualization

15.3.1 Bringing Together Data Providers and Data Users

There exist many proprietary Web-based geographic information systems (WebGIS) that allow management, retrieval, synthesis and visualization of spatial data. However, most of these systems are complex and expensive, and require that users have considerable technical background [311; 439]. These systems typically integrate "as-much-as-possible" information for "as-many-as-possible" user needs. EnviroMap has followed a different approach, namely the "as-simple-but-exciting-as-possible" combination and visualization of maps that are relevant to the environment. This approach for the integration, retrieval, analysis and presentation of environmental data to the public takes into consideration that "speed" and "simplicity" of data retrieval are at least as important qualities for the usefulness of such systems as is "completeness" or "integration" [131].

Summarizing the above, the EnviroMap system it not a WebGIS but a rapid and easy map upload and display system that serves as an interactive electronic atlas. To achieve this simplicity and speed, the spatial representation is designed as a raster system with a standard spatial projection.

The EnviroMap system links data holders and data users through the World Wide Web. It is an open system that matches data seekers with data providers. Thus it can act as a marketplace for spatial environmental data (Figure 15.1).

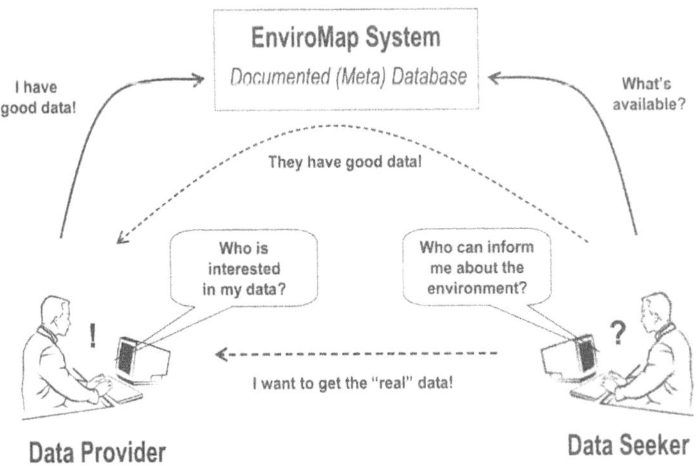

Figure 15.1: EnviroMap as a marketplace for environmental maps

Experience in Austria (and also elsewhere) has shown that digital spatial environmental data exist in a variety of formats in many different institutions. Yet only a few institutions are able to provide their data treasures to the world for combination with data from other users: obstacles are sometimes technical (e.g. the inability to bring the different data formats into a uniform presentation), financial (e.g. the high costs of geographic data management systems), or institutional (e.g. the need to protect the value of the data sets).

Thus one of the aims of the EnviroMap system is to support holders of spatial environmental data to advertise their data holdings to a larger audience. To do this, they can upload a simplified raster version of their original data to the EnviroMap together with information about quality and availability of the original (native) data. This feature would at the same time allow data owners to protect the value of the original data as they would only make available a simplified version of their original data. The system allows upload of data by data holders as long as these are formally registered, and as long as their data conforms to the predefined standard format requirements.

Data users have access to all data on the system. For most users - particularly from the interested public - the simplified raster data provides sufficient information to assess the spatial dimension of environmental parameters; expert users who need access to the detailed original data can get a preview of the available data and they can contact the original data holders.

As EnviroMap users can retrieve information about Austria's environmental situation with the possibility of combining data to create their own maps, it is an attractive tool for "playing" with environmental data. Expert users who need the maps for specific applications would have a reference to the original data holders, where they could turn to get access to the detailed original data.

15.3.2 Simple Standards

The EnviroMap system is based on a standardized data format. This facilitates the combination of data from different sources and with different contents. Standardization includes spatial representation, geoprojection, data description, and data format.

15.3.2.1 Standard Spatial Representation

The standard spatial representation of EnviroMap is a grid (or cell raster) system. Referring to the systems objective, there are a number of advantages. First, there are thematic reasons: a raster representation is appropriate for environmental data as these are mostly ubiquitous and continuous with no clear-cut and defined boundaries. Often, vector or polygon data force contents to be related to statistical (administrative) spatial units so that users get an impression that environmental pressures or states are discrete units, which is not the case.

Furthermore, raster data are better suited to data integration: integration of different vector data systems would need detailed data format transfer routines and much effort for integration of vector data attributes that would not allow concentration on the thematic side of a Web atlas. Line and point data cannot be presented on a small scale unless they are superimposed. This is not a problem for raster data: in a raster map system all features are related to at least one raster cell, not requiring superimposition within a rather large scale-range. The size of the raster cells is selected as a compromise between providing sufficient detailed spatial information and data management efficiency. For Austria, the system is based on a 250 × 250 m grid system, which is equivalent to about six million raster cells. This data size allows detailed spatial information with fast data transfer combination and visualization.

Finally, raster data can also be used to protect data confidentiality. There are cases in which national privacy regulations prevent detailed publication of environmental pressures. The raster format allows consideration of these aspects by integrating confidential data through building sum-, density-, or mean-indicators.

15.3.2.2 Standard Geoprojection

In order to locate the proper positions of the raster cells, the system must have a standard spatial projection. For the application of EnviroMap in Austria, we have defined the Austrian Lambert conformal conic projection as the projection standard. Of course, any system could be defined as standard because the projection is mainly necessary for harmonized visualization of maps.

15.3.2.3 Standard Data Description

As EnviroMap allows dissemination of data from different data holders to many users it is necessary to provide a common level of information quality. Acquiring data within the framework of environmental research is often time-consuming and expensive. To secure optimal data application and analysis it is important for the data holders to provide all information that is necessary for the map data users to review the suitability and applicability for the intended data analysis as "metadata" information. There are several internationally recognized standards for documentation of spatial data sets. However, most of them require a very detailed description. For instance, the European Geographical Data Description Directory (GDDD) that is related to the ISO 19115 standard requires information about 14 metadata categories with more than 100 parameters [265].

Again, the EnviroMap system focuses on the "as-simple-as possible" approach. The EnviroMap data description is an XML-based metadata system which takes into account the important elements of international standards, but has only a few obligatory data fields. The reason for not using a complete international data documentation system is to make it attractive and simple for data holders to upload and document their data. The EnviroMap metadata standard was reduced to a necessary minimum (layer title, content, meaning, purpose, limitations, source, ownership, access restrictions, descriptors, temporal and spatial coverage, spatial accuracy, attribute variables, units and data uncertainty) in order not to chase away users who want to "upload" data to the system because of too high requirements for data documentation. The reduction of metadata requirements is tenable as EnviroMap only provides secondary (lower resolution) raster data and (usually) not the original data sets.

15.3.3 Data Quality Requirements

In order to ensure applicability and reliability of the data there are a number of requirements for spatial data sets to be uploaded to the system. These refer to data interoperability, comparability, and harmonization [121].

15.3.3.1 Spatial Data Interoperability

Spatial data interoperability is "the ability to freely exchange all kinds of spatial information about the Earth and about objects and phenomena on, above, and below the Earth's surface; and to cooperatively, over networks, run software capable of manipulating such information" [77]. Interoperability is a crucial issue if different data sets are used in combination. Six elements of spatial data quality are important [121]:

- *Lineage* describes the data set production history, the data sources, the methods of derivation, and the operations and transformations performed.
- *Positional accuracy* is a measure of the closeness of data compared with its location in reality. For EnviroMap this is of minor relevance but it has to be stressed with respect to the original data sets.
- *Generalization quality* describes the simplification of a geometric data set in the sense of its abstraction of reality in a scale- and purpose-adequate manner.
- *Attribute accuracy* refers to the quality of the thematic property information. This is particularly important because in order to achieve satisfactory data harmonization it is necessary to have a consistent understanding of the data content.
- *Logical consistency* ensures that data are neither contradictory nor inconsistent. This requirement involves reporting the results of logical tests of validity run upon the data set.
- *Completeness* defines whether all necessary real-world phenomena related to a certain analysis or presentation purpose are integrated into the map.

15.3.3.2 Data Comparability

Integration of data from different sources involves the problem of compatibility and comparability: data are often not compatible and thus lack comparability due to differences in data formats and data definitions.

This has been recognized in many assessment and monitoring systems [303]. With respect to a raster cell system this means that attributes which contain absolute numbers or density information must refer to a spatial reference that is appropriate for the respective cell size.

15.3.3.3 Data Standardization and Harmonization

Standardization proper refers to data generation rather than to data management. An ex-post standardization of data that were generated through non-standardized monitoring is not possible. Consequently, the EnviroMap system which is maintained through bottom-up data upload by different users, cannot achieve the standardization of data attributes. Thus the emphasis of the EnviroMap system is on data harmonization, which is integration of attributes to secure thematic comparability of the data. This means that new data should match those that are already integrated in the system. Data providers can use the system metadatabase to describe their data sets and to offer the users advice on relating new data sets with existing ones.

15.4 Implementation

The EnviroMap system was established and programmed in 2002. Since late 2002 a prototype version has been operational with unrestricted access for data users [698]. A detailed description of the EnviroMap system which summarizes the most important features has recently been published [401]. Figure 15.2 shows how the system structure links data holders and data users. The two main system elements, namely the "upload tool" and the "map tool", are described in more detail below.

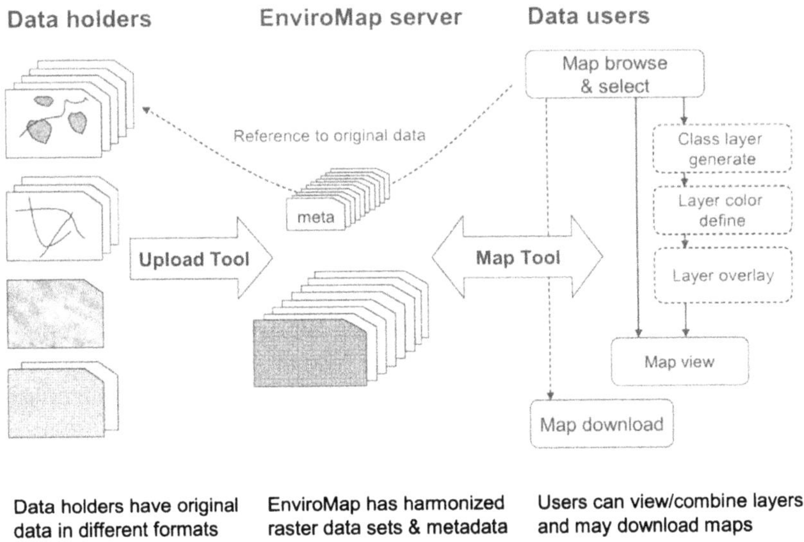

Figure 15.2: EnviroMap system structure and functionality

15.4.1 Maps and Layers

15.4.1.1 EnviroMap "Maps"

In EnviroMap terminology, a "map" is a raster file in standard format that contains a value in each raster field. The two possible types of data need different consideration.

- Continuous Data: In this case, data in the raster cells refers to a numerical value in a continuous scale of one attribute. For such maps, there is only one unit to which the values of all raster cells refer (e.g. mm of precipitation per year). There is virtually no limit to the different values. In order to produce meaningful maps, such data must be aggregated into a finite number of classes (e.g. an aggregated class that refers to annual precipitation of 50–150 mm per year).
- Discrete Data: Data in raster cells may also be coded, i.e. the values of the cells refer to certain properties that are defined externally. There is a limited number of (numerical) codes, and each code signifies a distinct and potentially dif-

ferent property. To produce meaningful maps, there must be a thematic reference for each of these codes. Such maps are often used for properties that cannot be directly represented by simple numbers (e.g. for land use categories, the value "1" could signify "forests", and "2" could indicate "water").

15.4.1.2 EnviroMap "Layers"

In the context of EnviroMap the term "layer" is used differently than in most proprietary GIS systems. Map layers are subsets of maps with one defined property, such as a category in maps with predefined discrete classes, or a defined class containing a value range in maps with continuous values. Layers are at the core of the EnviroMap user interface, and allow the user to combine elements or property ranges from different maps to generate "new" ones.

15.4.2 System Elements

EnviroMap consists of two main application system elements, namely an upload tool (for data providers), and a map tool (for data users).

15.4.2.1 Upload Tool

Data holders are usually GIS experts who have experience in transforming their original maps into the standard raster grid format. Before uploading, the data holders must convert their original map data into the EnviroMap standard format, i.e.

- a spatial projection that conforms to the system standard (for Austria this is the Lambert conformal conic format with 13°20' central longitude meridian, 46° and 49° standard parallel latitudes, 48° central meridian);
- a 250 × 250 m raster data set in the industry-standard ESRI ASCII grid export format with a header that contains georeference information (this conversion is a routine operation for most advanced GIS systems).

The upload tool supports data providers to document the resulting EnviroMap raster map for upload and optionally to document the (detailed and valuable) content of and access to the original data set, and to store this information as an XML metadata file.

This tool also allows to define a default color table for data classes that can also be uploaded together with the map. The tool allows to preview the map data set before uploading, and to upload both the map data and the map description. To facilitate and speed up uploading, the ASCII data are compressed to a 16-bit binary code. During upload the user registration information is attached to the file information. After receiving both the map data and the map description the system checks the formal validity, re-converts data to the original values, and publishes them on the server.

The upload tool is a Java program that can be downloaded by a data provider and will then reside on the data provider's local system. Thus it is possible to prepare and edit all data offline before connecting to the system.

15.4.2.2 Map Tool

The map tool is a Java application in the browser window. It is the main tool for data users. The map tool contains a map and a layer window (Figure 15.3). The map window allows users to retrieve and select available maps from a list, and to display metadata information about the map and its layers. Furthermore, the map window allows users to preview a thumbnail image of the selected map.

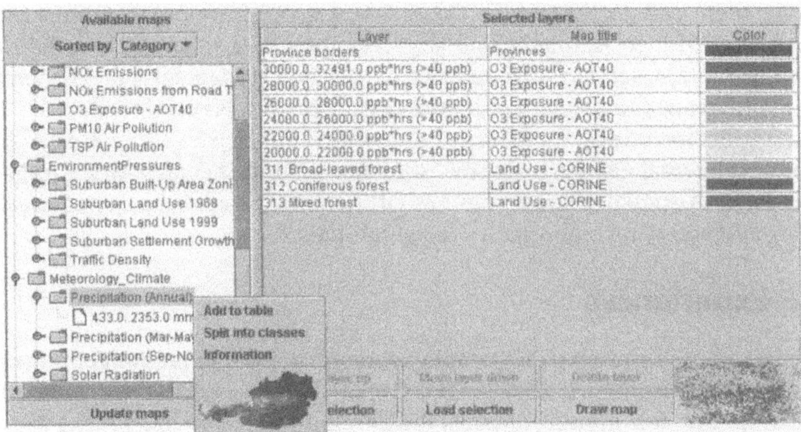

Figure 15.3: Features of the map tool. The map window (left) lists all available maps and their layers in a standard folder structure. The layer window (right) shows the layers selected by the users; in this example, the selection consists of ten layers taken from three different maps. All layers can be assigned user-defined colors, and their order of appearance can be rearranged.

The layer window allows users to select predefined layers of maps that contain discrete classes, or to generate self-defined layers from maps with continuous values: this can be done by splitting continuous values into self-defined discrete classes. For all layers, colors can be defined by the users, who can rearrange the order of selected layers from different maps, choose their own color preferences, or create color-ramps (here named color transitions) by simply clicking the colors of the first and last class for which a color-ramp is to be calculated. In summary, users can interactively "play" with the maps.

Through this flexibility of selection and rearrangement of (opaque-colored) layers and individual color selection, users can produce new maps interactively, combining the stored data sets. They can then display these maps on the screen, zoom into regions of interest, and download the map as an image to their local system.

15.4.2.3 Map Download

Users can display their map selection as standard image files in full color resolution. They can download and save these images locally in standard image formats (bmp, png). The EnviroMap team had planned to allow download of EnviroMap raster data to registered users, but implementation of this functionality has been postponed until all legal and copyright issued are resolved.

15.4.3 Applications

The system can be used as a platform on which holders of environmental data can publish a version of their data and data seekers can get access to these data. The intended users of the system are, as already described, data holders as well as data users.

Data holders who are interested in data upload might include public administration, public research organizations or private consulting firms who wish to provide their data to the public. Other target groups are organizations or individuals who wish to announce and advertise their data treasures by publishing a coarse-resolution version of their data on the EnviroMap system. Through this, they might get in contact with potential customers of their original data.

Data users include (a) members of the general public who are interested in spatial environmental data, e.g. schools, students, environmentally aware public and (b) experts in spatial planning and environmental issues who are interested in finding and accessing high-quality "original" data.

15.5 Experiences

Since late 2002, a prototype EnviroMap server has been operational on the Internet at www.enviromap.at [698]. About 30 maps are available for public users. During the first few months of 2003, the development team has sought user feedback in order to further improve the system.

15.5.1 System Performance

In the first year of operation, the system has been stable and fast. Even with a very basic Pentium-2 with an "archaic" speed of 266 MHz hardware with 128 MB memory, the system runs sufficiently well when powered with a Linux operating system. A satisfying performance is important as the response speed is a crucial element that determines the usability and public acceptance of Web-based environmental information systems [350]. For instance, the typical upload time for a 10-megabyte ASCII raster data set is about one minute; installation of the file on a data server takes another 10–20 seconds. For users, response speed is not only determined by the server but also by the performance of the Java Applet on the local host; however, typical loading times for a display of several layers from several maps would take about 10 seconds.

15.5.2 Acceptance by Data Users

As the availability of the EnviroMap Web site has not yet been advertised to the general public, only a limited number of users have given feedback to the development team up to now. In general, all users were surprised by the power and speed of the system. However, several shortcomings were reported via user feedback and via a professional usability study, where a sample of test users was confronted with the EnviroMap system and asked to pass six to eight exercises within 90 minutes [486]:

- The system offers too many possibilities and thus is difficult to understand for users who are not familiar with map-handling principles. Many novice users have (within eight minutes of being first confronted with the system) not understood some operational functionalities (such as how to split ranges of continuous values into classes).
- The Java-style graphical interface contains unusual elements that differ from the typical "Windows" layout: e.g. shapes and positions of buttons and folders.
- Map outputs lack elements for orientation, such as names and borders of major cities and municipalities. This was felt to be particularly important for the interpretation of zoomed maps.
- The possibility of combining many layers from many maps to derive user-defined maps is too demanding for many non-expert users or too time-consuming for impatient users. What such users expect from a Web atlas is not necessarily to produce (and subsequently to interpret) their "own" maps, but to retrieve useful maps for predefined purposes together with an explanation of these maps in simple language.

Users from the European Union, Austrian Federal Agencies and private planning institutes who have been guided during one-hour presentations have found the system easy to handle and valuable for retrieving and combining exercises.

15.5.3 Acceptance by Data Providers

In the few months of operation, there has been very little usage by data providers. One of the reasons for this was that the prototype system was not advertised. However, there are three serious intrinsic causes for the poor usage:

- Many data providers are willing to make public a raster version of their data, but they are not willing to put extra work into format conversion and proper documentation of their data sets. Even with the very basic documentation requirements of the EnviroMap system, they feel too stressed to upload data.
- A number of data providers are reluctant to publish their data. This has to do with the "value" of the data that some providers do not wish to share publicly, even if the data are to be shared in an aggregated raster format. However, one federal institution has provided an additional 30 data sets which have been uploaded for test purposes into a secure area that is not accessible to the public.
- Several data holders cite legal aspects as obstacles for uploading data: Who actually owns the data on the system? To whom should references be made when data are used, even if it is in the simplified EnviroMap raster resolution? Furthermore, some institutions have data that were synthesized from other "original" data and therefore they are unsure whether they "own" the data and have a right to distribute to the public.

15.6 Discussion and Outlook

There are various approaches to visualization of environmental data via Internet. However, these are often sophisticated GIS-linked systems with many features high server system requirements. Examples of excellent environmental Web atlas systems are the Baltic Environmental Atlas [656] and the National Atlas of the United States [746]. These systems have the advantage of high functionality for viewing and using vector data in different map scales; the drawbacks are that handling these systems needs some cartographic experience and that they do not provide features for map upload.

In our view, the advantages of EnviroMap are that the system is better suited for easy up- and download of data, that users can edit their "own" maps, and that data storage is simple and rapid. The EnviroMap prototype has demonstrated the usefulness and the limitations of the system design. In general, the system concept of providing an open marketplace for standardized environmental spatial data has proven useful.

The most important conclusion from the prototype test is that requirements for an appropriate user interface have been grossly underestimated. Most users consider the range of technical functionalities too advanced and confusing. Thus the EnviroMap team plans to improve usability through introduction of a separate, simplified user interface for non-expert users, where only predefined maps and selected layer-combinations from maps that contain map explanations in simple language can be viewed.

The success of the system will also depend on the attractiveness of EnviroMap for data providers. Their reluctance to use the system is mainly due to the additional work for documenting the data, and to fears of giving away data "value". Development of the EnviroMap system has in the past heavily relied on technical solutions. To make the system more attractive for data providers, future work will require more emphasis in terms of organizational development.

Part III
Corporate Sustainability

Sustainability and Social Responsibility
Environmental Reporting

Chapter 16
System Architecture and Maintenance of the Ecoradar Web Portal

Rudolph Braun • Werner F. Schulz • Helmut Krcmar • Markus Russ
Sebastian Maute • Benjamin Hermann • Valentin Nicolescu
Martin Kreeb • Henning Kolb

Abstract. *The complexity of Web portals requires the implementation of integrated maintenance management. Ecoradar represents a portal in the domain of sustainability and environmental management. After presenting the system architecture and general components of this portal, we have a closer look at the domain-specific application framework: EcOObjects, as a software class framework for developing environmental management software, EcoAvatar, as an instrument for a multimedia environmental communication and EcoWizards, as a computer-aided and interactive completion of checklists in environmental management. A description of the portal maintenance tasks follows. This chapter provides conceptual background information for persons responsible for Web portal management, especially for portals in the domain of environmental management. It can also be used as a blueprint for portal planning and implementation.*

16.1 Introduction

The portal for sustainable management "Ecoradar" [695], established in the context of a project promoted by the Federal Ministry for Education and Research, has the objective of sensitizing decision-makers in small and medium-sized enterprises (SMEs) to the issue of sustainable management [313].

In particular, neither established jobs for environmental tasks nor extensive budgets for sustainable management projects are usually available in SMEs. The Internet can reduce expenditure on environmental tasks, especially in the process of collecting information. In order to find case-based data in the Internet, usually more than the designated time is spent for the investigation, because of the amount of overhead data.

Thus, the supply of direct and simple information to visitors must be mastered. The Ecoradar portal consists of eight ecological radar screens (Figure 16.1). Each of these radar screens includes a surrounding and an operational radar. The surrounding radars ("macro level") provide information for managing the tasks of the relationship between the company and the stakeholders. The operational radar ("micro level") focuses on the task of environmental management inside the company [313]. The portal will also be translated into English.

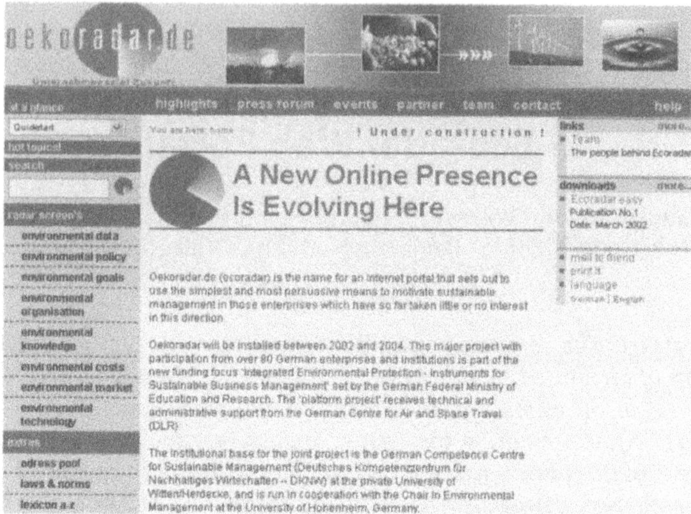

Figure 16.1: The homepage of the Ecoradar portal

Ecoradar offers problem-oriented support, enabling the user to obtain its problem solution (e.g. strategies for introducing EMAS, the Eco Management and Audit Scheme) with the help of the system. In addition, several tools and other services are available [68]. To manage all the content, tools and services, an integrated portal architecture is operated [312].

Like in a theatre, a visitor can see only a small part of the system organizing the show. In the following paragraphs, we will lead the reader behind the curtain to introduce the technical backstage of the Ecoradar portal.

16.2 System Overview

16.2.1 Layered Portal Architecture

The portal consists of multiple components which interact with each other. The portal system and its components can be divided into six layers (Figure 16.2): hardware, operating system, database and application providing, application, presentation, and security. Several management tasks can be divided between these layers. This is necessary because of the complexity of the jobs connected with each layer. The task of the portal system is not only to present a Web site to visitors, but also to provide a platform for teamwork and collaboration.

The first layer is the hardware. Next is the operating system layer consisting of Red Hat Linux and Microsoft Windows. This dual-platform strategy allows the implementation of a wide range of applications which often require a certain platform. It also allows the programming team to use a wide range of implementation languages for different operation areas. On layer three, next to the relational database backend which enables data storage and retrieval, reside further software packets delivering the required functionality like a Web server which delivers the HTML pages and an application server which enables the application framework

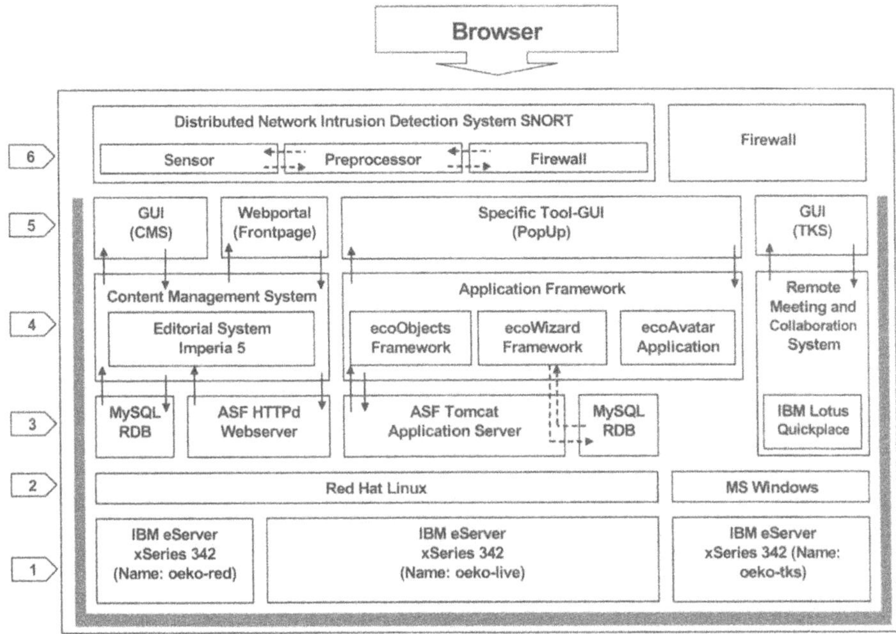

Figure 16.2: Overview of the portal's layered architecture

While Ecoradar is a geographically distributed project, it maintains editorial collaboration through a remote meeting and collaboration system based upon Lotus Quickplace. The look-and-feel of the implementation is the same as the Corporate Design and Corporate Identity standards as described in the Ecoradar portal [356].

Layer four is the core of the system where the intelligence of the environmental portal is conserved. Next to the "Content Management System", an "Application Framework" is developed to carry the ecological or environmental tools or services. EcOObjects is a class framework where environmental tool developers can use a ready-to-implement software code. EcoWizards combine the principals of known checklists (e.g. checklists to determine which environmental representative a company needs) with easy-to-use Web applications, so that users can use checklists comfortably. At the end of each EcoWizard, the results are immediately visualized in charts and enriched with explanations.

The search mechanism is supported by an EcoAvatar, in the shape of a human being, who supports the user in the search process. The virtual human being can identify information about the visitor with the help of an attached chatbot by creating a user profile. These profiles can be compared with the profiles of articles which are updated daily by semantic analysis. Thus the avatar presents the best contributions covering the specific user's problem. EcOObjects and the EcoAvatar are explained in detail in Sections 16.2.2 and 16.2.4.

Looking at the fifth layer, we can clearly see the division of content and presentation. Using a content management system, the content is stored at the third layer, processed at the fourth layer and presented at the fifth layer.

Access to the fifth layer is filtered through the sixth layer. This layer implements a firewall. All traffic passing through the firewall is monitored by several sensors of the signature-based intrusion detection system and processed by a so-called pre-processor to analyze and detect signs of malicious behavior [358].

16.2.2 A Closer View of EcOObjects

As well as the Ecoradar portal objective, providing portal users with cost-free appropriate content for environmental management, there is also the goal of supplying users with interactive aids. These aids are small Internet-based applications which should encourage users to come in contact with "computer-aided environmental management" and should lower the initial costs of implementing environmental management. In this context, such aids are called "EcoTools" (Ecological Tools) or simply "tools". Tools should not been seen as competitors to established environmental management systems. The tools support users in the performance of small but regularly occurring tasks of environmental management. Tools represent known instruments or methods of environmental management or environmental controlling such as the ABC-Analysis of dangerous goods, instruments for generating an environmental report, methods for the structured handling of environmental indicators or environmental benchmarking. EcoTools are developed according to the rules of object-oriented programming. Thus the functionality of the tools can easily be extended, and the code can be reused for the development of similar problems.

To identify appropriate objects of environmental management, text scanning can be used [1]. After searching for an article related to the domain "Environmental Management", the subjects of the text have to be extracted. After consolidating the list of subjects (only retaining subjects with a close relationship to the topic of environmental management), an accumulation of candidates for so-called EcOObjects (Ecological Objects) is created. The two OO's indicate the development as defined by object-oriented programming. After consolidation and returning to the site of discovery (the sentence in the text), the verbs belonging to the concerned subject are identified. These verbs become candidates for the operations of the EcOObject and are supplemented by the attributes of the object. Figure 16.3 shows the EcOObject "Environmental Performance Indicator" (EPI) as a UML-diagram.

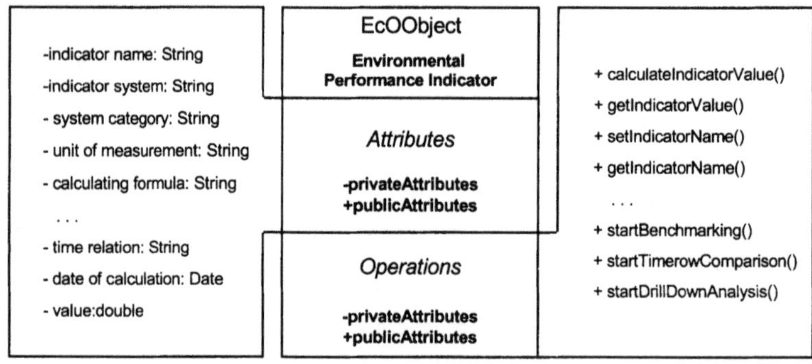

Figure 16.3: Overview of the EcOObject EPI

Each EcoTool consists of 1 to *n* EcOObjects. There is a hierarchic relationship between the EcOObjects. Each EcOObject consists of 1 to *n* classes of a software code. Commonly needed functions for several tools, such as generating diagrams, are defined as EcoFeatures. Figure 16.4 shows the structure of the EcOObjects and EcoFeatures.

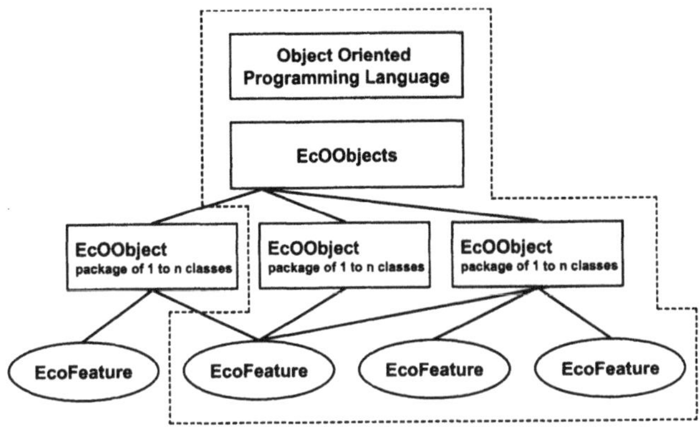

Figure 16.4: The structure of the EcOObjects

It is necessary to implement an extensive technical tool platform to guarantee the interoperability of the EcoTools. A three-tier platform architecture turned out to be a suitable structure. Figure 16.5 presents the tool system architecture and its objects. Next to the EcOObjects there are other object frameworks, such as objects for handling the presentation on the users' client (Internet browser), system objects, which manage client requests for example, and database objects, which handle the read and write processes.

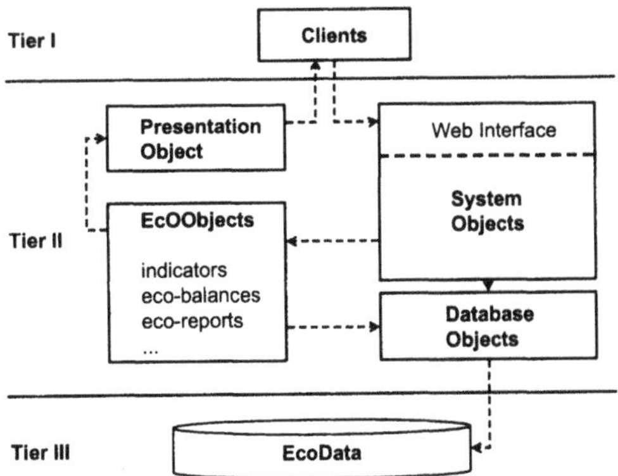

Figure 16.5: Overview of the tool system architecture

16.2.3 Application Scenario

Based on the EcOObjects and the tool architecture, the Environmental Indicator -nd Benchmarking Tool (short: EPI-Tool) was developed to support employees in environmental management department of SMEs in handling daily tasks and -pport them in their decision processes.

The tool offers a variety of different features helpful for different problems (see Figure 16.6, right column). It can be used to plan and control environmental parameters in the enterprise, so this tool can be regarded as an environmental controlling instrument.

The aim was to develop an open tool which can be used on different operating systems and which is suitable for different industry branches. In a first step, breweries were considered to test the functionality of the tool. In a further step, other industries will be integrated.

First of all, the user has to register and login. Then the user receives a password secured account which has to be filled with company-specific environmental indicator values. This can be done manually or semi-automatically.

The compatibility and cooperation of the EPI-Tool with standard office software were regarded as a very important in creating an integrated system in which environmental indicator data can be exchanged and used easily. This exchange is supported by import and export functions for MS Excel, enabling exchange of data with ERP (Enterprise Resource Planning) software like Navision or SAP R/3.

The user can enter and modify all the environmental indicator data in his account via the Internet browser. So the EPI-Tool can be seen as a first environmental management module which enlarges the functionality of a company's ERP system. The entire hosting and analysis of environmental indicators in SMEs can be enabled by using the EPI-Tool.

In the following some features of the EPI-Tool are described in detail. The benchmarking feature allows the comparison of a company's indicator value with the indicator value of other companies of the same industrial sector. The user of the tool can choose for instance the indicator "heating oil consumption" which can be compared with the best and the worst value of other enterprises of the same industrial sector. The results are shown in a user-friendly illustration.

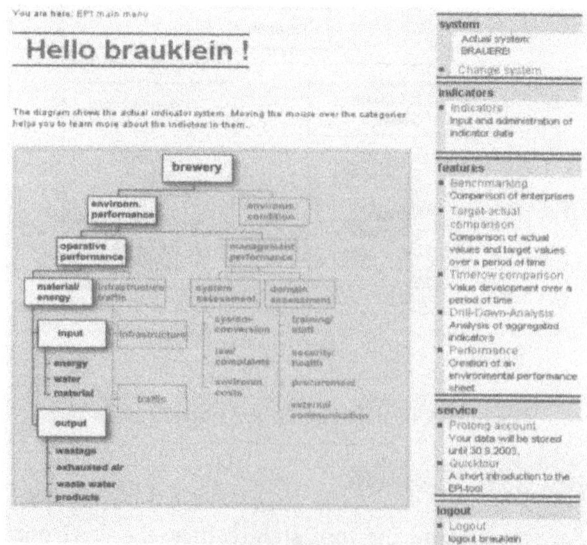

Figure 16.6: The initial screen of the EPI-Tool

Thus this benchmarking feature can help to identify ecological strengths and weaknesses in the company. The consumption of material and energy in particular holds great potential for cost reduction [477]. Furthermore, the management can concentrate on the most urgent ecological problems.

Trend analysis is another interesting environmental feature. With this instrument an indicator's value can be monitored over a period of time. It is possible to recognize certain trends in the development of an indicator value. In addition, the trend of the branch's best value is drawn in contrast to the trend of the user's own value. This can help decision-makers to initiate appropriate activities for improving the environmental performance of their own enterprise. Thus the eco-efficiency of the operational activities can be improved.

16.2.4 A Closer View of EcoAvatar

The EcoAvatar system was developed to simplify and accelerate access to the constantly growing quantity of scientific papers on the Web site www.oekoradar.de [695]. An avatar, which represents generally the graphic representation of an artificial or natural intelligence, has the task of establishing a more personal relationship with the user by its human appearance [621]. In the case of Ecoradar, it is a young woman named Gaia, who embodies the role of a receptionist or a librarian.

Apart from the function of purely graphic representation, Gaia has the possibility of communicating with visitors. This is managed by the chatbot software (AliceBot Program D), which permits natural-language communication on a textual basis. These chatbots, which go back to the idea of the Turing test, have been used since the 1960s to examine the social interaction of humans. Thus the first chatbot "Eliza" was developed as a virtual psychiatrist [609]. The combination of chat and the animation of a graphic figure arouses, despite better knowledge, the impression of being in a video conference, i.e. communicating with real humans [507]. This synthesis is accomplished by playing one of approximately 30 animations in GIF format by the chatbot depending upon discussion contents and the mood of the avatar.

The knowledge base of the chatbots plays an important role in the simulation of real humans. The number of existing rules is highly correlated with the degree of humanity that the avatar seems to have. Gaia's set of rules can be divided into five domains. The standard rules contain contents, which are necessary for small talk about general topics. The general Ecoradar contents give information about project data, the process and the state of development. The Eco-tools contain questionnaires and other sequential rule constructions, which are to be put into Gaia's mouth. The Eco-lexicon contains data from the Web site's own encyclopedia, in order to address questions about certain technical terms directly to the avatar. The range of the Eco-articles contains the categorization data of all scientific papers presented in Ecoradar.

The first three domains are maintained manually. The Maintenance of Eco-articles, Production and Actualization of Eco-lexicon are automated by the proprietary software "Ecoradar content analyzer". The encyclopedia of the Web site is converted into an XML-based format readable by the chatbot, while a semantic ontent analysis takes place for the categorization. Each article is examined on the is of a great number of groups of synonyms, in order to provide an article profile for each contribution. The characteristics of the profile contain for example values such as industries, regions and departments for which this contribution could be of interest. The characteristics, as well as the steps that lead to their extraction, are represented in Figure 16.7.

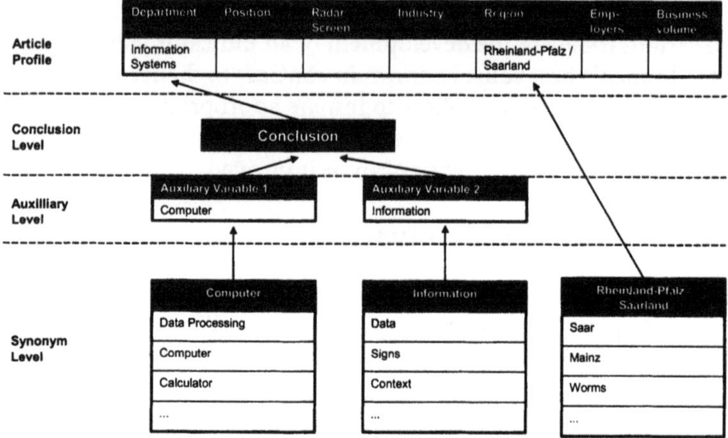

Figure 16.7: Scheme of article profile extraction

Parallel to the extraction of article profiles, which runs daily, the avatar tries to collect and arrange information about the visitor to create an equivalent user profile by passive shaping [78]. This enables a comparison between user and article profile, in order to suggest to visitors the most suitable contributions in the fastest way. In order not to confront visitors with contributions that have already been recommended to them, the avatar stores the suggestions for each user to avoid repetitions. The EcoAvatar thus provides a quick way to find suitable information on Ecoradar, also for laymen [312]. The interaction of the EcoAvatar system components is shown in Figure 16.8.

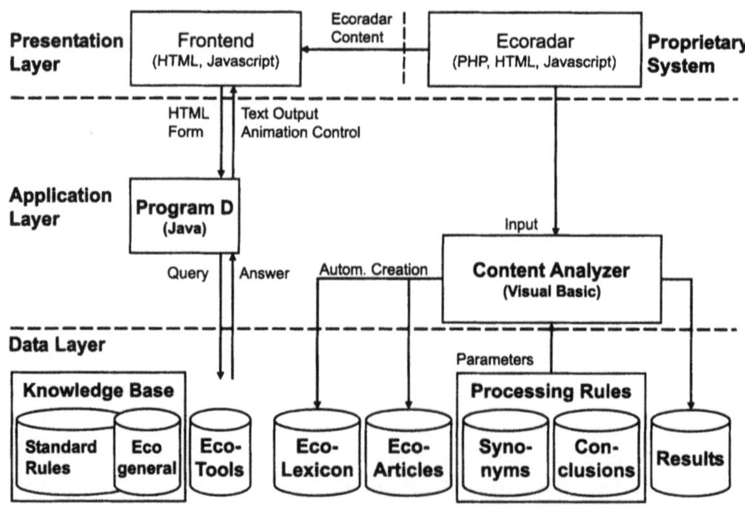

Figure 16.8: Overview of the EcoAvatar system architecture

16.2.5 A Closer View of EcoWizards

EcoWizards combine the principals of known checklists (e.g. checklists to determine which environmental representatives a company needs) with easy-to-use Web applications, so that users can use checklists comfortably. At the end of each EcoWizard, the results are immediately visualized in charts and scales, enriched with explanations. The EcoWizards can be viewed from the user's perspective and the developer's perspective.

16.2.5.1 The User's Perspective

A Wizard presents itself to the user as a particular program sequence with a specialized GUI (Graphical User Interface). Functioning on the same principle as the Windows installation wizard the user is led step-by-step by a program and is supported by numerous hints. There is no point during the program sequence at which the user doesn't know how to continue or is overtaxed with the required information.

The EcoWizards' central objective is to challenge the user intellectually through the GUI and increase his interest in environmental management. Nevertheless, EcoWizards are designed as be as simple and informal as possible to make their use easy for inexperienced users. Consequently, an EcoWizard primarily consists of several questionnaires (Figure 16.9). The user is led step by step and is supported by toolboxes and explanation texts. Moreover, background information and important explanations are given.

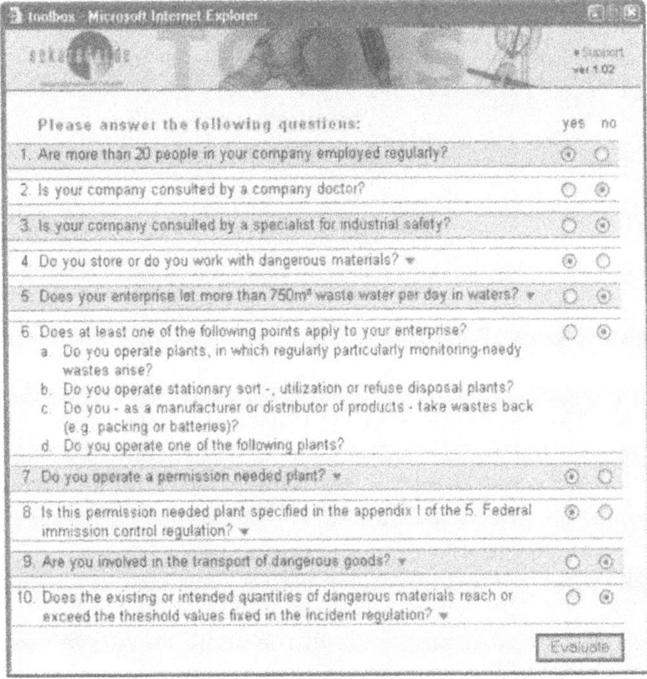

Figure 16.9: EcoWizard Beto start screen

The motivation for using the Wizard is that the user needs almost no preparation time. It is also possible to work with inaccurate or estimated data. The inexperienced user explores the tool continually by filling out the forms constantly. At the end of the program sequence, the user gets individually produced results including explanations and export possibilities (Figure 16.10).

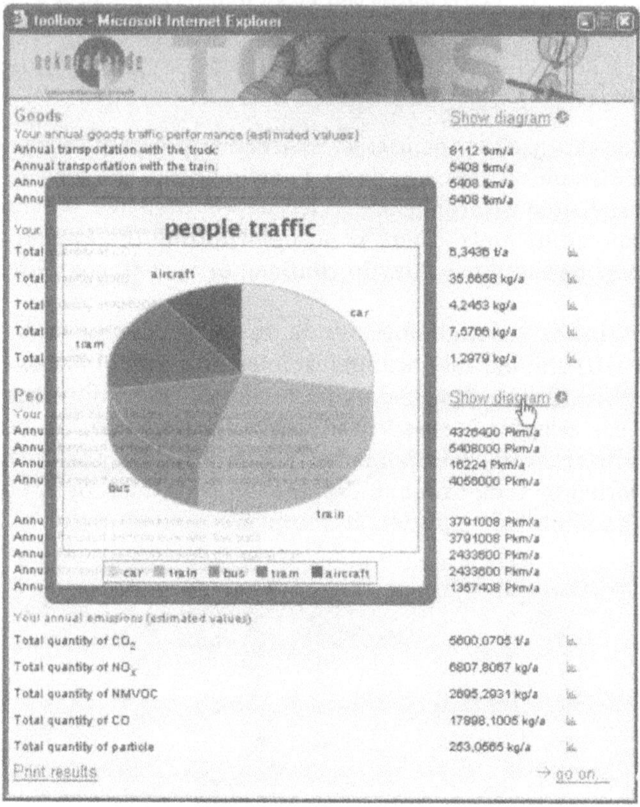

Figure 16.10: EcoWizard Veto result screen

16.2.5.2 The Developer's Perspective

Java Server Pages (JSP) and client-side JavaScript are used for the EcoWizard system technology. The development of the EcoWizards follows the rules of object-oriented programming [235]. EcoWizards do not require browser plug-ins or dedicated client software. Moreover, the wizards work on all current browsers. The standards of the W3C-DOMs are taken into account, particularly the use of client-driven JavaScript [594].

Through object-oriented programming the developer gains a high degree of modularity. This provides several advantages. First of all, the EcoWizards can be administered quickly and enhanced rapidly. In addition, the software code of one EcoWizard can be reused for another.

The data are held on the server during a tool session. No database is required for these low data quantities. This means that porting and installation of these tools are extremely simplified. Furthermore, the EcoWizards are a fast way to record data used for the EcOObjects-based tools. The transfer of the values can take place in pre-defined parameters. For example, the EcoWizard Veto presented in Figure 16.10 helps the user to collect company traffic data and converts them into ecologically relevant values, such as the overall quantity of CO_2 that the lorry traffic had caused. These and further values can be defined as interface parameters, so that data collected by the user can be transferred easily from the EcoWizard Veto to the EPI-Tool. Now the user of the EPI-Tool can analyze the company traffic value with the help of the tool features presented in Section 16.2.3.

16.3 Managing the Maintenance of a Portal

16.3.1 Hardware Layer

The hardware is rack mounted and located in a separate server-room where the climate is controlled. Thus we can ensure physical security through physical access control. To decrease the risk of data loss, the hard disks are hot swap capable with RAID-5 (Redundant Array of Independent Disks, Level 5) and one hot spare disk. This means that we can replace defective hard disks on power with real-time data recovery. In addition, an external tape backup is done on the third layer. Roaming administrative access on the hardware is made available through integrated maintenance components.

To keep the hardware up and running, it is necessary to secure the servers against physical manipulation. As long as a dust-free, temperature-controlled and fire-secured environment is provided as recommended by the hardware manufacturer, we ensure the physical network connection through a second backup connection. An uninterruptible power supply and power load distribution enable a clean shutdown if power is not restored in time. Thus, the hardware layer needs minimal maintenance once properly installed.

16.3.2 Operating System Layer

Unlike the hardware layer, the operating system only requires maintenance on demand. Ecoradar uses the Linux standard distribution (Red Hat Linux 7.3 Valhalla), which is upgraded manually with the software packages needed. Thus there is no need to pay license fees for enterprise editions or commercial products. A regular update on the basis of the Red Hat Network is cost free and levels up security through updated packages available via this service. This means that the software packages are installed and updated through the Red Hat network automatically. For security reasons such updates are crucial. Thus for system administrators it is only necessary to monitor log files, system load and watch system parameters. These tasks may be automated to a certain degree. Windows is maintained automatically through Microsoft Windows Update. The other servers monitor the availability of the windows system and manage the system through NMP (Simple Network Management Protocol).

The log files are processed through automated filters and reporting mechanisms and finally the reports are delivered by e-mail. On Linux, the system load is monitored through a cron job running system tools. If the system load increases, the system parameters such as kernel or file system settings are optimized to improve response time. This optimization must also be done on layers three, four and six.

16.3.3 Database and Application Providing Layer

At layer three, the content is stored in a relational database system. In addition, this layer provides a Web and an application server which both deliver functionality required for software running on layer four. Ecoradar uses Apache as Web server and Jakarta Tomcat as application server.

As most applications are implemented in the PHP programming language, the Web server needs a module which enables processing of PHP (here Apache's dynamic shared object technology is used). Unlike the Web server, the application server needs no further modules but a Java Runtime Environment, because this is the implementation language of the application framework.

Nevertheless, the main maintenance tasks are the clean configuration of both the Web and application server as the project evolves and expands with new functions. Like any other software, the installed software may have security holes or may need bug fixes. Thus an update must be done. These updates may result in performance gains.

Regarding the database, the administrator must ensure data integrity and performance, less for the static content but more for the application framework (see "application framework" layer number four in Figure 16.2), because the framework stores and retrieves data from the database in real-time.

16.3.4 Application Layer

The Ecoradar-specific software resides at the application layer. The project provides a content management system for the editorial work to enable the editor to focus on content and not on the technical conversion. Content management systems store information separated from the layout. Using the capabilities of both a tele-cooperation system and content management system, a third-party expert can contribute content to the portal in direct collaboration with the editorial staff. Both systems together provide a flow of continuous feedback and review cycles. So a new article must pass different stages until it is put online. Figure 16.11 presents an overview of the workflow.

First and second level support are installed for maintenance tasks. The first level support means that a user help desk for the content / tele-cooperation management system is provided. Here, the users receive answers to questions of system handling. All procedures are documented with the corresponding solutions in the tele-cooperation system. The second level support includes on-site support for the users. This also means bringing together the people participating in the content production workflow to solve system-wide problems. Occasionally, it is required to debug and patch the software.

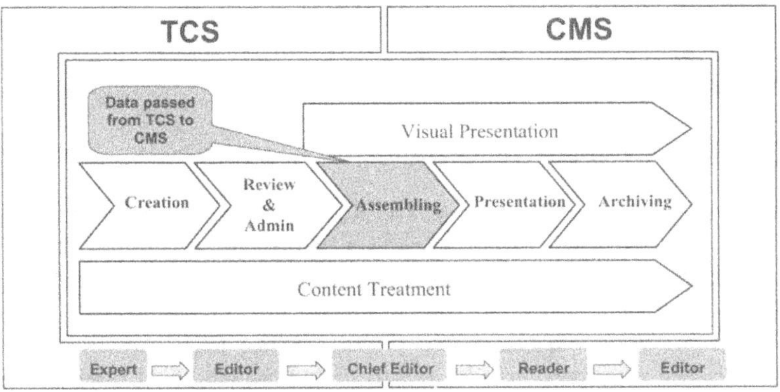

Figure 16.11: Content production workflow of Ecoradar (TCS = tele-cooperation system, CMS = content management system)

16.3.5 Presentation Layer

The presentation layer is only specialized for the content layout. For the easy-of-use principle of Ecoradar, the presentation has to be close to standard software ergonomic guidelines.

The idea of sustainability may only be transported by easy, quick and fault-tolerant access to both content and tools. Thus an environmental appointee can retrieve the needed information in less than no time. On the other side a brand is to be established: Think of environmental sustainability, think of Ecoradar. Thus the project defines a detailed guideline how to present content. This means that a corporate identity has to be established which also has to be maintained regularly.

Many cooperations evolve in Ecoradar. Our experience shows that cooperating enterprises and associations wish to provide either content, references like a logo, or a Web presence with the corporate identity of Ecoradar. Thus the project provides everything required in a Web Development Kit, which also has to be supported and tended [355; 357; 358]. As mentioned before, the process of continually improving the usability of the portal software is crucial.

16.3.6 Security Layer

The first and uppermost layer is a filtering layer. This layer provides the security, privacy and anonymity of users. Ecoradar implements a centralized data storage and analysis through the application framework. These data must be secured. Considering the tasks of security in the information technology, this also means that the security layer is responsible for keeping the whole system running. Malicious behavior is recognized and blocked. No attack should bring the system down or corrupt stored data, both personal and enterprise. The demand for secured data is deduced from legal regulations and the best-practice examples as seen by well-known Internet portals and service providers [28].

In technical terms, the implementation consists of an intrusion-detection system and a firewall. The firewall filters non-standard network traffic and blocks all unnecessary ports. Traffic passed through the firewall is then processed through a processor to be finally recorded by the intrusion-detection system's sensors.

The processed traffic is analyzed to identify malicious behavior. The signatures are updated automatically using a Perl script. The result is a secure platform which can be used without any concerns.

16.4 Summary and Outlook

Until the end of 2003, each radar screen will be filled with appropriate content. One of the next steps will be to transfer Ecoradar in an international context. There are already several international organizations interested in cooperating or contributing to the content.

Based on the knowledge acquired through using EcOObjects and its system architecture, the next steps will be the implementation of an open source community and the creation of EcoWebServices which could integrate the EcoTools functions easily into the landscape of the companies' information systems. The method of identifying ecological objects has to be refined.

As well as the EcoWizard system technology, the EcoAvatar will be developed further. There is a high potential that avatars can animate and motivate people with regard to sustainable management.

Chapter 17
Environmental Investment Sites: Sector Analysis and Development of GreenMoney.at

Arno Scharl • Irene Pollach • Michael Pieber • Horst Treiblmaier

Abstract. *Web sites pertaining to both traditional and socially responsible investment (SRI) have emerged in recent years, seeking to help investors make informed decisions. This chapter illuminates how such information systems accommodate the particular information needs of their audiences and unveils ideological differences in the arguments put forward. For this purpose, a sample of five SRI sites was investigated qualitatively and quantitatively and juxtaposed to an equally sized sample of traditional investment sites. The main results of these investigations were used to refine an independent investment site (EcoInvest.org), which was implemented using an open-source content management system. In a joint project with the Austrian Society for Environment and Technology, the prototype was then integrated with the content of GreenMoney.at, an established site that targets the Austrian investment market.*

17.1 Introduction

Private sector investment has significant implications for environmental sustainability, in terms of both the ecological footprints of specific investments [595] and the development trends they reinforce [636]. Investors increasingly consider and favor "green" companies and mutual funds [60; 335; 501; 544]. Accompanying this trend, the concept of *Socially Responsible Investing (SRI)* arose in the 1970s. An array of investment opportunities was introduced in the 1980s, before the movement gained mainstream access in the 1990s [70]. Historically, SRI funds lagged behind traditional equity funds in terms of their return on investment, but recent studies confirm that current SRI funds perform as well as traditional funds [448].

SRI integrates social and environmental values into the investment process, grounded in the observation that the corporate sector fundamentally influences both communities and ecosystems. SRI strives to deploy capital in a manner that is consistent with the needs of society and the natural limits of ecosystems [70]. Investors buy shares of companies whose practices reflect their personal values and withdraw capital from companies not sharing these values. Relying on such reward-and-punishment mechanisms to create incentives to become better corporate citizens, SRI follows three main strategies [70; 501; 544; 636]:

- *Screening.* Avoidance screening excludes organizations whose practices or products are perceived to harm the environment, such as nuclear or armaments companies. Affirmative screening deliberately seeks out leaders in environmental performance that also offer solid financial returns.

- *Community investment* targets organizations operating at the grass-roots level of social and environmental change; it helps them generate resources and opportunities for economically disadvantaged groups.
- *Shareholder advocacy* initiates environmental change within corporations by dialoguing on issues of concern, insisting on corporate disclosures, or filing proxy resolutions.

Although most business activities produce a wide range of economic, social and environmental impacts, most corporations tend to emphasize the demands of their current and potential investors and neglect ecology-oriented frameworks such as *bioeconomics* [202; 374] or *thermoeconomics* [117]. These interdisciplinary frameworks concentrate on complex interactions that occur at the interface between organizations, individuals, and natural ecosystems. Once investors become aware of the close relationship between investment and ecological outcome, they embrace the idea that the right kind of business activity yields not only economic but also social and environmental benefits. A new awareness is dawning that sustainable portfolios might translate into better economic performance as well [230; 441].

This chapter deals with the development of GreenMoney.at, an environmental investment Web site maintained by the Austrian Society for Environment and Technology. Originally converted from print material, the content has been enriched and adapted to the hypertext environment. The following sections present a study on investment Web sites conducted to gain insights into online investment communication with a view to refining the communication strategies of GreenMoney.at. The second part of this chapter documents the conceptualization of the site, its history of development, major components of the current system, and potential future extensions.

17.2 Capturing and Analyzing SRI Communication

More effective communication and the heightened interest in environmental investing provide companies that embrace sustainable business practices with a better financial standing, as they are typical investment objects of socially responsible investors. Yet many investors still shy away from SRI, even though its nonmonetary gains for society add to the profits generated by the investments. Seeking to help investors make more informed decisions, Web sites pertaining to SRI address this reluctance. This section identifies starting points for enhancing SRI communication via Web-based channels by studying the communication strategies employed by both environmental and traditional investment Web sites.

17.2.1 Methodology

A judgment sample chosen on the basis of site scope and depth was examined quantitatively and qualitatively. It includes five popular sites focusing on environmental investing and five traditional investment sites, representing both mutual fund companies and independent advisors.

- *Environmental investment sites:* Domini Social Investments [685], Environmental Investors Network [699], Green Century Funds [712], SocialFunds.com [773], Social Investment Forum [772];
- *Traditional investment sites:* CNN Money [672], Invesco Funds [724], InvestorWeb [725], Julius Baer Group [731], SmartMoney [771].

The quantitative part of this survey employs an automated method for extracting raw textual data from the Web in order to investigate common themes and linguistic characteristics such as textual richness, available languages, and interactivity. The framework for the qualitative analysis is functional discourse analysis, as put forward by Halliday and Hasan [222], Fairclough [172; 173], and Stillar [537]. Examining the ideational, interpersonal, and textual functions of Web sites, discourse analysis does not require the systematic coding of texts but focuses on language characteristics and word usage.

In this chapter, the ideational function focuses on the sites' attitudes toward traditional investing and the persuasive appeals used to present SRI as an attractive form of investment. The interpersonal function investigates how the Web sites present themselves as trustworthy sources of advice and how they include their audiences in the discourse linguistically and through hypertext features. The textual function examines the structure of the messages conveyed, which in the case of Web-based communication includes hyperlinks and other navigational features.

17.2.2 The Ideational Function

17.2.2.1 Argumentative Patterns

Since trust is a key issue for investment decisions, investment sites need to present themselves as reliable sources of information and advice. The traditional investment sites emphasize their financial expertise, publishing articles on investment, taxes, mortgages, loans, and the like. Their primary goal is to help their audience maximize their income by offering advice on numerous financial issues. Traditional investment sites typically need not convince their audience of the benefits of investing.

By contrast, SRI sites are often stand-alone undertakings without bricks-and-mortar outlets and therefore face the challenge of building trust solely via their Web sites. They must make a case for engaging in socially responsible investing. The discourse analysis revealed an array of arguments that SRI sites use to accomplish their persuasive goals.

- *The bandwagon effect:* SRI sites argue that a growing number of people base their investment decisions on social and ethical criteria and that SRI is rapidly gaining popularity among investors. By engaging in SRI, therefore, investors join a current trend that must be worthwhile because everybody is doing it. Examples of such claims include "by every measure, the SRI field is growing", "one of the most optimistic emerging trends in the financial world", and "10,000 new social investors each month".

- *Numerical evidence:* When stating past returns of ethically screened funds to suggest that environmental investing is as profitable as traditional investing, SRI sites often provide unreferenced and thus unverifiable figures.
- *A better world:* Environmental investment sites claim that SRI is personally rewarding, as it provides for "a better future for everyone". Appealing to emotions, they assign responsibility for the world's future to their audience in statements like "remember, together we can make a difference", or "to make the environment a better place for ourselves, our children, and our communities". The justification for this argument is that SRI funds encourage companies to raise their social and ecological standards in order to be considered for investment.
- *The long-term perspective:* SRI sites emphasize systems thinking and the long-term aspects of investment decisions. An argument found repeatedly was that socially responsible companies are more successful in the long run than companies that do not integrate social or environmental considerations into their business strategies. The latter often face costly litigations, for example, or might be more sensitive to price fluctuations of non-renewable resources.
- *Contingency:* Another pattern found on SRI sites is hedging claims with vague utterances that can be both true and false at the same time. This is reflected in the elastic quality of statements such as "social investing is working for a lot of people", "several recent studies support the notion that", or "some analysts suggest". Since quantifying determiners like *some*, *a lot* and *several* do not express totality, they render it likely that the opposite is true as well in some cases.

17.2.2.2 Lexical Diversity

Type token analyses provide insights into the Web sites' vocabulary. The standardized type token ratio (TTR) is a syntactical index that divides the number of distinct words (types) by the total number of words (tokens), computed as a running average based on consecutive 1,000-word chunks of text (see Section 4.4.1). High values indicate texts with a heterogeneous vocabulary [326; 561]. For an investment Web site, a lower TTR seems preferable if the target audience includes novices who prefer simple terminology and linguistic structures of moderate complexity. Still, the results of the type token analysis can only be meaningfully interpreted by supplementing them with qualitative research [254], e.g. keywords in context to account for phenomena such as synonymy (words with similar meaning) and polysemy (words with multiple meanings).

The spider-web graph in Figure 17.1 shows the types, tokens and TTRs for each Web site in the sample. A comparison of 06/2003 data shows that the environmental sites have a slightly higher TTR of 0.381 (sd = 0.068) compared with the traditional sites with 0.353 (sd = 0.060). By comparison, the Web sites of Fortune 1000 companies achieved a mean TTR of 0.358 (sd = 0.053).

Since TTRs say nothing about the homogeny of the vocabulary across the two samples, it is worth carrying out a consistency analysis to identify recurring lexical items [492]. A ratio indicating consistency in vocabulary divides the numbe of types appearing at least once on all sites in the sample by the number of ty in the sample. These ratios are 7.27% for the SRI sample, 5.53% for the traditional-investment sample, and 2.25% for the total sample. The low level of lexical consistency in the total sample and the considerably higher ratios of the individual samples indicate that SRI sites and traditional investment sites deal with different topics and concepts. The consistency ratio of the SRI sample is higher than that of the traditional investment sites, suggesting that the latter cover a broader range of topics, whereas the SRI sites focus on a particular market segment.

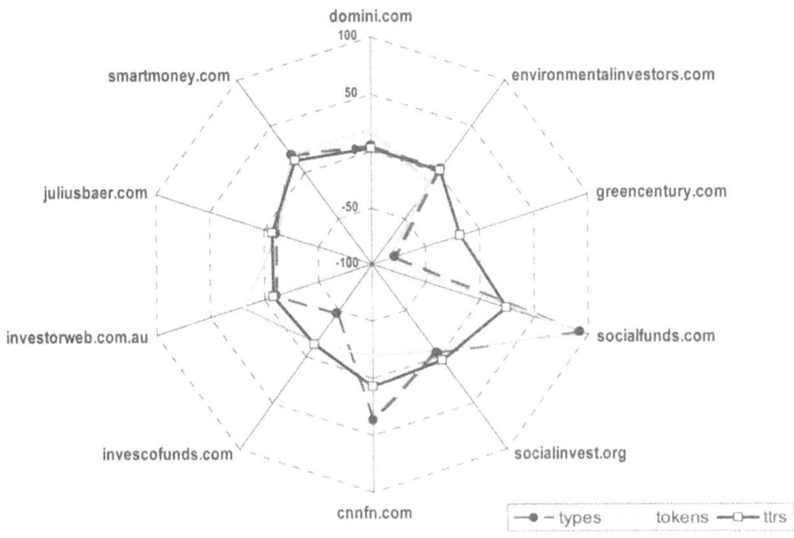

Figure 17.1: Types, tokens, and standardized type token ratios (deviations from mean values in %)

17.2.3 The Interpersonal Function

17.2.3.1 Source Credibility

The Web sites on traditional investment team up with major players in the financial-news sector, e.g. *Money Magazine* [672], *Fortune Magazine* [706], *Barron's* [658], or the *Wall Street Journal* [787], and frequently post articles from these media. These alliances add credibility to their financial advice. SRI sites, which typically do not benefit from the reputation of well-known partners, use an array of credibility appeals to enhance the acceptability of their claims:

- *Humanization:* Almost all SRI sites provide names and e-mail addresses of their key staff, sometimes supplemented by their vitae and pictures, in order to give the content a human touch and to demonstrate their staff's expertise. This strategy was not found on traditional investment sites.
- *Objectivity:* They draw on testimonials by third parties to evoke trust in the site. SocialFunds.com, for example, lists favorable mentions in media such as *International Herald Tribune* [722], *Reuters* [764], *Time Magazine* [781], *Newsweek* [750], *Money Magazine* [672], and *Business Week* [664].

- *Lexical prominence:* It is characteristic of the discourse of SRI sites to use attitude markers when describing themselves or the information they provide. They often use adjectives in the superlative form, like “most comprehensive”, “first”, “largest”, “oldest”, “the nation’s premier”, “extensive”, or “unique”. Traditional investment sites do not exhibit this lexical prominence in their self-descriptions and, in fact, do not describe themselves at length at all.

17.2.3.2 Audience Communication

The discourse on traditional investment sites turned out to be more factual than that of SRI sites. A good indicator of factuality is the use of first-person personal pronouns (“we”, “us”, “our”), which reduce the factuality of a text, indicating that the author communicates beliefs rather than facts [84]. The first-person pronoun “we” occurs more often on SRI sites than on traditional investment sites. On average, “we” achieves relative frequencies of 0.23% on SRI sites and 0.12% on traditional investment sites, which suggests that traditional investment sites are more inclined to communicate facts than SRI sites. The first-person pronouns “our” and “us” have lower indicative value in this respect and might distort the results. Prominent in repetitive features (e.g., menu items entitled “Contact Us”), they were thus excluded from the count.

An indicator of how well a site includes its audience in the discourse linguistically is the relative frequency of direct addresses in the form of second-person pronouns (“you”, “your”). Traditional investments sites outnumber SRI sites in terms of direct audience addresses with a relative frequency of the pronouns “you” and “your” of 1.4% and 0.54%, respectively.

Community building is a strategy more common to SRI sites, promising member benefits such as “networking opportunities”, “access to latest research”, “up-to-date news”, or “informative discussions”. Eager to engage their audiences in a dialogue, SRI sites encourage feedback through emotive statements, e.g. “we would love to hear from you”. Further, they offer subscriptions to newsletters and encourage (paid) membership, which gives the site a touch of exclusiveness and conveys a sense of belonging together.

Additional interactive features for registered members of SRI sites include online conferences, discussion forums, and chat rooms. These interactive features actively involve the audience in the discourse. The traditional investment sites, in turn, largely fail to offer community-building features but offer investment-related features such as savings calculators, stock screeners, portfolio trackers, or retirement planners.

17.2.4 The Textual Function

Text cohesion occurs lexically through repetitions and synonyms or grammatically through pronouns, substitutions, and conjunctions [222]. On the World Wide Web, text cohesion takes on a broader meaning, including visual consistency (e.g., color schemes, text format, text headings) and different types of hyperlinks [49]. While internal links are defined as relative or absolute references to resources that belong to the current host, external links identify resources locate at hosts different from the one that provided the link. Anchor links are und stood as (internal) hyperlinks referring to anchor tags embedded in a document. They structure the current document or establish semantic connections to specific paragraphs of other documents.

The discourse of SRI sites tends to adopt a narrative format, using lexical and grammatical cohesive devices extensively. The long flows of text without separate headings found on these sites are not optimized for online environments and should be broken into shorter sections to make reading from screens easier [388]. Traditional investment sites make better use of hypertextual cohesion than SRI sites by providing annotated links, which serve as cohesive devices by structuring text and assigning visual prominence to the list items. Lists also render transitional phrases unnecessary, which makes the text less verbose [174]. Thus, traditional investment sites respond far better to their audiences' online reading strategies by choosing annotated links and short paragraphs over narrative text.

A quantitative analysis of the hyperlinks on each Web site assessed the significance of hyperlinks as cohesive devices. Figure 17.2 displays the number of internal links, external links, and anchor links on each site. The diagram shows that the Web sites examined make little or no use of anchor links as cohesive devices.

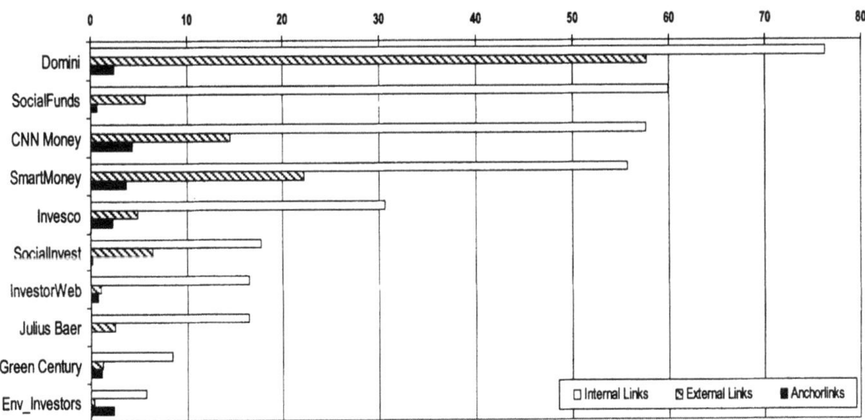

Figure 17.2: Breakdown of internal, external, and anchor links per document

17.2.5 Implications for Conceptual Site Design

The findings of the preceding sections suggest that socially responsible and traditional investing are separate movements communicating with different *clientèles*. The results demonstrate that organizations of both categories seek to accommodate the particular information needs and interests of their audiences. The study has unveiled ideological differences in the nature of the arguments put forward.

The original content of GreenMoney.at was based on print material and has been redesigned to enhance its readability and to avoid the missionary zeal that other sites use to promote SRI. To tailor its content to the hypertextual environment more effectively, paragraph lengths have been shortened and more subheadings have been introduced. Some of the community-building features found environmental investment sites, e.g. newsletter services and discussion fo-s, have also been implemented on GreenMoney.at.

17.3 Development of GreenMoney.at

The findings presented above guided the development of GreenMoney.at in a joint project between the Austrian Society for Environment and Technology [655] and the ECOresearch network [696], both of which operate sites with similar goals at GruenesGeld.at [714] and EcoInvest.org [693], respectively (the term "Gruenes Geld" is the German translation of "Green Money"). Essentially, the current version of GreenMoney.at [713] is a merger between these two environmental investment sites, seeking to provide factual content and retain credibility as an independent source of information. The official launch of the integrated site is scheduled for the first quarter of 2004, concurrently at GreenMoney.at and the other two addresses.

17.3.1 Conceptualization and Prototyping (EcoInvest.org)

At the time of EcoInvest.org's initial conceptualization, only a few dedicated environmental investment sites existed, none of which addressed Austria's emerging SRI market. Investors interested in SRI had access to two categories of Web sites. Environmental sites, on the one hand, provided a broad spectrum of topics ranging from general environmental problems to specific issues of social responsibility. These sites did not address questions of where to buy shares of mutual funds or how to profitably invest money in certain industries. Financial sites, on the other hand, published company profiles, reports on international financial markets, and the composition and performance of mutual funds – typically neglecting social and environmental impacts of investment decisions. EcoInvest.org combined aspects of both categories in its attempt to meet the specific needs of private and institutional investors, mutual fund companies, sustainable capital seekers, and environmental non-profit organizations. The goal was to create the leading Austrian environmental investment site.

Based on an analysis of the strengths and weaknesses of related projects, students of the Vienna University of Economics and Business Administration originally implemented EcoInvest.org under the supervision of two of the authors, Pieber and Treiblmaier. Major milestones of the first project phase included analyzing the competition, developing a strategic marketing plan, and implementing a prototype Web site. A content management system was the logical choice for this prototype, ensuring rapid implementation, affordable maintenance, and intuitive interfaces for content providers and users. After evaluating available alternatives, the open-source platform PostNuke was chosen [762]. Like most content management systems, this package offers the following core functions [333; 591]:

- Interfaces to database management systems for storing, retrieving, and tracking content; EcoInvest.org used mySQL [745], an open-source database management system as the central content repository;
- User registration and login management to handle the read, write, and execute permissions of different (classes of) users;
- Templating components to separate design, content, and programming logic from each other;

- Graphical editors to create, change, and annotate documents via standard Web browsers;
- Components to preview and publish documents (including the option to specify when articles should appear and expire);
- Advanced search functionality covering the site's major components;
- Plug-in management to incorporate modules and content elements.

17.3.2 Site Integration (GreenMoney.at)

GruenesGeld.at had a well-known domain name, an extensive knowledge base, and the capability to provide value to a broad range of visitors. A content manager edited the newsletter and updated the site on a regular basis. The technical platform of the site, however, was unable to serve dynamic content. EcoInvest.org incorporated a database-driven content management system, but faced difficulties updating content and providing news due to changing development teams.

In the second phase of the project, therefore, a synergistic strategy aimed at combining EcoInvest.org's technical foundation with the content structure of GruenesGeld.at pursuing the following goals:

- Leveraging an open-source content management system to simplify and accelerate content creation and maintenance;
- Incorporating interactive site features and emphasizing usability and reliability without increasing operational costs;
- Providing an executive summary in English for international users;
- Preserving the independence and credibility of the original site.

17.3.3 Content Structure

GreenMoney.at consists of six main sections with several subsections each as shown in Figure 17.3. The *SRI Information* section defines the site's intentions, since the term "green" means a variety of things to different people, especially fund managers [448]. It sketches the SRI movement and explains major investment vehicles such as shares, investment funds, bonds, life insurances, direct investments, and savings instruments. In order to shore up the credibility of GreenMoney.at, the entry page of this section states the main supporters of the Web site: the Austrian Society for Environment and Technology [655], the Austrian Federal Ministry of Agriculture, Forestry, Environment and Water Management [654], and the ECOresearch network [696].

The *Background* sub-section employs some of the argumentative patterns typical for environmental investment sites, as delineated in Section 17.2.2.1. Arguments in favor of SRI aim to convince the reader of the benefits of environmental investing. As the mission of GreenMoney.at is to promote environmental investments in Austria founded on unbiased information, the copy of the site's other sections adopts a more neutral and factual tone.

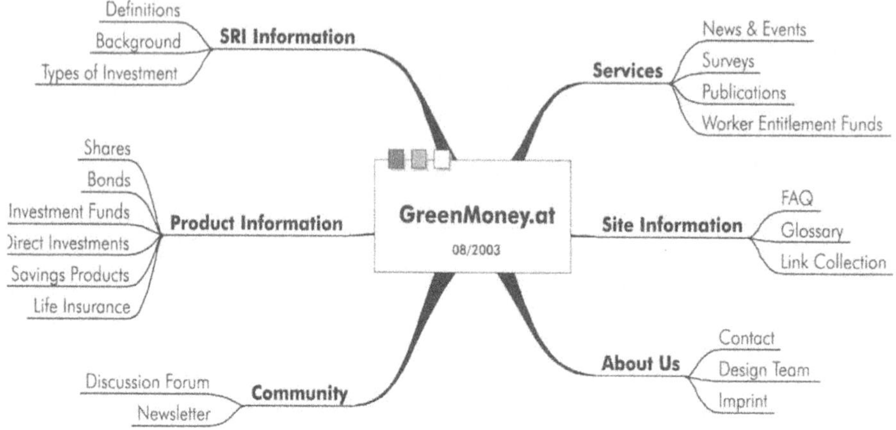

Figure 17.3: Content structure of GreenMoney.at

The *Product Information* section presents investment products available to Austrian and international investors. The information on mutual funds includes their names, their identification numbers, paying agents, inception dates, performance histories, value added since inception, and premiums. GreenMoney.at uses this numerical evidence to demonstrate that SRI products perform as well as standard investments, although return is not necessarily the only criterion for SRI investment decisions. Demand for ethical investments is remarkably stable, as ethical investors often refrain from redeploying their capital in case of temporary losses [335].

The site's *Service* section posts SRI-related events (lectures, seminars, etc.), Austrian surveys, and an extensive bibliography for novices and experts alike. This section also provides information on SRI criteria for worker entitlement funds as provided for in Austria's labor legislation.

About Us lists the contact details of a number of Austrian SRI experts who support the site as members of the "Platform for Socially Responsible Investing". The credibility and reputation of the institutions these experts represent helps GreenMoney.at build trust among users and recognition as a provider of accurate and reliable information.

Some of the interactive features found among the sample of five environmental investment sites have also been implemented on GreenMoney.at. In the *Community* section, registered users can share opinions and experiences with other members in a discussion forum. Visitors can also subscribe to an electronic newsletter to receive regular news on events, legislative changes, and research findings.

To make the site as user-friendly as possible, a glossary explains technical investment terms. This glossary is either accessible from the *Site Information* section, or by selecting a term on the screen and clicking on the "Explain" button in the right frame.

Finally, the most important parts of GreenMoney.at have been translated into English to give international users an overview of environmental investment opportunities in Austria. This should promote Austria, generally a country with good environmental track record [395], as an excellent place for green-min investors.

17.3.4 Launch and Promotion

The launch of GreenMoney.at should benefit from recent bankruptcies and scandals linked to managerial misconduct, especially in the energy sector. Private and institutional investors alike have concerns about corporate governance, realizing that good corporate conduct is vital to financial wellbeing. Financial and environmental debacles have led to an unprecedented level of shareholder scrutiny and boosted investment in socially responsible mutual funds. In 2002, SRI funds saw US$ 1.5 billion of net inflow, while US$ 10.5 billion were withdrawn from diversified US equity funds [60; 448].

After its final release in the first quarter of 2004 (the screenshot in Figure 17.4 shows the most current version at the time of writing), the site will be promoted through different channels, e.g. flyers, advertisements in specialized print media, articles in online newspapers or journals, banners on topic-related Web sites, search engine submissions, and presentations at universities.

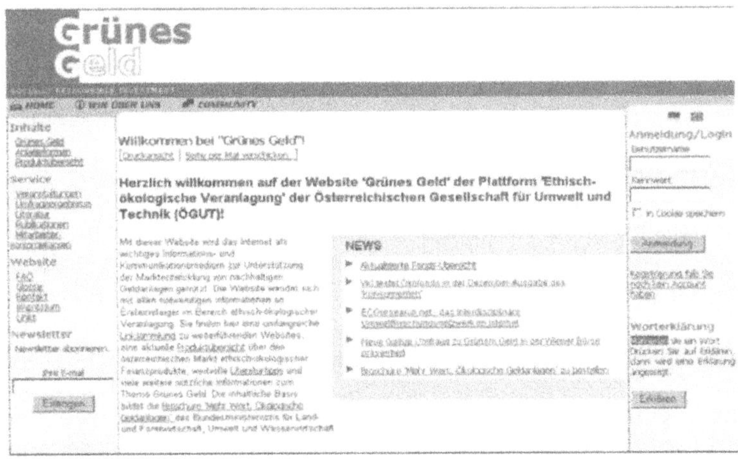

Figure 17.4: Screenshot of GreenMoney.at as of December 2003 (the logo in the upper left corner represents the German translation of the term "Green Money")

Many financial institutions underestimate the usefulness of independent sites in customer acquisition, product matching, and selling. Mutual benefits can be reaped by engaging in co-branding, outsourcing, and co-selling partnerships with successful organizations [470]. The Austrian Ministry of Agriculture, Forestry, Environment and Water Management has sponsored the Austrian Society for Environment and Technology and its Web site GruenesGeld.at from the beginning. Platform members come from a wide variety of institutions, e.g. the ministry, banks, independent financial consultants, labor unions, and consumer advocacy groups. This ensures a broad audience and reinforces the credibility of GreenMoney.at.

Since July 2002, all Austrian employers have to contribute 1.53% of their payroll to worker entitlement funds in order to provide for future redundancy payments [61]. Seven out of nine institutions responsible for these funds have de-[illegible]red their intention to invest in a socially responsible way [396], which should [illegible]late SRI interest among the Austrian population.

17.3.5 Additional Functionality

As new trends evolve, they tend to have intense informational and educational requirements. Know-how transfer and community building can be supported effectively by conveying the main issues in a user-friendly way. Distributed content such as academic reports, market analyses, company profiles, and documents by regulatory authorities will complement the current offering. In the long term, such a repository could be developed into a brokerage platform for educational materials on environmental investing and related concepts.

Additional extensions to the site might include real-time market information and a transaction system for buying and selling securities, both of which would require cooperation with a third party. To preserve the independence of the site, however, no such cooperation is planned at the moment.

After the launch, the site will concentrate on distributing financial news, similar to the traditional investment sites analyzed in Section 17.2. Since many scientists from universities and independent research institutions are involved, the publication of scientific content provides a competitive advantage. Further Web assessment, both qualitative and quantitative in nature, will enable GreenMoney.at to track current developments in online media and other investment sites.

17.4 Conclusions and Outlook

Due to increasing global environmental damage and ongoing educational efforts to promote sustainability, ecological awareness is growing. Nevertheless, many organizations that promote sustainable business models are still undercapitalized, especially in Central and Eastern Europe and developing countries [132]. Environmental investment sites address this capital shortage and the resulting restrictions.

The findings of the discourse analysis described in the first part of this chapter have been used to refine GreenMoney.at, an investment portal originating from a joint effort of the Austrian Society for Environment and Technology [655], the Austrian Federal Ministry of Agriculture, Forestry, Environment and Water Management [654], and the ECOresearch network [696]. Future efforts will focus on enhancing overall site usability, optimizing the site's navigational structure, and adding interactive content modules. Most importantly, an audience survey will be conducted as soon as the site has been promoted to determine whether it accommodates the information and communication needs of Austrian investors interested in SRI.

Chapter 18
How to Provide Customized Environmental Reports Properly

Ralf Isenmann • Jorge Marx-Gómez

Abstract. Customization is vital for environmental (online) communication, especially for corporate environmental reporting. Meeting target groups' information needs strongly influences reporting success, perhaps whether users pay attention to reports at all, how they assess the value of these documents, and to what extent they are willing to make use of these communication vehicles. In contrast to its wide acceptance in guidelines however, current practice shows significant room for improvements. Hence, we propose a framework on how to provide customized environmental reports properly, described in terms of three conceptual components: (i) stakeholder analysis and information requirement analysis, (ii) XML-based document engineering, and (iii) ICT-architecture of an environmental reporting system. Customization is a real step forward to the benefit of all groups involved in or affected by environmental reporting: managers, accountants, employees, financial analysts, standard setting institutions and organizations focused on benchmarking, rating or ranking.

18.1 The Challenge of Customized Environmental Reporting

Since its inception in the late 1980s and early 1990s, corporate environmental reporting has increasingly entered the business mainstream. Today, reporting on environmental performance is part of companies' daily affairs, and thus Marshall and Brown [352] feel that it is merely a question of *how* to report on environmental issues, no longer whether to report at all. Regardless of nationality, this is not just true for environmental pioneers and sector leaders, but also for global players, multinationals and an increasing number of small and medium-sized enterprises (SMEs) whose activities either result directly in high environmental impacts or at least are suspected of causing them. Examples abound – among others – in the pharmaceuticals, chemicals, mining, transport, electronics and automotive sectors [176; 223; 305; 306; 307; 308; 317; 616].

Within several industrial sectors, there is empirical evidence that corporate environmental reporting truly has become of *competitive relevance* [181] and *strategic importance* [323]. Companies that have learned their lessons are clearly benefiting from environmental reports used as green management tools [304], e.g. for decision-making with internal and external value [158], especially in terms of cost reduction [367], resource controlling, accounting, material and waste management [535], reputation [50] and compliance [3].

Marshall and Brown's introductory note may answer, at least to a certain extent, one of the critical questions stated early in the field: "Will corporate environmental performance reporting be a 'passing fashion', like social reporting some years ago, or will it increasingly enter the business mainstream?" [568, p66] Now that companies are making more use of environmental reports, a great number are moving away from obviously outdated, very early stages simply providing "green glossy brochures", usually prepared as "one size fits all" documents on print media, towards a fully ICT-supported environmental reporting system that provides customized reports and other fine-tuned communication vehicles on different media and in various presentation styles, while meeting the requirements of several regulations, guidelines and stakeholders' different information needs [263].

The reasons why there is an increasingly demand for *customized* environmental information can be seen in the fact that interested parties, target groups and other stakeholders are generally becoming more critical of companies' business. Moreover, as environmental reporting matures, reports are analyzed by more competent and professional users with higher expectations. Thus, not only employees, customers, local authorities, but also investors and financial analysts want fine-tuned vehicles and they expect reporting instruments that truly meet their specific needs in content, form, media and distribution principles. Environmental communication only via uniformed reports on print media may hardly fulfill emerging requirements in information supply, stakeholder dialogue, feedback and interactivity.

Corporate environmental reporting supported by Internet technologies and Internet services – in particular the World Wide Web – is a rapidly emerging and increasingly popular method. The rationale why more and more companies are using the Internet can be seen in its *unique capabilities* in form and content and other technical benefits provided by this computer-based method, e.g. opportunity to address multiple issues in flexible depth, ease of access, great potential to reach a wider audience or disseminate target group tailored vehicles to specific stakeholders, many facilities to effect two-way and even one-to-one communication, ease of updating, great potential for supporting the whole workflow – including administration, preparation, distribution and presentation – and finally the outstanding opportunity to produce a number of customized, target group tailored, individualized or even personalized communication vehicles in an automated, effective and cost-saving manner [261; 354].

Compared with traditional methods on print media, Internet-based environmental reporting embraces a broader range of *beneficial characteristics* for smart presentation, e.g. combining text, still and moving images, sound, feedback, interaction, dialogue, integration of different contents, etc. [261]. Because of its overall *added value* creating nature, the Internet is ultimately seen an indispensable tool [276; 553], suitable for any forward-looking approach to corporate environmental reporting [263; 338]. At present, the Internet is already used by several reporting companies and target groups as the *pivotal platform* to provide access to information on environmental performance and other related issues, surely influencing the development the field will probably take in the future.

The limits of an exclusively paper-based practice are closely linked with diff culties involved in using print media for communication for which they are c poorly suited. In the words of Mach: "An organization needs to send the right messages through the right distribution channels to the right audiences. To accomplish this, it may need a variety of communications vehicles – not just a single report. One size doesn't fit all in today's Internet world of mass customization" (cited in: [345, p248]; compare [505; 619]). More to the point, a "report designed to appeal to everybody may end up serving nobody's real needs" [257, p6]. According to conceptual contributions [331] and empirical findings [263; 306], however, in many cases it is clear *target group tailoring* that is still lacking in the field. In order to improve current practice, extend reporting success and multiply the number of users actually reached, companies are going to customize their reports, intending to provide a variety of fine-tuned communication vehicles, not just a single report, from uniformed instruments and standardized ones towards a more advanced approach also providing individualized or even personalized reports.

Although more and more companies are using the Internet as an accessible distribution channel, despite the fact that it encourages paperless information, regardless of its favor for a digital world of online communication and though some reporting companies are slimming down the size of printed reports while some others again are publishing exclusively on the Web, many users still like to be provided with paper-based reports. Even now a considerable number prefer print media, perhaps to read reports later on. According to Scott and Jackson [493, p196] "abandoning paper-based reporting altogether is a risky strategy" and probably not a successful way to perform future reporting requirements. Companies doing so may tend to overestimate the usefulness of the Internet in terms of reaching target groups and providing information in a format useful to all.

18.2 Framework for Customized Environmental Reporting

The question of how companies are reporting [352], in particular how to provide customized environmental reports properly, involves a number of crucial aspects, perhaps depending on nationality and industry sectors, but also closely linked to companies' information management, communication strategy, image profile, stakeholder relations, organization, staff and ICT infrastructure. Due to various strategic implications, companies should consider these aspects carefully because they influence the overall reporting capabilities. Furthermore, these aspects may also influence the facilities to design customized environmental reporting instruments and other fine-tuned communication vehicles in form and content.

For example, companies need to ponder the question of which media to use for managing environmental reporting, particularly along the reporting workflow and its underlying core processes, i.e. preparation, administration, distribution and presentation. Should these be predominantly print media or computer-based media like the Internet and CD-ROM? Do the Internet and its associated technologies such as the emerging markup language XML carry any unique design and content capabilities that facilitate tailoring of reports according to different information needs and reaching the target groups addressed? No less importantly, what is the role of ICT applications when companies are moving away from providing simple standard reports on print media towards a forward-looking approach, perhaps in the sense of "cybernetic reporting" [619, p6]?

The cutting-edge approach in the rapidly developing field of corporate environmental reporting seems to be an *Internet-based reporting system*. Such a comprehensive and fully ICT-supported approach offers a variety of added value creating features compared with early environmental reporting stages. For example, an Internet-based reporting system provides a set of:

- Important contents (primarily environmental issues but more and more linked with financial and social issues and their mutual interrelations) that comprise the core themes for corporate sustainability [112; 257; 378; 553],
- Different media (print media, Internet, CD-ROM etc.),
- Corresponding distributing principles (push, pull),
- Various presentation styles (media-specific, target-group tailored).
- In technological terms, the reporting system is fully supported by an underlying ICT infrastructure that has its basis in using the Internet and employing XML.

For customized environmental reporting it is characteristic to take into account the requirements of several standards, guidelines and different needs of a number of users and then to produce reports exactly meeting all these requirements and needs. There is consensus that such customization or target group tailoring is vital for the success of environmental reporting [261; 505; 523]. In contrast to its wide acceptance in concepts and guidelines, however, current practice shows significant room for improvement, even for the best reporters.

In summary, customized reporting remains rather *unrealized* and still contains challenges, perhaps managed in the near future but clearly lacking up to now. Hence, throwing more light onto customization is argued to be a *real step forward* in the field of corporate reporting, be it environmental reporting, financial reporting [234], social reporting or its integrated fashion as sustainability reporting [260]. Approaching this goal, however, is not as simple a process as it may appear at first glance. On the contrary, such an enterprise represents a challenging and multifaceted problem requiring both identification of relevant stakeholders and clarification of their needs, and also a pool of report contents that companies are willing to disclose, preferably arranged in a specific structure appropriate for automated machine processing through ICT applications.

In this chapter, we propose a framework for a customized environmental reporting system, described in terms of three conceptual components, while fully exploiting the capabilities the Internet may offer for environmental reporting on the whole:

- *Component 1:* Stakeholder analysis and information requirement analysis, representing the demand for environmental information.
- *Component 2:* XML-based document engineering for corporate environmental reporting, representing the contents that a company is willing to disclose.
- *Component 3:* Environmental reporting system, representing a suitable ICT architecture for cross matching offer (component 2) and demand (component 1), intended to provide truly customized reports, not just smartly polished versions of a uniformed report.

The framework mentioned above may serve as a *step-by-step-approach* to developing from early environmental reporting stages towards an advanced and more sophisticated one, with special emphasis on providing customized environmental reports properly. Without the direct link to ICT, progressing in the field and finally approaching customized environmental reporting is seen to become quite difficult, as moving away from orthodox stages is a complex task.

While its early incarnations have been focused primarily on single, free-standing environmental reports, predominantly produced on print media and usually prepared as "one size fits all" universal documents, in future, as companies learn their lessons and practice mature, the focus will likely become cross-media, customized and more comprehensive, also disclosing how environmental issues are linked with financial and social aspects, concurring to the triple bottom line approach and finally leading to sustainability reporting [260; 263]. In any case, current reporting requirements as well as future challenges in the field may hardly be met without truly taking into account users' needs and preferences on the one hand (customized environmental reporting) and addressing latest trends in ICT (Internet-based approach) on the other.

18.3 Component 1: Stakeholder Analysis and Information Requirement Analysis

Any corporate environmental reporting system focused on customization should be based on two analyses that need to be carried out prior to the system's implementation: stakeholder analysis and information requirement analysis.

Stakeholder analysis identifies the primary users and typically asks: Who are the key target groups (including the critical ones) that should or want to be addressed through environmental reporting, inside and outside the company? Surveying current literature in the field, Lenz [330] identified twelve key target groups for corporate environmental reporting and arranged these in four clusters because of their comparable information needs (Figure 18.1).

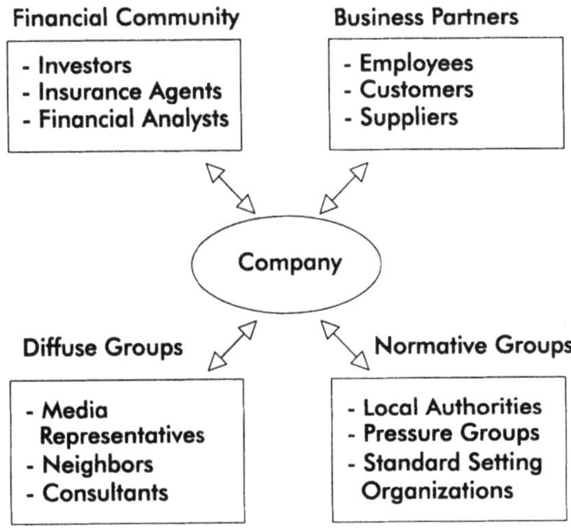

Figure 18.1: Target groups involved in or affected by environmental reporting

In addition to stakeholder analysis and the identification of primary users, *information requirement analysis* is performed to find out and study stakeholders' specific information needs and other preferences expected to be met in a more detailed fashion. Such an analysis closely related to the former one usually leads to the question: What relevant contents do target groups expect, and what preferences do they want to be fulfilled, perhaps regarding form, layout, design, media and distribution channel? At present, very little work is being done to conceptualize users' information needs, especially concerning media-specific aspects like distribution channels and presentation styles. Thus, van Dalen [572] complains about a lack of profound insights in this area. With this in mind, Lenz [330] surveyed a number of empirical studies [31; 105; 257; 488; 592] on user needs (Figure 18.2).

For example, with growing general environmental awareness, *employees* are interested in the environmental performance of their employers and companies. Further, they want to understand how companies are seen by local community groups, wishing to see their employer's business as a going concern.

Establishing supply chains and other manufacturing networks implies extensive environmental communication and proper channels along the whole supply chain or network [339]. *Suppliers* and other *business partners* need environmental information mainly regarding resource efficiency, regulatory compliance, new product and service opportunities, especially in terms of extended product stewardship and other environmental liabilities [479; 639].

Investors, including institutional and private shareholders, financial analysts and investment consultants, are increasingly interested in environmental issues and their financial interrelations since they are aware that environmental reports make good business and environmental sense [59] and they expect that environmental performance influences financial performance and shareholder value [473; 583]. For example, in November 2000 a group of 39 leading financial investors, managing combined assets in excess of US$ 140 billions, sent a letter to the CEOs of the 500 largest US companies urging them to encourage the provision of integrated environmental reports [515].

The information requirement analysis clearly demonstrates that employees, customers, suppliers, local authorities, legislators, neighbors, consultants, financial analysts, investors, insurance agents, media representatives and members of rating and ranking organizations have *heterogeneous information needs.* These different needs cannot be fully satisfied or easily be met just by "reporting as usual" through orthodox practice, via one universal document (on print media), mostly produced as a "one size fits all" report. Users increasingly expect customized, target group tailored, individualized or even personalized reporting instruments.

In summary, the insights gained from both analyses described above can be used to develop user profiles where information needs and other preferences of key target groups are stored (component 2) and then prepared ready for machine processing in an environmental reporting system (component 3).

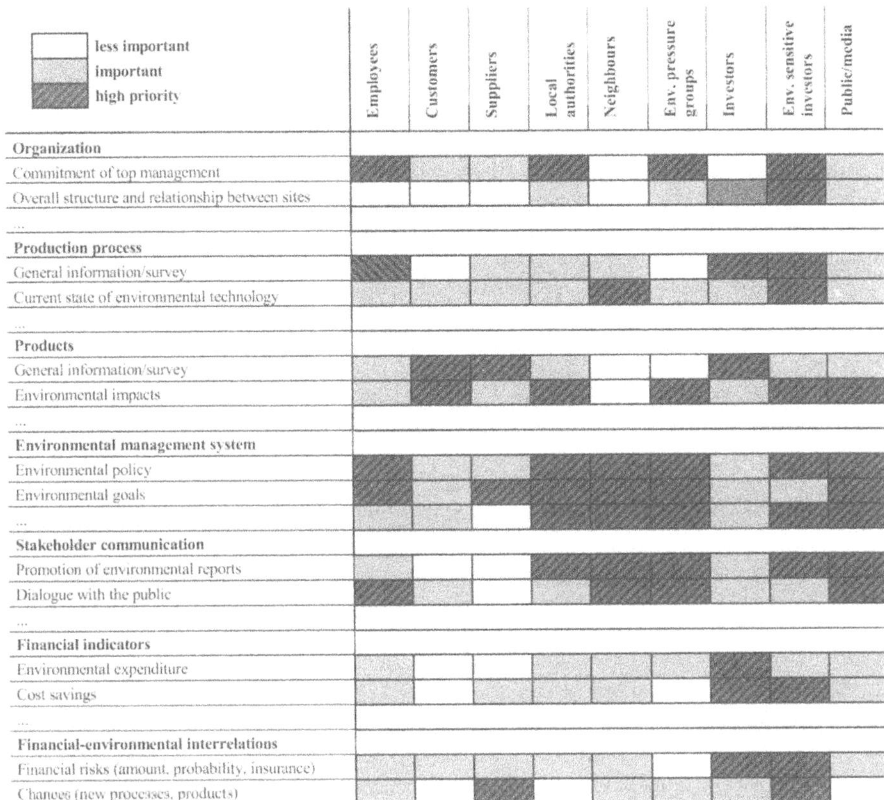

Figure 18.2: Environmental information needs of key target groups, extract [330, p232]

18.4 Component 2: XML-based DTD for Environmental Reporting

The results of stakeholder analysis and deeper insights of information requirements are used for XML-based document engineering. This ICT-laden area is where possible report contents and the basic structure of a certain group of documents are defined, e.g. for corporate environmental reports. The main task of document engineering is developing a proper Document Type Definition (DTD). Such a DTD represents the heart of any XML-based ICT application, and of a (Web) content management system that may be used for maintaining companies' environmental Web sites. Within a DTD the overall pool of possible report contents are stored in a basic structure. From this pool of structured contents, a number of customized reports can be prepared in an automated fashion, just by machine processing [264; 354].

Developing a comprehensive XML-based DTD for environmental reporting involved a sophisticated effort because a great number of different requirements ave to be taken into account. Methodically, its development rests on a process-nted model initially proposed by Schraml [485] and adapted by Lenz [330]:

- First, the main target has to be defined, i.e. developing a comprehensive DTD for XML-based environmental reports.
- Next, a multitude of resources has to be analyzed [2; 166; 167; 194; 200; 215; 568] to extract contents from regulations, standards, guidelines, available reports and users' needs and preferences, finally identifying the pool of possible semantic components a report may contain.
- Then, from this pool, the relevant semantic components are selected and arranged in a catalogue. The result is a total of 115 semantic components (Figure 18.3).
- Subsequently, the document type model could be designed. Therefore, all selected components are organized in a hierarchy, typical for XML documents. Finally, the document type model is implemented, i.e. noted according to XML and transformed in a DTD. Today, the process of standardizing and harmonizing this DTD is still ongoing [262]; see Chapter 19.

When employing such an XML-based DTD, it is argued here, the reporting company will be in a position to carry out its tasks of environmental reporting in a forward-looking way, intended to improve its information management, support its processes along the whole workflow while exactly meeting requirements proposed by certain guidelines, and communicating with its users in a meaningful and target group tailored manner.

ID	Description			r/o	Source	Generic identity
1	foreword			o	future 6.1, II	foreword
2	organization			r	EMAS II, A III, 3.2	organization
3		organization description		o	instances	orgDescription
4			corporate culture	o	users	corporateCulture
5			relationship to parent organization	o	EMAS II, A III, 3.2	parentOrg
6			sites	o	future 6.1, I	sites
...	...	...	...	...	...	...
98		economic-environmental interdependences		o	users	econEnvInterdep
99			financial risks	o	users	financialRisks
100			financial chances	o	users	financialChances
101	formalities			r	EMAS II, A III, 3.2	formalities
102		imprint		o	instances	imprint
103			publisher/author/originator	o	DIN 33922, 5.6, instances	authorOriginator
104			publicationDate	o	instances	publicationDate
105			reporting period	o	DIN 33922, 5.6	temporalCoverage
106			date of next report	o	future 6.1, X	nextReport
107			responsibility and participation in environmental reporting	o	future 6.1, III	reportTeam
108			contact	o	DIN 33922, 5.6	contact
109		verification		o	UNO 5, I, 11	verification
110			verifier name	r	EMAS II, A III, 3.2	verifierName
111			verifier accreditation number	r	EMAS II, A III, 3.2	verifierAccredNo
112			verifier address	o	DIN 33922, 5.6	verifierAddress
113			verifier statement	o	future 6.1, X	verifierStatement
114			verification date	r	EMAS II, A III, 3.2	verificationDate
115		additional information		o	future 6.1, X	additionalInfo

Figure 18.3: Catalogue of 115 relevant semantic components [330, p243]

18.5 Component 3: Environmental Reporting System

Based on the insights of stakeholder analysis and information requirements analysis as one component and built upon a suitable XML-based DTD for environmental reporting as another component, the next step is to combine these components to work together in an environmental reporting system, performed by an underlying ICT architecture. The main task of such a reporting system is to cross-match demand and offer of environmental information appropriately, to provide customized reports and other fine-tuned communication vehicles that exactly meet users' needs and preferences or that fulfill specific requirements of certain guidelines, in an automated manner via machine processing.

Based on a generic classification of customization in terms of user modeling and systems adaptation proposed by [330], here we describe an *adaptable* environmental reporting system producing *individualized* reports; users could modify and interact with the system to a certain extent. For example, every single user may create his own "report à la carte", while choosing contents, structure, layout and distribution channel according to his individual needs and preferences, just selecting with keywords, clicking on preferences on a menu or choosing a certain guideline – perhaps creating an environmental report according to DIN 33922 or an environmental statement according to the Eco Management and Audit Scheme (EMAS) at one's fingertips. Then the document is generated by the reporting system in an automated manner. An anonymous user model is stored for any user, but the user may always control system behavior by changing the profile.

At the heart of the ICT architecture lies Cocoon, a Java-based, modular structured, open source publishing software [650], able to perform XML-based DTDs, and suitable for providing individualized or even personalized environmental reports. Users just need to submit personal information, e.g. name and address, and the adaptable environmental reporting system provides personalized reports, truly able to realize one-to-one-communication on the fly. Companies are then in a position to make progress in customization step by step, perhaps starting with stereotyping and next moving towards an individualized or personalized environmental reporting system.

The reason why we have employed Cocoon lies – among other benefits – in its powerful and sophisticated application capabilities. The modular application components can be arranged rather flexibly, perhaps serially grouped in so-called pipelines where different reports are then created dynamically on the basis of an XML-based DTD, exactly meeting users' different needs. At present, this customized environmental reporting system is realized as a prototype, but soon it will be implemented in a number of German SMEs that then may be regarded as pioneers in the field of corporate environmental reporting, at least in terms of customization and using the Internet for environmental reporting properly.

18.6 Conclusions

Corporate environmental reporting is increasingly entering the business mainstream all over the world and thus it has become almost part of companies' daily ffairs. Since its initial incarnation in the late 1980s and early 1990s, companies made considerable progress. Consequently, companies are moving away from obviously outdated, very early stages simply providing "green glossy brochures", usually prepared as "one size fits all" documents on print media, towards an advanced reporting stage while paying more attention to target groups' different information needs, willing to provide customized reports and fine-tuned communication vehicles on different media and in various presentation styles. Now that the field has matured and companies have learned their lessons, environmental (online) communication via uniformed reports in print media or on the World Wide Web may hardly fulfill emerging requirements and satisfy current expectations in information supply, stakeholder dialogue, feedback, and interactivity.

Hence, more and more companies are using the Internet because of the unique media-specific benefits provided by this computer-based method. Its great potential for supporting the whole workflow and the outstanding opportunity for producing target group tailored, individualized or even personalized communication vehicles in an automated, effective and cost-saving manner, however, seems to be hardly exploited, as yet. Customization in the field of corporate environmental reporting is still at a premature stage. In order to improve current practice, we have presented a framework on how to provide customized environmental reports properly, described in terms of three conceptual components, while fully exploiting the capabilities the Internet may offer on the whole: (i) stakeholder analysis and information requirement analysis, (ii) XML-based document engineering, and (iii) ICT architecture of an environmental reporting system.

All in all, throwing more light onto customization is argued to be a real step forward in the field, to the benefit of all groups involved in or affected by corporate environmental reporting: managers, accountants, employees, members of the financial community, standard-setting institutions, or organizations focused on benchmarking, rating and ranking.

Chapter 19
Harmonizing Document Type Definitions for Corporate Environmental Reports

Jorge Marx-Gómez • Ralf Isenmann

Abstract. *Corporate environmental reporting using the Internet is a rapidly emerging and increasingly popular method. Despite the considerable progress companies have made recently, exploitation of the whole range of benefits of this computer-based method is still at a premature stage, especially in terms of automated production and customized provision of environmental reports. Employing the eXtensible Markup Language (XML) and using a harmonized Document Type Definition (DTD) for these leading reporting vehicles offer an array of benefits, finally elevating Internet use beyond the status of a mere online distribution channel of electronic duplicates of hard copy reports. In this chapter we present the harmonization of three XML-based DTDs that have been proposed for corporate environmental reports. On the basis of a harmonized XML-based DTD, companies are in a position to provide a customized environmental reporting system, prepared by machine processing in an automated manner.*

19.1 Using the Internet and Employing XML for Corporate Environmental Reporting

In corporate environmental reporting, using the Internet and employing XML are some of the latest and fashionable trends in the field. Today, environmental online reporting has become part of companies' daily affairs, and thus a growing number make use of certain capabilities that this computer-based method provides: reports, brochures, leaflets, newsletters, slides, presentations, audio sequences, video clips, etc. are accessible on the Web, e.g. several documents via download in PDF and/or online in HTML (MP3 and MPG for audio and video, respectively), prepared for being pulled or automatically disseminated via e-mail or other push technologies.

Any company improving its practice through proper Internet use will probably come across the area of XML-based document engineering, in particular when the focus is on structuring and standardizing communication vehicles in the form of documents as in the case of environmental reports. XML-based document engineering represents a crucial task for establishing an ICT-supported reporting approach. The conceptual core when exploiting the whole range of media-specific benefits of the Internet and XML lies in the development of a comprehensive and harmonized XML-based DTD for environmental reports.

A DTD is a declaration defining the overall structure and possible contents of a ·ain group of documents. Within the whole variety of documents and other environmental communication tools, an environmental report is seen as the pivotal and leading vehicle [369] because of its unique claim to credibility, the reliance target groups ascribe to it and its balanced nature including quantitative and qualitative data. Companies use such reports for communicating their environmental performance, usually addressing a wide range of target groups, often produced as single, free-standing documents and issued for a certain period of time.

Employing an XML-based DTD for environmental reports represents a truly forward-looking effort, intended to facilitate companies' underlying reporting workflow, from automated generation and comfortable administration towards customized distribution and target group tailored presentation, finally to the benefit of all groups involved in or affected by environmental reporting, inside and outside the companies. For example, on the basis of a harmonized XML-based DTD, companies are then in a position to provide a customized environmental reporting system, prepared by machine processing in an automated manner (see Chapter 18). In the following, we give an outline of how to harmonize three XML-based DTDs for environmental reports that have been proposed in the field of corporate environmental reporting.

19.2 Three XML-based DTDs for Environmental Reports: Kaiserslautern, Magdeburg, Berlin

All three XML-based DTDs proposed for environmental reports have been prepared autonomously and published more or less simultaneously. The first DTD was developed by the Department of Business Information Systems and Operations Research, Kaiserslautern University of Technology [331]. The second DTD was presented by the Institute for Technical and Business Information Systems, Otto-von-Guericke-Universität Magdeburg [316]. The third DTD was proposed by the Institute of Information Systems, Humboldt University of Berlin [25].

Throwing more light onto the harmonization of environmental reports helps to elevate Internet use beyond the status of a mere online distribution channel of electronic duplicates of hard copy reports. Hence, the research groups mentioned above started a still ongoing process of harmonization, embedded into efforts towards shaping a unifying Environmental Markup Language (EML).

19.3 Harmonization of XML-based DTDs

The overall harmonization is based on a sound methodology, i.e. a sophisticated DTD design process proposed by Schraml [485]. First, the main target of the DTD has to be defined. Next, a pool of possible semantic components should be identified and structured. Then, from this pool, the actually relevant semantic components are selected and arranged in a catalogue. Finally, the document type model can be designed. As the outcome, a basic proposal on the top level is suggested. The next step is to harmonize the approaches in a more detailed manner. However, harmonization on the second level is not as simple as it may seem at first glance, because of rather different design profiles underlying certain DTD developments (Figure 19.1).

	Kaiserslautern	Magdeburg	Berlin
Main target	– Academic approach – Standardization – Sustainability reporting – Stakeholder dialogue – Target group tailoring – Multiple media publishing	– Case study: Hasseröder brewery – Automation – Life cycle analysis – Target group tailoring – Integrated communications – Multiple media publishing	– Meeting EML-requirements – Using metadata
Restrictions to be considered	– EMAS II – EN ISO 14001 – DIN ISO 33922 – National guideline of future e.V. – International guideline of UNEP	– EMAS I – DIN ISO 33922	– DIN ISO 33922
Information needs taken into account	– Employees – Customers – Suppliers – Government/local authorities – Neighbors – Environmental pressure groups – Investors – Journalists – Public	– Customers – Employees – Government/local authorities – Neighbors	
Document instances	– Print media – Computer-based media	– Print media – Computer-based media	

Figure 19.1: Design profiles underlying DTD development

Based on the insights above, any XML-based DTD for environmental reports primarily used in Europe should fulfill at least several basic criteria [262], i.e. recommendations of the EML initiative [26], and clarification of crucially important aspects for defining the DTD's underlying design profile. Concerning the main target, we propose a general, flexible and comprehensive approach. Regarding the identification of possible semantic components, we recommend the inclusion of the European Eco-Management and Audit Scheme (EMAS II), the ISO standard 14001 on Environmental Management Systems, the German standard 33922 Environmental Reports for the Public, widely accepted guidelines such as Company Environmental Reporting proposed by UNEP or Preparation and Distribution of Environmental Reports and Statements by future e.V. and the Institute for Ecological Economy Research, as well as the forthcoming ISO standard 14063 on Environmental Communication. Further, we recommend analyzing document instances on print media and on the Web, perhaps also to identify logical components like heading, paragraph, abstract, chart, etc. To select relevant semantic components out of the pool of possible components, we strongly recommend employing Schraml's methodology. All these requirements are met by the current proposal for a harmonized DTD for environmental reports (Figure 19.2).

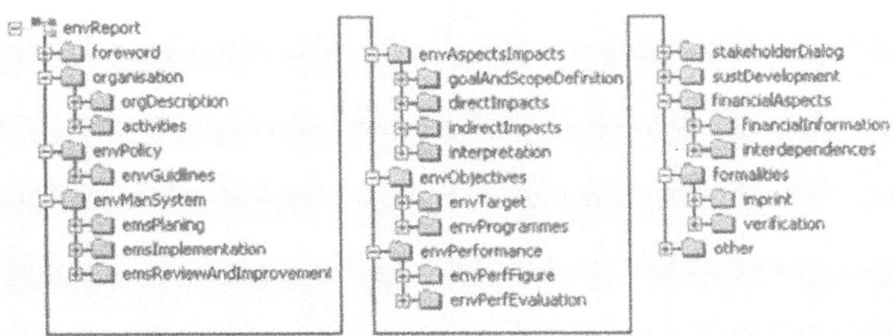

Figure 19.2: Proposal for a harmonized DTD for environmental reports

To upgrade the provisional result to a globally applicable approach, now we are going to take two further aspects into account: inclusion of the GRI guideline that will likely become the *de facto* standard for advanced environmental reporting worldwide, and analysis of similar trends in financial reporting, e.g. the efforts to establish a commonly accepted Extensible Business Reporting Markup Language (XBRL).

19.4 Conclusions

From an academic's point of view, harmonizing XML-based DTDs for environmental reports is thought to be a crucially important effort in the field of corporate environmental reporting. Such an enterprise may promote initiatives towards shaping a unifying markup language within the area of environmental informatics, perhaps in the sense of an EML. From a practitioners' perspective, a harmonized XML-based DTD helps companies to use the Internet properly and to fully exploit the advantages of XML for environmental reporting compared with other data formats. When employing an XML-based DTD in an Internet-based environmental reporting system, the company will be in a position to carry out its tasks of information management well, using its human and organizational resources more efficiently, and communicating on environmental issues in a meaningful way, e.g. providing a number of target group tailored environmental reports, prepared by machine processing in an automated manner.

Chapter 20
Environmental Information Practices in the Chemical Industry

Corinne Faure • Thomas Hillenbrand • Susan Röver

Abstract. *In this chapter, we systematically study the extent to which leading chemical companies are using online communication to raise awareness about environmental issues and answer environment-related questions linked to their activities and products. We analyzed the corporate Web sites of twelve leading chemical companies, distinguishing between two types of online environmental information: (1) corporate information about the company's environmental activities and (2) product-related environmental information. Our results indicate relatively good (but unequal) practices in corporate environmental online communications and relatively poor product-related online environmental information in the chemical industry. Our analysis allows us to identify the companies with the best practices for each criterion considered.*

20.1 Introduction

Chemical production and the use of chemicals are often associated with potential environmental damage. Air, water, ground, animals, and people can be affected by hazardous chemical substances. In fact, the chemical industry is one of the main targets of environmental interest groups. In response to these image problems, the chemical industry has voluntarily committed itself to an environmental self-regulation effort, the Responsible Care Program [253]. Through this program, each member company commits itself among other measures to educate the general public as well as potential users of their products about environmental issues of relevance to their business area. The complexity of the production chain in the chemical industry makes it particularly difficult to deal with environmental issues in this industry: many steps generally occur between the production and the end usage of a chemical product, including many different intermediaries and intermediate products. This complexity leads in turn to significant information gaps about the materials comprised in chemical products and about the properties of these materials. This situation makes environmental risk management in the chemical industry particularly challenging; so far, no satisfactory solutions to this issue have been found. These difficulties often lead to governmental interventions in the form of regulations. For instance, the European Union is currently working on implementing a new chemical policy. At the core of the policy is REACH: an integrated system for the Registration, Evaluation, and Authorization of CHemicals. REACH will compel companies which produce, import and use chemicals to assess the risks arising from their use and to take the necessary measures to manage any risks they identify [53; 163]. One necessary condition for the successful implementation of this new EU policy for the chemical industry is a significant improvement of information flows between producers, finishers, and users of chemical products about the properties of the products [163]. While in the past environmental regulations for the chemical industry were mostly focused on the direct environmental effects of manufacturing and distribution, the new regulations will bring new challenges. The results of our analysis can provide helpful indications for the implementation of these new regulations.

20.2 The Internet as Medium for Environmental Information

Among the different media available to fulfill these new information flow commitments, the Internet appears particularly attractive because of its media richness and its accessibility. Thus, corporate Web sites can be used as a medium to fulfill these commitments.

Online communication is uniquely attractive because of the richness of the medium. For most communication media, content and form are a matter of trade-off: when form is put forward (for instance on television), then some content has to be left out; when content is put forward (for instance in paper communication), then the form suffers. Online communication in contrast allows us to optimally combine form and content (or in Veen's terms, form and function), so that, as necessary for a successful communication [452; 584], both feed from one another but are not made at the expense of one another. Thus, the Internet allows us to present a wide depth of information in an attractive manner [227]. Online communication can include in a visually attractive environment documents such as white papers or position papers for download (which would otherwise be cumbersome to diffuse), search tools, or interactive documents, thereby better fulfilling the needs of the recipients [227; 549], particularly in the growing business-to-business markets [114; 190].

Accessibility is the second characteristic of online communication. From the company's viewpoint, accessibility can be further broken down into ease of access, immediate access, and controlled access. For relatively low investments, companies can broadcast corporate information through this medium [452]. Access too is immediate; online information can be updated almost instantaneously, thereby offering the possibility to react quickly to changes in environmental regulations, company policies, or product characteristics. Companies also have controlled access to the information diffusion; unlike other public relations activities, where the company depends on the goodwill and interpretation power of the press, the information presented on corporate Web sites is not exposed to third-party interpretation and can be controlled by the company. Even further, the company can selectively control who gets access to which information, thereby for instance controlling which information is made accessible to competitors, partners, customers, or government agencies [452].

From the recipient's viewpoint as well, accessibility is an important characteristic of online communication. Unlike other media, online communication is accessible to all Internet users at all times, irrespective of their geographical location [452; 549]. The global nature of the Internet as a communication medium is particularly attractive for the chemical industry, where many of the transactions place at an international level, and where customers and suppliers are widely spread throughout the world [721]. Recent studies show that the Internet is particularly used as source of information in business-to-business transactions [130].

Finally, global accessibility also signifies increased transparency; best practices in terms of environmental policy and of environmental communication become available to all. Customers, suppliers, competitors, and government agencies can evaluate the extent to which each company in the chemical industry fulfills the environmental information flow requirements they committed themselves to.

Such an evaluation is precisely the purpose of our research: we explore whether and to what extent the chemical companies actually make use of the online medium for environmental communication. In this article, we systematically study the extent to which leading chemical companies are using online communication to raise awareness about environmental issues and answer environment-related questions linked to their activities and products.

20.3 Method

20.3.1 Evaluation Criteria

To conduct this analysis, we first developed a list of evaluation criteria. We distinguished between two types of environmental information available on corporate Web sites: (1) corporate information, which helps a company to generate a positive image about its environmental activities and targets either the general public, or environmental and governmental organizations, and (2) product-related information, which is directed at customers, employees, suppliers, and finishers, and whose purpose is to reduce misuses or incidents when handling the products. Each of these two categories was evaluated on two main criteria: content (range and depth of information provided) and form (quality of presentation of information) of the environmental information provided on the Web site. The form criteria were developed on the basis of the literature on online communication (adapted to the specific context of environmental communication); the content criteria were developed based on the literature and the authors' expertise in environmental issues in the chemical industry.

20.3.1.1 Criteria for Corporate Information

For the content of corporate information, we evaluated the following criteria: (1) availability of online (HTML) versions of corporate environmental reports, which describe the company's overall environmental strategy and environmental activities – such online versions allow the user to directly go to parts of the reports they are interested in; (2) possibility for ordering or for downloading environmental reports – this possibility is more convenient for users desiring a printed version of the documents; (3) information about environmentally friendly activities of the company and description of these activities; (4) information about industry-wide environmental activities; (5) information about historic developments (for instance through availability of past reports) and about future plans in terms of environmental activities; and (6) availability of links to external environment-related activities and organizations (such as Responsible Care).

For the form of corporate information, we identified the following criteria [170; 375; 388]: (1) direct link to environmental information from corporate homepage; (2) organization of content/ease of navigation; (3) presence of content lists, titles, and internal links on individual documents; (4) presence of summary graphics; (5) date of last update, to inform users about the currency of the information; (6) availability of contact details, allowing users to request further information or provide feedback on the Web site.

20.3.1.2 Criteria for Product-Related Information

The objective of the product-related criteria was to evaluate whether the Web sites provide potential users of their products with enough information to ensure that these products are handled in a safe and environmentally friendly manner.

For the content of product-related environmental information, we identified the following criteria (note that providing information especially about the third criterion is usually required by law): (1) composition and properties of chemical substances; (2) information about applications and prohibition of usage of the substances; (3) classification of chemical substances, for instance as a hazardous material; (4) material safety and technical data sheets, which not only reveal all information related to environment, health and safety issues, but are also needed by customers when purchasing the materials; (5) handling instructions; and (6) technical characteristics of the chemical.

The criteria for the evaluation of the form of the product-related environmental information were as follows [297; 528]: (1) structured product catalogue; (2) search engine to directly access desired products; (3) ease of navigation for the product Web sites; (4) comparison tool of different products and of their specifications; (5) last information update, particularly important for Data Sheets which have to adhere to the latest regulations; and (6) contact details.

20.3.1.3 Overall Evaluation

Each Web site was evaluated on these 24 criteria by the authors and received a pass/fail grade for each criterion, resolving disagreements through discussion.

20.3.2 Sample

To conduct our analysis, we first identified the top 20 chemical companies of the year 2002 [500]. In order to focus the analysis clearly on the chemical industry, only companies with more than 60% of their income stemming from chemical products were included in our sample (thereby eliminating Mitsubishi Chemicals and petroleum companies such as Royal Shell or BP). Also, to ensure a meaningful comparability of the companies in the sample, we only included specialty or basic chemical producers with a diversified product range, thereby eliminating from our analysis Air Liquide [646]. Using these criteria, our final sample consisted of twelve companies: Akzo Nobel [647], BASF [659], Bayer [660], Dainippon Ink and Chemicals [679], Degussa [681], Dow Chemical [686], DuPont [687], Huntsman [717], Imperial Chemical Industries [719], Mitsui Chemicals [743], SABIC [766], and Sumitomo Chemical [778]. The analysis of the twelve Web si was conducted in August 2003.

20.4 Results

To obtain an overall view of the aggregate performance of the Web sites, we totalled the number of criteria achieved for each Web site studied. Figure 20.1 represents the number of companies meeting a certain number of criteria (out of twelve possible respectively for corporate information and product-related information).

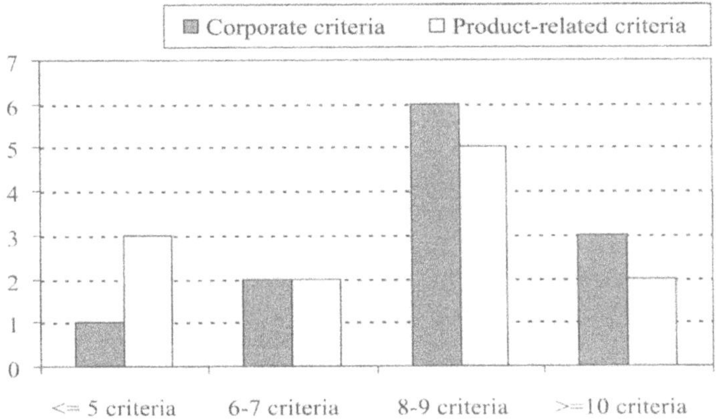

Figure 20.1: Number of companies meeting the evaluation criteria

20.4.1 Corporate Information

Overall, our results indicate a positive picture for environmental corporate information in the chemical industry, with nine of the twelve companies passing at least eight of the possible twelve criteria. However as these criteria are not truly additive, a more detailed analysis appears necessary. Such an analysis reveals large variations between the companies studied, as well as some areas where improvements could be made across the companies studied.

20.4.1.1 Content

While the majority of the companies provided information about their environmental activities (past and present) and access to reports for download, only a minority provided these reports in HTML format or provided links to external environmental activities or organizations. Further, even though all companies stated their concerns about the environment, the level of detail provided varied greatly. While some companies precisely informed about the entire environmental strategy, its activities and principles, others only briefly stated that environmental care is important without providing further details. Similarly, even though two thirds of the companies in the study provided some general information about Responsible Care or industry activities, the extent of information provided was, with the exception of the Web sites of BASF and Dow Chemical, very limited. Table 20.1 summarizes the evaluation of the content of the corporate environmental information.

Table 20.1: Evaluation of the content of online corporate environmental information

Criteria	# Companies	Comments
HTML version of environmental reports	5	
Possibility to download or order reports	11	
Information regarding the Responsible Care Program and other industry environmental activities	8	Different level of detail
Importance of environmentally friendly activities for the company and description of activities	12	Different level of detail
Historic developments and future plans	9	
Links to external environmental activities or organizations	5	

20.4.1.2 Form

Generally, the form of the information was evaluated positively, leading us to conclude that the basic principles of user-friendly Web site design are now almost standard. Nearly all companies provided a link from the homepage to the environment or sustainability section. The environmental Web sites were in general well structured, providing content bars which enabled an easy and effective navigation through the often extensive amount of information. Ten of the twelve Web sites offered titles and internal links on the individual Web site. Most Web sites also effectively used graphics, such as charts or pictures, to help clarify the content. Negative aspects were the lack of contact details for questions related to the environment and lack of information about the date at which the content was last updated. Table 20.2 summarizes the evaluation of the form of the corporate environmental information.

Table 20.2: Evaluation of the form of online corporate environmental information

Criteria	# Companies
Direct link to the environmental Web site from the corporate homepage	11
Well organized content / easy navigation	11
List of contents, internal links and titles	10
Summary graphics	10
Last information update	3
Contact details	4

20.4.1.3 Best Practices

Our analysis of corporate environmental information identified the Web site of the company BASF as best practice. This was the only company passing almost all criteria on both content and form. BASF offered many details through their online version of environmental reports as well as general information about environmental activities in the chemical industry. The BASF Web site also described current activities of the different BASF locations around the world related to health, safety and environment; finally, many presentations and speeches emphasize the importance of environmental management. Second best were the companies Dow Chemical and ICI for the content of their Web site on corporate environmental information. Dow Chemical presented detailed information about objectives and environmental philosophy and distinguished itself by offe many links to external organizations. By contrast, ICI presented in detail its own activities regarding sustainability, offered useful information through downloads of brochures and reports, but somewhat neglected external information.

20.4.2 Product-Related Environmental Information

Whereas the corporate environmental information was in general evaluated positively, a different picture was found for the product-related information.

20.4.2.1 Content

The quality of content varied. While almost all companies mentioned the uses and applications of their chemical substances, only a few drew the attention of the user to the applications these substances should not be used for. Information about the composition of the chemical and its properties, although necessary for the user to choose the most environmental friendly and most appropriate product, was only provided regularly by six of the twelve analyzed companies. Moreover, this information was often located in technical data sheets and not on the Web site itself. The same applies to technical properties as well as to advice regarding handling and storing the product, which were mentioned by seven of the twelve companies. The classification of the chemical substance was also found on only a few Web sites. The lack of information about the classification, the chemical properties and handling and security recommendations on the companies' Web sites is alarming, as laws require that the companies inform about these aspects. Material safety data sheets were available on half of the analyzed Web sites. Note that these data sheets were often secured by a password. As these data sheets are usually only intended for customers or employees that use the product, it is however legitimate for a company to restrict access to selected users in order to avoid misuse by competitors. Another negative aspect identified by our analysis was inconsistency of information across products of the same company. For instance, Inkjet Ink and Softack P.O., two separate product lines of Dainippon, were evaluated radically oppositely: whereas Inkjet Ink passed the majority of the criteria, Softack P.O. failed most of them. Similar examples were found across all the Web sites analyzed. Therefore our analysis considered a criterion as "passed" when the majority of the product Web sites of a company passed this criterion. As a consequence, even criteria indicated as passed for a single company do not mean that the criterion was indeed passed for all products of the company in question. Table 20.3 summarizes the distribution of the factors used to evaluate the content of the product-related information.

Table 20.3: Evaluation of the content of product-related environmental information

20.4.2.2 Form

The form of the product-related information on the Web sites studied was evaluated more positively than its content. Of the twelve companies analyzed, eight had a logical product catalogue, where the user could look for a specific product by various factors, such as the industry of the application, the generic product name, or the brand name of the company. Product search options and the general structure of the information were also usually adequate. Product comparisons were only very rarely offered (only four of the companies studied offered such a tool, and then not for all product types). As was the case for corporate information, information about last updates or about contact possibilities was often not available. This appears particularly problematic for product-related information which tends to be even more time-sensitive, and where users are more likely to have specific usage questions.

Finally, we should mention that the form of the product-related information, as its content, varied significantly within single companies, with individual product Web sites often having a different appearance within a single company, thereby reflecting a lack of corporate identity (see as an example the navigation and design differences of the product lines Solvents and Biodegradable Plastics at BASF). Table 20.4 summarizes the factors used to evaluate the form of the product-related information.

Table 20.4: Evaluation of the form of product-related environmental information

Criteria	# Companies	Comments
Structured product catalogue	8	
Product search machine	9	
Comprehensive structure of the information	2	
Comparison of the products	4	Usually not for all product groups
Last information update	2	
Contact details	7	

20.4.2.3 Best Practices

While none of the companies studied obtained perfect marks, three companies could be identified as best practice companies for product-related online environmental information.

Dow Chemical stood out for the good standardization across (almost) all its product sites. These Web sites offered the same structure, thereby facilitating information search for customers buying multiple products. In each Dow Chemical product site, an information sheet summarized the main characteristics of the product, including information about environment, health, and safety. A material safety database was also available on the corporate Dow Chemical Web site, where data sheets for all products could be obtained.

The second best practice company was BASF. Although its product information was not always presented consistently, the main factors such as applications, chemical properties, handling and technical advices were typically covered, often complemented by a detailed technical data sheet.

Finally, Bayer offered an original tool. Even though its product information was generally not as detailed as the one offered by Dow Chemical or by BASF, Bayer made up for it somewhat with a feature called the 'Bayer Eco Check', which evaluates online the appropriateness of a given product for a given application, also from an ecological point of view.

20.5 Conclusions

In this study, based on publicly available information, we assessed the performance of twelve companies on corporate and product environmental information for the form and the content delivered, thereby providing a current picture of the environmental online communication practices in the chemical industry in August 2003. Generally, our analysis showed relatively good (but inconsistent) practices in corporate environmental online communications for the chemical companies studied. In contrast, product-related environmental information was relatively poor. This shortcoming is somewhat surprising given that providing such information is required by law from chemical companies, and that the Internet should be an appropriate medium to fulfill these requirements. Obviously, the decentralization of Web site management within companies leads to very unequal respect for the law by individual product managers. This is certainly an area where companies in the chemical industry still have much catching up to do.

From the twelve companies studied, the Web sites of BASF and Dow Chemical were identified as best practices for both corporate information and product information, ICI for corporate information, and Bayer for product information. These four companies thus should serve as standards.

Our results reflect well the environmental activities of the chemical industry up to now; companies have put much effort into reducing the direct impact of their activities in production and finishing - partially under legal pressure. Not surprisingly, considerable information about these efforts was found on the corporate Web sites. The flows of information about environmentally relevant data about the products of the chemical industry between producers, finishers, users, and end consumers are in contrast still lacking. Providing such information is however necessary to enable industrial users of chemical substances to actively search for the least harmful substances that fulfill their technical requirements, which will be mandatory in the near future [7]. Thus, improvements in product-related environmental online communication appear necessary to enable efficient risk management of chemical substances.

A major contribution of this chapter is the development of a framework to identify best practices in online environmental communication in the chemical industry. We certainly hope that this comparative study will be used by chemical companies to raise their performance in this respect. Thus, beyond providing a current snapshot of online environmental practices in the chemical industry, our framework can also be used at different points in time, enabling us to monitor the performance of single companies as well as the performance of the entire chemical industry in this respect over time.

Clearly, one limitation of our analysis is the fact that we could only include formation available to the general public. It is possible that some of the compa- included in our study provide more online environmental information in some privileged spaces that are only accessible to registered users, such as for instance Extranets for their customers. These Extranets usually provide extensive product information, and often also material safety data sheets. Our analysis however indicated that only a few of the companies analyzed had such a facility. The fact that we did not get access to these Extranets should not have affected the validity of our results, as we purposely focused only on publicly available information (consistently with the Responsible Care Program objectives as well as with the REACH program of the European Commission).

Further research in this area could certainly further refine the evaluation criteria proposed in this chapter to develop true benchmarking metrics. A similar framework could also be applied to other industries where online environmental communication is also of strategic relevance (for instance refineries, electronic equipment, textile industry, automotive engineering, or paper industry) because of either production or recycling needs (material efficiency).

Part IV
Networks and Virtual Communities

Online Collaboration, Trust and Credibility
Knowledge Management

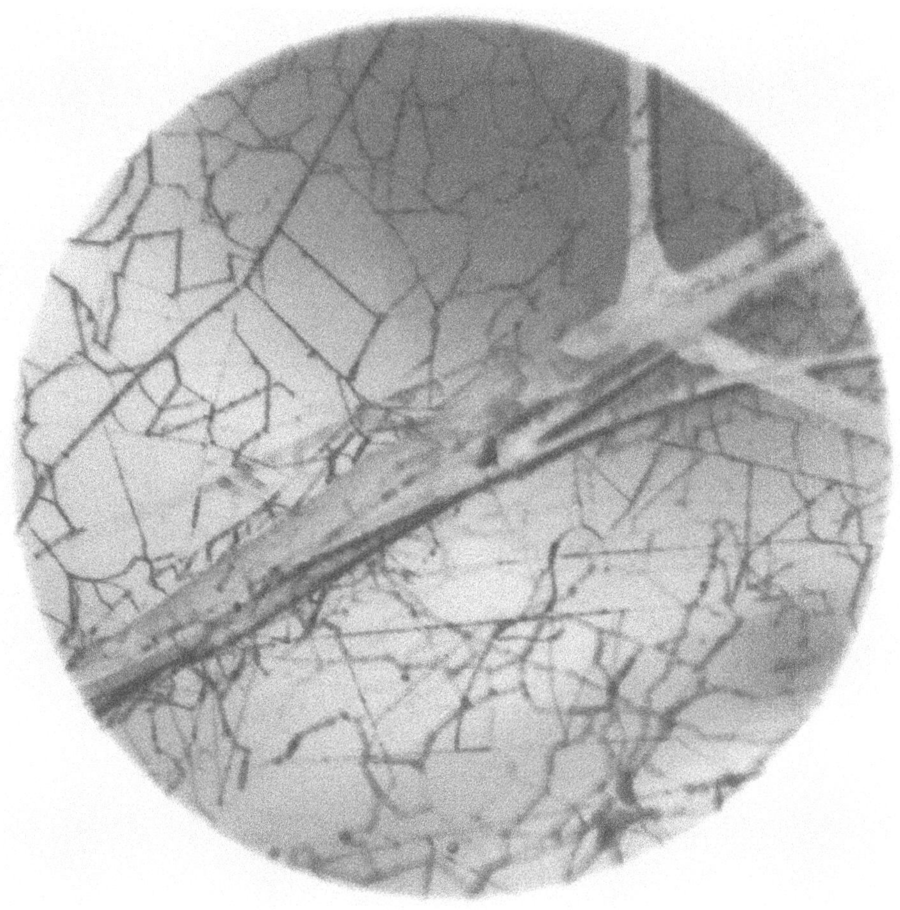

Chapter 21
Epistemic Communities and Online Environmental Data Systems

Nancy A. Van House

Abstract. This chapter attempts to understand some of the social and organization problems that confront online environmental communication in two areas: credibility of content, and the design and management of multidisciplinary systems. This chapter approaches these both as knowledge work, where trust and credibility are critical. Using the perspectives of science studies and science and technology studies, I suggest that it is useful to see tensions in online environmental systems as resulting from crossing epistemic cultures.

21.1 Introduction

Understanding the complexities of environmental conditions over time often requires combining, collating, analyzing, and comparing data of many kinds from many sources. The purpose of environmental online communication is to make more available and combinable a variety of useful information: raw data, analyses, models, projections, and so forth. Data from unexpected or unusual sources is particularly valuable. Amateur observers may provide detailed on-the-ground information. Old data may provide needed baseline information; some of its value and difficulties stem from its "historic" nature.

Online communication makes it easier than ever to collect and combine information generated over time in multiple places, rooted in a variety of disciplines and understandings, for differing purposes, using varied methods and formats. There are, of course, problems. Some are technical. Other, serious problems have to do with metadata and how the world is conceptualized and described [65]. Yet others are social and organizational.

This chapter attempts to understand some of these latter problems using the understanding of science studies [236] that knowledge work is situated, distributed, and social. It also draws on the insight of the related field of science and technology studies (STS) that the meaning of technology is flexible [271]: multiple groups may participate in the development of technology or a technological artifact without agreeing on its meaning or significance [579]. The argument of this chapter is that the organizational and social stresses observed in online environmental systems are not necessarily signs of trouble, but are inevitable and manageable (although difficult) results of the boundary-crossing nature of the work, and that science studies and STS can be useful resources in understanding the nature of boundary-crossing knowledge work and technological innovation.

This chapter examines tensions in two areas in particular, which, I will demonstrate, are closely related. The first is credibility and trust in the content of online environmental information systems. The second is the design, management, and policy decisions in such systems. This discussion is rooted in both empirical and analytical work. The empirical work, reported in more detail elsewhere, consists of interviews and observations in two areas of environmental work in California: water planning [478; 575; 580; 581] and the creation of a system of botanical data [576; 578]. The analytical bases are understandings of knowledge work and technology rooted largely in science studies and science and technology studies [579], and investigations of trust and credibility in a variety of disciplines. This chapter is not about California environmental information *per se*, but uses the empirical work to illuminate the analysis of environmental work as knowledge work.

21.2 Empirical Findings

The empirical base of this chapter comes from two related projects. The UC Berkeley Digital Library Project [785] has been funded since 1995 by the US National Science Foundation as a research project. It consists of a mix of computer and information science research and the construction and maintenance of a testbed of documents, images, databases, and geographical layers. Much of its content is non-copyrighted reports, including government documents, and data about the California environment. My colleagues and I conducted interviews early in the initial project with representatives of state, federal, and local agencies and non-profit organizations engaged in water planning about their needs for online environmental data [478; 575; 581].

My more recent work [576; 577; 578] has addressed CalFlora [665], a project that began in collaboration with the Digital Library Project and is now independent. CalFlora provides online scientific information about California plants for research, conservation, and education. CalFlora consists of a Species Database with geographic distribution, habitat, and lifeform data for the approximately 7,600 vascular plants native or naturalized in California, based on published literature and documented observations; and an Occurrence Database of over 800,000 observations of plants within California, from about 19 public agencies, herbaria, private organizations, and individuals. Calflora is also linked to the UC Berkeley Digital Library Research Project's database of more than 29,000 photos of California plants. Its users and data contributors include state, federal, and local agencies, non-profit organizations, and individuals. It is funded by government agencies, foundation grants, and private donors.

The following section summarizes observed problems, concerns, and stresses. Then I consider the factors that promote trust and cooperation. The following section presents a more complex analysis of trust in knowledge work drawn from sciences studies and science and technology studies (STS). The final section presents conclusions and suggestions specific to online environmental data.

21.2.1 Concerns about Data Sharing

In our observations and interviews with people engaged in environmental work, we found both enthusiasm and concern about environmental online communication. The concerns can be classified into those of information users and information owners or providers. More details can be found in previous publications [478; 575; 576; 577; 578; 580].

21.2.1.1 Information Users: Quality of Information

Information users were deeply concerned about the quality of information. "Quality" is a complex concept. One concern was the methods by which data were collected and analyzed; different disciplines have different methods, and users were concerned about whether the appropriate methods, by their standards, had been followed. Another concern was the expertise of the data contributors. For example, the botanical community has to consider an observer's ability to accurately identify taxa, especially for field observations for which identifications cannot be independently verified. Information users were also concerned about how the interests and biases of the information providers may shape not only interpretations and analyses, but the data themselves. Representatives of non-profit environmental organizations, for example, were suspicious of data from resources extraction industries. Some were equally suspicious of data from government agencies whom they saw as under political pressure to promote certain projects.

21.2.1.2 Information Providers: Misuse of Data

Information owners reported a variety of concerns; here I focus on their worries about misuse of data. One concern was out-of-date information. Many databases are dynamic, but even owners of "static" databases updated their databases as errors were discovered and corrected. Anyone who downloaded data from their database might be working with outdated information.

Information owners were often concerned about the possibility of information being misunderstood or misused. For example, water forecasters worried that their critics, unfamiliar with errors in water supply data and the ways in which errors were corrected, would use the supply data to dispute their forecasts. Some members of the botanical community feared that identification information about endangered taxa could promote their destruction rather than preservation. For example, some were afraid that photos of endangered species would be used by landowners to identify and destroy specimens of endangered taxa in order to avoid the resulting legal constraints.

21.2.2 Tensions in System Creation and Management

Another area of stress uncovered in our empirical work had to do with creating and maintaining environmental information *systems*. Many online systems are the result of collaborative work by people from a variety of disciplines, with different understandings, interests, and goals. One common problem, in this area and others, is that funding is available for research, not for operating systems. Funders and researchers are often interested in innovative, unproven technology, while the users are looking for stable, working systems [606].

Another area of tension was between people who were more technically-oriented and those who were more closely aligned with the user community, as they debated needed functionality and the cost versus the value of various design choices. A third area of tension was among people in environmental work with different purposes for the data and different institutional mandates. One area of friction, for example, was the trade-off between comprehensiveness and quality control of the contents.

Finally, a system like CalFlora, whose value is in coordinating data from a large number of sources, requires that participants be willing to make those data available and to put in the effort required to do so, which is generally not a priority with the contributing organizations.

The creation of an online environmental system like CalFlora requires a large number of decisions of varying levels of detail about its vision and purpose, policies, funding, content, presentation, functionality, and operation. It also requires continual cultivation, coordination, and accommodation to the needs and priorities of participants and contributors. Much of this work is invisible. Star and Strauss [532] write about "articulation work", the often-invisible work needed to keep things on track. Bowker and Star [66] demonstrate that infrastructure disappears when it works; similarly, creating and maintaining online systems is only apparent when it fails. As a result, observers often underestimate the amount and difficulty of this work, especially, ironically, when it is done well.

21.3 Environmental Work as Knowledge Work

Although these may seem like different classes of problems – those related to the data, and those related to the creation and management of online systems – they have commonalities rooted in their nature as knowledge work that crosses boundaries of disciplines, institutions, political orientations, and values.

21.3.1 Trust, Knowledge Work, and Online Data

Knowledge work is always collaborative, directly or indirectly, as one uses the work and understandings of those who have gone before. Collaboration of any kind requires trust. In online environmental systems, those creating and managing the system must trust one another and their work. Information users must trust the capability of information sources. And information owners must trust that users will not misuse their information.

With networked information, people are faced with increasing amounts of information from unknown sources, divorced from the context that makes it understandable and warrants its accuracy. This raises questions about what has been called epistemological trust [124], epistemic authority [208], or cognitive authority [625]. These judgments are not, in themselves, new. What is different is that information technology and networked information are making them more prevalent, and changing the means by which they are made and the evidence available. Designing systems to support knowledge work requires understanding both pre-existing practices and criteria of trustability, in general and in specific knowledge domains, and their translation into a networked world.

21.3.2 Epistemic Authority

A number of disciplines address the topics of trust and epistemic authority, including philosophy, psychology, economics, and political science. A central question is the conditions that promote (or undermine) trust, and the processes and criteria by which people judge one another. Here we review these discussior briefly, because they indicate some of the criteria and processes that need tr translated to an online environment, although understanding knowledge work requires much more than this. Many investigations into factors that promote trust and credibility attempt to identify relevant characteristics of the truster, trustee, or situation. One major shortcoming with most of this literature is the focus on one-to-one relationships, not communities or networks of trust [35].

A person's capability is generally a function of his or her knowledge or expertise, indicators of which include experience, credentials, and reputation. Capability, however, is a potential. A physician, for example, needs to be knowledgeable, but also to pay attention to a patient's symptoms and complaints and to conscientiously bring that knowledge to bear on the current situation. Good work is a combination of knowledge, capability, and propensity to appropriate behavior.

Another characteristic of the trustee that may be relevant is interests or motives, perceived or inferred [389]. Since trust is about future behavior, past experience is an imperfect indicator; motives may be more reliable determinants of future behavior.

The relationship between trustor and trustee is also important. Trust often builds over time, with repeated interaction. People who expect to have future interactions (or expect information about their behavior to be available within a shared social network) have incentives to be trustworthy [48].

Another key element is shared culture, which may provide culturally-specific clues concerning trustworthiness [193]. It increases the chances of the parties sharing important values. Shared culture also can provide a basis for solutions to newly-emerging problems. Shapin's [498] impressive study of trust in seventeenth-century English experimental science, for example, demonstrates that the newly emerging field adopted pre-existing gentlemanly (sic) practices to solve the problems of credibility and trust.

A final set of influences has to do with the larger social context. Groups and societies tend to differ in their trustingness and whom they trust, e.g. family members versus others [193].

In sum, while indicators of expertise are important, they may be difficult to interpret and may be insufficient to predict whether a person will behave in ways that others consider appropriate. People who know one another, have information about one another, and/or are members of the same community or social network are likely to have a greater propensity to trust one another. Expectations of future interaction create incentives to behave in trustworthy ways. Shared culture, values, interests, propensities, language, and practices will increase the likelihood that each will behave in ways that the other will find acceptable, and that participants will be able to (or believe that they are able to) interpret clues about others' behavior, interests, motives, values.

Online interaction may be with unknown as well as known others. The ability of the Internet to facilitate access to data heretofore unavailable also means interaction with unfamiliar people and institutions. The Internet has paradoxical effects in creating connections between people who would not otherwise know one another. People use information from others whom they don't know, but they also come to know people who are otherwise distant. Environmental information that crosses disciplines, space, and time is of great value; but it raises questions of loss of context, lack of knowledge and interaction between source and user, and lack of shared culture, values, interests, and experiences.

21.3.3 Trusting Online Information

In asking how people determine the trustworthiness of online environmental information, we can look at the evidence about people's use of computer-mediated communication and online information in general. Computer-mediated communication among stable, distributed work groups has been the subject of considerable research. Among the factors found to promote trust in computer-mediated communication are continual and frequent interaction [252], supplementing online interaction with informal and face-to-face interaction, and use of richer communication channels, e.g. video as opposed to text [399]. Olson and Olson [399] describe the value of what they term common ground: knowledge that participants have in common, and shared experience and norms. This is based in part on general knowledge about one another, and in part on specific knowledge gained from interaction. Another source of evidence is research on the credibility of Web resources. While normative discussions stress the importance of the credentials of the information source or sponsor [11; 217] and, experts in an area tend to assess a source's credentials, non-experts pay more attention to the superficial aspects of a site, such as design [185; 527].

Overall, then, information about the source of data, personal and/or institutional, is important in assessing the credibility of information. While personal interaction – face-to-face or computer-mediated – may be reduced, other information may be easier to acquire. Transactions and reputation may be more visible through reputation mechanisms at sites like eBay and ePinions. It may be easier to link behavior across time and place with an identity, e.g. some CalFlora participants reported that reviewing all the observations contributed by one person helps them to judge that person's experience and expertise [577].

However, identity can be a problem in a networked world. Online identities may not be linked to offline identity, history, reputation, and accountability, or to other online contexts. Information may not even have an identifiable author or source; if there is no author, who is accountable [389]?

In sum, making judgments about the credibility of information requires knowledge about its source – direct and indirect. Online interaction makes some information – including some information about potential partners – more available, and some less so. However, this approach to trust is largely concerned with one-to-one trust, and doesn't ask how a trustor is able to make judgments based on the information provided.

21.4 Epistemic Communities

Further insight into knowledge work, collaboration, and trust, as well as the creation of information systems for sharing data, comes from science studies [236] and STS [271; 579]. Science studies is concerned with the content of scientific knowledge (and, by extension, other kinds of knowledge); STS is concerned with the relationship between technology and social systems. Most of science studies and STS takes a situated, social constructionist perspective that insists that there is no "view from nowhere" – knowledge and understanding are always situated in a place, time, conditions, practices, and understandings. This approach contends that most knowledge claims are underdetermined, that is, most of our obs

tions of the world are open to multiple interpretations. What we believe to be true and whom we believe are not determined (solely) by nature or reality, but by our interactions with others and the understandings of our community.

This stress on the local, accomplished nature of knowledge give an important place to *practice* [436; 475], people's actual, daily, embodied activities, including skills, tacit knowledge, and presuppositions, as well as their interaction with others and with material and other resources, in the construction of knowledge. Artifacts, tools, and technologies are integral to the practical accomplishment of work and of knowledge [110]. Among the artifacts of knowledge work are representations, including images, graphics, all kinds of documents and texts, databases, and metadata. Part of becoming a member of a community of practice is learning to create and use representations. The work of representation gets "deleted" [530], which makes invisible, after the fact, the choices that are made and by whom, and what gets left out. Bowker and Star [66], for example, describe the work that classification does in ordering human interaction, and how categories and the work of classification tend to be invisible.

The *community* is critical to this located, historicized, practice-based approach to knowledge. Lave and Wenger [324; 613; 614] developed the influential notion of "community of practice", which emphasizes the mutuality of community, knowledge, activity, and social practice. People learn to construct meaning and to create and interpret representations in the course of becoming members of communities of practice. Similarly, practice, artifacts, history, and community are central to Knorr Cetina's [300] notion of epistemic cultures, "those amalgams of arrangements and mechanisms – bonded through affinity, necessity, and historical coincidence – which, in a given field, make up *how we know what we know*. Epistemic cultures are cultures that create and warrant knowledge" [300, p1]. She is interested in "the machineries of knowing composed of practices" [300, p10], which are both technical (e.g., scientific instruments) and social (e.g., how decisions are made). A critical part of her argument is that epistemic cultures and machineries are diverse, not just between science and non-science, but even within science. She demonstrates similarities but also differences between the two lab sciences of high energy physics and molecular biology.

In both Lave and Wenger's and Knorr Cetina's approaches, to become a member of an epistemic community is not simply to acquire knowledge but to have one's identity transformed, to be shaped by the community and its practices, to see the world in terms of these practices, understandings, and orientations. In both approaches the epistemic subject, the knower, is not the individual but the group. Issues of trust and credibility in shared knowledge and in collaborative work go well beyond the questions of what people know about one another and indicators of their expertise. The knowledge community and its shared history and practices are deeply implicated in the creation of knowledge. Within a community, members share methods of creating and using representations (including data, databases, and images).

Of significance for this analysis also is the STS idea of the "interpretive flexibility" of technology. Different groups may have different understandings of a technology or a technological artifact. Online environmental data systems are likely to understood very differently by these groups – resulting in conflicts over key sions about design, policies, and operations, and about what "works".

21.5 Trust, Epistemic Communities, and Online Environmental Data

The empirical findings described above re-examined through the lens of epistemic communities gives us a picture of the world of environmental work as consisting of multiple, overlapping, interacting, interpenetrating, and sometimes antagonistic epistemic cultures. Any one epistemic community is held together by understandings and values, shared work practices, as well as common artifacts and representations, and uses and understandings of them. People who share, not only practices, but values, interests, meanings, and understandings are more likely to trust one another, and to feel confident in their ability to assess one another's capabilities. They are more likely to be able to work together to design and maintain shared systems, and to be willing to use one another's data.

But the worlds of environmental work and online environmental data are multiple, disparate, and varied. Networked information – and the creation of networked systems – not only facilitates but requires the crossing of social and technical boundaries between epistemic cultures. This may call into question that which is taken for granted within communities, challenging the established cognitive order. Shared practices and understandings can no longer be assumed; the indicators of credibility may be questioned or missing; inscriptions and data may be used in ways not envisioned (or accepted) by their creators. Online information systems potentially destabilize the processes and the hierarchies of credibility and power relations.

Innovations in networking and information technology also enable new practices, representations, and functionality. Such innovations are at least initially usually based on pre-existing practices. For example, Bijker [56] includes a picture of an early bicycle that looks like a horse, complete with artificial horsehead and saddle. But the boundary crossings enabled by online communication can raise conflicts about which pre-existing practices will govern. For example, in combining data from a variety of sources, CalFlora has had to reconcile differing policies that originate in the different goals of the participating systems. One arena of prolonged negotiation was the choice of the appropriate level and methods of quality control on submissions. Some participants' work requires a high level of confidence in the data. Others prefer a less-certain but more extensive database; for them, erroneous observations are a small price to pay for comprehensiveness.

The argument here is that the networking of digital information has made these epistemic machineries more visible, in part by instantiating them in system design, in part by facilitating the flow of information – and representations – among epistemic cultures, and in part by requiring that groups work together in new ways to create online information systems as well as to use one another's data. The power of an online information system is in its integration of heterogeneous elements, its flexibility, and its usefulness to multiple professional communities. But these same factors may destabilize the processes and standards of knowledge creation by mixing participants and information from multiple epistemic cultures with differing goals, practices, values, criteria, and understandings

21.6 Resources for Understanding Collaborative Work

Conflicts and misunderstandings – among data providers and users, and among the multiple groups required to build and maintain online environmental systems – may begin to seem inevitable. However, science has always required collaboration and data sharing across various kinds of boundaries. Science studies also offer resources for understanding why and how collaborative online environmental work may succeed. In this section, I briefly describe three useful approaches from science studies, both as useful in themselves, and as illustrating the power of learning from other investigations of knowledge work.

In earlier work [578], I found Actor Network Theory [325] useful in understanding online environmental systems as existing at the confluence of interests among participating groups. Actor Network Theory (ANT) is concerned with the stabilization (always temporary) of socio-technical systems by the enrollment of allies by means of the translation of interests; when multiple groups see their different interests as served by a common solution, they cooperate. These disparate interests continually tend to pull apart, so the stabilization of the shared solution is always temporary. The implication of ANT for online environmental data is that, first, the tendency toward destabilization is continual and expected. It is not a symptom of trouble, particularly; but it does require constant attention. Second, it sensitizes us to the need for on-going translation and enrollment as a normal, not extraordinary, part of the work of maintaining systems.

Another concept developed to help understand knowledge creation and sharing in science and other fields is what Star [529; 531] has labeled "boundary objects". Scientific work has always required that groups share resources without necessarily agreeing in other ways. Star was seeking a more decentralized approach than ANT's. The term "boundary object" refers to artifacts and (notably for our purposes) information that can be used by multiple users and communities for a variety of purposes. Star describes boundary objects as both plastic enough to adapt to local needs and have different specific identities in different communities, and robust enough to maintain a common identity across sites, and be a locus of shared work. She identifies several types of boundary objects, one of which is repositories, "ordered piles of objects indexed in standardized fashion" [531, p408]. Star and Greismer's example is the University of California, Berkeley, Museum of Vertebrate Zoology, a collection of specimens with extensive, standardized metadata that required the cooperation of collectors, funders, and university administrators, as well as scientists.

A third concept is Galison's [196] trading zones, where two or more cultures can come together to trade objects that have radically different significance for the two groups, and communicate with one another by means of simplifications of their own languages. His example is theoretical and experimental physics, which use one another's work, discuss common problems, and coordinate their work without altering their different internal understandings.

The point of this discussion is that environmental work, like other areas of scientific and other kinds of knowledge work, requires sharing data and building shared information systems. ANT, boundary objects, and trading zones are all ways of describing and understanding work that crosses boundaries without removing those same boundaries, without eliding the differences across groups.

21.7 Implications

Online environmental systems are loci of knowledge work of two kinds: the work of creating and using *information*, including data collection, representation, and analysis; and that of creating and maintaining online data *systems*. Problems in these two areas are often seen as local problems; I contend that they signal *systemic* stresses that are not only likely, but perhaps inevitable. If so, local attempts to reduce conflicts and resolve problems are likely to succeed, at best, temporarily. Such tensions can be self-perpetuating if they are seen as aberrant rather than inherent in this kind of work. For example, participants may be suspicious of others who seem not to share their understandings, and wonder whether cooperation is possible or desirable, unless they see that such conflicts are predictable given the nature of the work. Understanding online environmental work as necessarily embodying boundary-crossing tensions may, ironically, reduce these tensions.

I argue that it can reassure participants to see difficulties not as local failures and individual differences, but as characteristic for boundary-crossing knowledge work. As solutions are not entirely local or idiosyncratic, applications can learn from one another. Understanding online environmental systems as boundary objects, trading zones, or processes of translation and enrollment are solutions from other areas that are useful here.

This analysis also cautions us to be sensitive to the variety of existing practices of knowledge creation. A successful environmental information system has to articulate with participants' hierarchies of credibility and processes of establishing trustability, which differ across scientific fields and other disciplines engaged in environmental work. This analysis also suggests that the environmental community – as varied and divisive as it may be – must take the lead in designing online environmental data systems, and not rely too heavily on technologists.

These findings are not unique to environmental data. Suchman's [550] reflections on the construction of technological systems are useful here. Crossing boundaries, she says, means "encountering difference, entering onto territory in which we are unfamiliar and, to some extent, therefore, unqualified" – and uncomfortable [550, p25]. Useful system design requires the on-going creation of situations for "the meeting of different partial knowledges" [550, p25] in "an increasingly dense and differentiated layering of people and activities, each operating within a limited sphere of knowing and acting that includes variously crude or sophisticated conceptualizations of the others" [550, p30]. This inevitably creates conflicts and misunderstandings across groups, but is necessary if online environmental communication is to make a difference in environmental work.

In summary, networked information systems of all kinds, but especially online environmental systems, are the loci of webs of relations among different epistemic communities, including varied suppliers and users of data, technologists, and others engaged in system design and management, people from various institutional bases, with differing agendas for the system and for the information that it contains and helps to create. Wherever people are working at the boundaries of knowledge communities we will see differences and disconnects, negotiations and assessments. The boundary-crossing work of designing and building systems, populating them with content, and keeping them operational requires work across uncomfortable boundaries, and the exercise of mutual trust.

Chapter 22
Communication Process Analysis in Virtual Communities on Sustainable Development

Rolf Kleef • Aldo de Moor

Abstract. *Collaboration on sustainable development involves stakeholders and experts from multiple organizations, who often work together in temporary or permanent communities of practice. The potential for joint work has greatly increased with the advent of networking technologies like the Internet. However, successful online collaboration requires a systematic analysis of the often very complex, dynamic and interdependent communication processes, while taking into account many organizational constraints. We present the extended Social Context Model that can be used to analyze and improve communication process support. We then apply the model to a case: the worldwide Friends of the Earth network of environmental organizations.*

22.1 Introduction

The problem domain of sustainable development is vast. Addressing the many problems related to nature conservation, environmental protection, and social justice involves more and more stakeholders from different backgrounds. The emerging networks and communities of stakeholders and decision-makers have to deal with increasing amounts of available and required data, information, and knowledge from many different domains.

Traditional knowledge management approaches are not sufficient, as they are based on a "commodity view" of knowledge, separating "knowledge" and "knower" [574]. This approach assumes a joint agreement that knowledge can be stored, replicated and distributed without further involvement of the person providing the knowledge.

However, in the area of sustainable development, the setting is different, as a proper distinction must be made between "external" and "tacit" knowledge [392]. Much of the available knowledge and its context of use is hard to document, and sometimes even impossible to codify (as in the case of a lot of "traditional knowledge"), making it hard to properly document "best practices".

Another complexity is that the value and power associated with certain knowledge (economic or cultural) inhibits "knowers" from being willing to lose control by having their knowledge externalized. Instead, these knowers should be brought together to discuss their positions and apply their expertise [282]. There is a need to mobilize, activate, and combine the experiences, insights, wisdom, and stakes of people involved in the issue at hand. High-quality communication between complex networks of stakeholders in different locations to exchange information and coordinate activities is becoming a critical success factor.

To unleash this potential, it is essential to provide balanced systems of information technologies to adequately support the subtle web of communication processes [127]. However, the "commodity approach" is propagated even into many Internet-based products to support "communities of practice" or "communities of interest". Although they usually provide ample facilities to store documents, case studies, best practices, and other "externalized knowledge", their functionalities are often rather generic, and not tailored to the specific purpose of the communication processes, or the politicized context in which they take place.

In this chapter, we present our extended Social Context Model, a diagnostic instrument to systematically analyze the quality of communication process support in virtual communities on sustainable development. In Section 22.2, we define what we mean by such virtual communities. Section 22.3 presents our Social Context Model for communication process analysis. Section 22.4 describes the Community Activity Cycle as a way to extend the Social Context Model. We apply the extended model to the case of Friends of the Earth International in Section 22.5 and end the chapter with a short discussion and conclusions.

22.2 Virtual Communities on Sustainable Development

Communication among stakeholders is not the result of random interactions. The actual work, and communication, is mostly done in communities of practice. A community of practice is a domain of knowledge which defines a set of issues, a community of people who care about this domain, and the shared practice that they develop to be effective in their domain [615]. In such communities, people interact socially for mutual benefit. They help foster knowledge development and creative interactions amongst highly specialized experts, and help to channel their efforts to where they are most needed [371; 511].

We consider the communities working on sustainable development as a special type of communities of practice. They consist of many experts and stakeholders, often with conflicting interests. The communities are very knowledge-intensive, under high time pressure, forced to work most effectively and efficiently because of a regular lack of resources, and have strongly evolving sets of goals, workflows, and organizational structures. They are often highly international, interdisciplinary, inter-organizational, and interactive.

Most communities on sustainable development are by necessity also virtual communities, as they are often dealing with global issues and policies, and involve participants from many different locations. Virtual communities describe the union between individuals or organizations sharing common values and interests using electronic media to communicate within a shared semantic space [487].

In virtual communities, the common space is provided by a suite of collaborative and communicative information tools, ranging from simple mail functionality to advanced Web applications [551]. This functionality comes in the form of standard tools, so that information systems development becomes more a process of tool selection than building whole new systems from scratch [471]. For virtual communities, the right ensemble of tools needs to be selected to serve their requirements. However, choosing the technology to use is not trivial, as there is a tremendous variety of design choice, and requirements may even vary with communities in different stages of their life cycle [209].

Given that communication processes are essential, yet complex to model and support, what we need is an analytical instrument for diagnosis of communication problems. The analysis should include key factors related to communication processes in their community context. The instrument should take into account the different phases in the development of a virtual community. The result of the analysis would be a systematic overview of communication process problems and opportunities. This overview could be used to make a set of recommendations on communication support.

22.3 The Social Context Model

To assess the effectiveness of tool-enabled communication processes, we need to know the context in which they operate. For example, it makes quite a difference if a discussion process is used for friendly conversations, or as a key element in an official document review. Once these communication processes are charted in their context, it is much easier to set priorities such as which tools to acquire and how to use them in the socio-technical system that is the virtual community. In this way, the socio-technical gap between community requirements and enabling technologies can be reduced [4].

In earlier work, while focusing on "authoring processes", we developed a Social Context Model (SCM) to help make these assessments of the context of communication processes [128; 129]. This model looks at communication processes along a process *context* dimension consisting of communication process layers, and along a process *structure* dimension describing the configuration of the elements of the communication processes.

22.3.1 The Communication Process Context

The SCM consists of four layers of communication processes, each higher-level process providing a context that embeds the lower-level processes. From high to low-level processes these are: collaboration, authoring, support, and interaction processes. *Collaboration processes* give purpose to the collaboration activities, discussions, and documents. *Authoring processes* produce the structured document or output of the collaboration processes. *Support processes* focus on the organization of the discussions between the participants, ensuring that they contribute to the creation and interpretation of the output. *Discussion processes* are the actual exchanges in which the argumentation between participants takes place. By embedding the basic discussion interaction process in three top layers, a systematic context analysis can be performed.

22.3.2 The Communication Process Structure

Each communication process, whether it is a discussion process or one of its embedding context processes, has a structure comprised of certain process entities. First, there are *process elements* (elements the process itself is made of). Second, there are processes constructed out of these elements. These we subdivide into *tions* (which constitute the actual communication process) and *change processes* (meta-processes in which the communication process can be adapted).

- *Process elements* can be divided into at least three types: goals, roles and objects. *Goals* define what the outputs of these processes should be focused on. *Roles* specify in what quality participants can act in processes. *Objects* are the inputs and outputs of processes.
- *Actions* define the workflows of the community and are composed of configurations of process elements. In actions, participants playing process roles generate more complex objects as outputs from simpler objects.
- *Change processes* describe how the evolution of the socio-technical system takes place. To ensure that communication processes and their supporting functionality co-evolve adequately, change processes need to be explicitly defined.

In both actions and change processes, there are *norms* that describe the acceptable behavior in the community by defining the authorizations of the participants in the process roles that they play. All communities have such norms, either explicitly laid down in charters and by-laws, or only implicitly defined, but no less strong in impact [443; 551].

22.3.3 The SCM Matrix

Combining both dimensions, we create the matrix in Table 22.1, with examples of context and structural elements, as an illustration of how the model can be used to organize the complexity of discussion processes in their context [129].

Table 22.1: The Social Context Model

	Process elements **Goals, roles, objects**	**Actions** **Workflows**	**Change processes** **Definition of socio-technical system**
Collaboration processes (Why is the discussion taking place?)	Collaborative roles (facilitator, judge, expert, keynote speaker), Stakeholder profiles, Policy reports, Conflict resolution priorities	Facilitation, Mediation, Conflict resolution, Debating, Political inquiry	Define goals of collaboration, Define rules of engagement, Set social norms for roles
Authoring processes (What is produced in the discussion?)	Authoring roles (editor, author, reviewer), Document structure elements (section, position, case, argument)	Editing, Reporting, Authoring, Reviewing, Publishing, Notification	Change authoring roles, Change authoring norms, Change authorizations, Adapt document structure
Support processes (How is the discussion organized?)	Support roles (moderator, technical facilitator), Discussant profiles, Message digests, Discussion summaries, Archives	Inviting, Reminding, Registration, Agenda-setting, Moderation	Set communication policy, Agree upon discussion planning, Change notification parameters
Discussion processes (How is the discussion conducted?)	Interaction roles (discussant, attacker, defender), Discussion elements (posts, replies, labels), Discussion objectives	Raising issues, Replying, Creating a position, Playing devil's advocate, Position taking	Assign attackers, defenders of position, Define discussion rules

22.4 The Community Activity Cycle

In this chapter, we extend the Social Context Model along the dimension of time. Virtual communities are not static, but continuously change their focus. They develop in a long and often difficult growth process, and over time, priorities with regard to the goals and structure of communication processes change [209]. Therefore, the stage of development or operations of a community needs to be taken into account in the analysis.

We build upon ideas inspired by the life cycle of communities of practice [387; 615]. Rather than taking the "community life cycle" as phases in a relatively linear fashion, we see a "community activity cycle" as consisting of activity aspects that often occur simultaneously in the community (Figure 22.1). In practice, however, a community often stresses different aspects in a linear fashion. For simplicity, we therefore assume a sequential model of stages here.

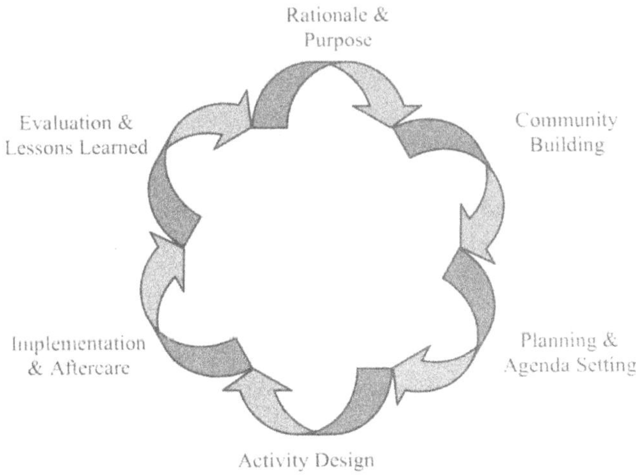

Figure 22.1: The Community Activity Cycle

Each of the stages in the Community Activity Cycle has an associated Social Context Model. In each of these SCM instances, different elements can be important. The configurations for these different stage models were obtained by analyzing project documents, and observing stakeholders in a number of virtual communities on sustainable development. In this way, best communication process-practices can be recorded and used for analysis, depending on the particular activity that has priority.

22.4.1 Stages in the Cycle

We briefly outline each of the six stages in the cycle. In our case study in the next section, we will show how the filled out social context model looks for the "implementation and aftercare" stage. Similar typical model configurations exist for the other five stages of the Community Activity Cycle. For lack of space, these will not be shown in this article.

22.4.1.1 Rationale and Purpose

This stage looks at the "raison d'être" of the community: what purpose does it serve, who will benefit from it, and how will its effectiveness be monitored? Most specifically, it needs to be very clear how intended or existing participants benefit from being active in the community (for instance, in terms of access to knowledge or political support, personal exposure and profiling, learning opportunities, having an "antenna" in other fields, etc.).

22.4.1.2 Community Building

This stage looks at ways in which the community holds together. What are the common grounds or stakes around which participants gather, how do community members gain trust in each other and the community, how do they communicate and collaborate? How can a new member become part of the community?

22.4.1.3 Planning and Agenda-Setting

In a "true community", the members set their own joint agenda of what they want to collaborate on (this of course may sometimes not completely overlap with the reasons for institutions to establish or support the community). Here we look at possibly existing agendas and ways in which they can be set. For temporary communities with a specific purpose, the agenda may be given beforehand.

22.4.1.4 Activity Design

Once an agenda and planning are set, individual activities can be designed. Important questions are: what should the process look like; how will the new activities be integrated with existing work methods, overall policies and processes; which roles are needed?

22.4.1.5 Implementation and Aftercare

Individual activities need to be carried out and supported. When dealing with online collaboration, this includes sufficient technical support and helpdesk facilities. Also, the roles defined in the previous stage need to be filled, and the overall process needs to be facilitated. There needs to be "aftercare" once the activity is over, to report, wrap up, and close down.

22.4.1.6 Evaluation and Lessons Learned

The last stage is to see how a community reflects on its own functioning, formulates criteria and derives lessons learned. The output of this reflection can lead to an adaptation of the original rationale and purpose, and feed forward into new activities to build the community, set the common agenda, etc.

22.4.2 Applying the Extended Social Context Model

By extending the SCM with the Community Activity Cycle, it becomes more [illegible]ful for real-world applications:

- We use the Community Activity Cycle to identify the stage at which the improvement of communication process support is most needed.
- For the identified stage, a detailed Social Context Model analysis is performed.
- Based on this analysis, we can then make a set of recommendations for communication process architectures and specific tools to deploy.

22.5 The Case of Friends of the Earth International

AIDEnvironment has advised and supported several organizations in various aspects of collaboration and change processes involving virtual communities on sustainable development. These practical experiences were mixed with critical investigation of the existing methodologies and with academic research in this area, resulting in the extended Social Context Model sketched in the previous sections.

In this section, we will describe one case to which we applied our model: the intranet of the network of Friends of the Earth International (FoEI). We will first sketch the project context in which we applied the model. Then we will describe the decision-making process of which we did a communication process analysis. We then apply the extended SCM. We end this section with some observations.

22.5.1 The "Strengthening the Message" Project

Friends of the Earth International is a network of 68 environmental organizations worldwide, each working in its own country with local, community-based groups active in civil society (around 5,000 local groups in total). It is a grassroots network, where decisions are made by consensus, and work and authority are decentralized as much as possible. In the over 30 years of existence, this has resulted in an extensive set of procedures and principles for decision-making and coordination.

The FoEI network forms a "micro-cosmos" in which many of the characteristics and problems around virtual communities on sustainable development occur. The network has grown considerably in the last few years, involving organizations of different sizes and with different resources available. They collaborate in different coalitions in campaigns and projects. The network has to deal with the influences of the "digital divide", North-South gaps, and barriers in language and culture.

For the past years, FoEI has been working towards enhancing the network's democratic global environmental grassroots character and simultaneously strengthening their capacity to respond quickly and effectively to global environmental challenges. The aim was to leverage their extensive community of experts and activists, and become more effective in their international campaigning efforts. As part of an internal project called "Strengthening the Message", FoEI analyzed their internal communication processes for improvement (this work was supported by the Ford Foundation under their "Governance and Civil Society" program).

One focus of analysis was the use of communication tools in decision-making coordination of international campaigns. A motivated choice of priorities in tools and activities to support the network was needed. Rather than comparing feature sets, it was imperative to find an appropriate match with the capacities and needs of the targeted community members and the existing culture and working practice of the network. Work on the project started in March 2002, with a first version of the intranet taken in use in June 2003.

22.5.2 Analyzing Decision-making Processes

FoEI and AIDEnvironment together identified the major decision-making processes related to international campaigning. We distinguished four key decision-making stages in the life cycle of a campaign:

- There is a continuous discussion in the network to identify "campaign issues", areas where FoEI should be campaigning. This discussion is fed by opinions and ideas from around the world. From these discussions, mostly conducted by e-mail, ideas should arise for specific campaigns.
- This leads to a "policy and work-plan" process in which one or a few members of the network will produce a campaign plan with specified objectives, goals, budget and funding. Part of the process is to create political support for the campaign in the network. The campaign plan can be approved in the Bi-annual General Meeting and become one of the international campaigns in the network.
- Within a campaign, the "coordination" process is done by one or more campaign coordinators, based on priorities and decisions made in a workgroup of all active members in the campaign. The campaign team basically forms a "sub-community" in the network.
- The formal accountability is done in a "work-plan and reporting" process in which progress reports are submitted for approval to the Executive Committee and the Bi-annual General Meeting.

In a workshop at one of the Bi-annual General Meetings of the network, participants were asked to further investigate the processes above, to come up with an analysis of the main problems and suggestions for improvements.

The main conclusion was that the best improvements could be made in the processes around the work-plan and reporting once a campaign was established. The general feeling was that increased transparency in the decision-making within a campaign, and, most of all, better accountability of the progress made in activities in the campaign would increase the effectiveness and reduce stress and distrust levels in the network.

Some general communication problems reported were that in between the face-to-face meetings, all of the communication processes are conducted over mailing lists, providing basic interaction opportunities, but suffering from poor support for more effective decision-making. There is no separation of management and content information, leading to obfuscation of the management process, and "e-mail overloads" of participants. Many groups, especially those with fewer resources and less staff, found it nearly impossible to follow the e-mail debates and as a result felt excluded from the decision-making process.

22.5.3 Applying the Extended SCM

In this section, we apply our extended SCM to the case. We first use the Community Activity Cycle to select the highest priority stage. We then analyze this stage using the SCM, use the results of the analysis for some design recommendations, and present some initial results and observations.

22.5.3.1 Community Activity Cycle: Identifying the Stage

The output of the workshop was matched against the elements of the Community Activity Cycle. In terms of "community", the FoEI network is well established, with many international activities, as well as an extensive handbook with general principles about the norms and values of the network. Regular regional and international meetings focusing on various aspects of the network specify the joint agenda, partly laid down in around 10–15 international campaigns.

Within campaigns, the campaigner community is basically set up during the "policy and workplan" phase when interested campaigners try to gain political support in the network, and define the initial focus and objectives of the campaign. This also leads to an initial agenda and planning for the campaign.

The main stage that required improvement was considered to be Implementation and Aftercare. In this stage, coordination of activities and synergy with the rest of the network, transparency of progress, and accountability with respect to the workplan are needed. This stage was considered to be most problematic by workshop participants, as transparency and accountability were perceived as weak. Having it organized better was considered a prerequisite for the final Evaluation and Lessons Learned stage to become more useful.

22.5.3.2 The SCM Model for Implementation and Aftercare

We illustrate the basic elements in the Implementation and Aftercare stage in the SCM matrix of Table 22.2. The italicized items were the primary focus in our implementation process, as these turned out to be the problematic ones. Note that we renamed the authoring layer of the original SCM into *production* to make this layer more widely applicable.

Table 22.2: The SCM for Implementation and Aftercare

	Elements	Action	Change
Collaboration	*Coordinator*, campaigner *Work-plan and reports repository*	Mediation *Conflict resolution* *Evaluating*	
Production	*Author*, *reviewer*, auditor *Work-plan*, *report*, budget	*Reporting* *Reviewing* Verification	
Support	Support roles (moderator, *technical facilitator*) *Discussant profiles* Discussion summaries	Reminding Moderation	Change notification parameters
scussion	Interaction roles (discussant, attacker, defender), Discussion elements (posts, replies, labels)	Raising issues Replying Creating a position Position taking	Assign attackers, defenders of position *Define discussion rules*

At the level of collaboration, evaluation of campaign progress was considered very important. As the work plans and reports were not readily available, campaigners developed conflicting opinions with respect to the status and progress of a campaign. This resulted in the tendency to over-define reporting and review procedures, creating a bureaucracy that overwhelmed the campaign coordinator.

At the production layer, there was only little support for the authors and reviewers to discuss the work-plan and reports to be delivered, and to monitor the progress and consensus-building in writing these documents.

To support improvements in these processes, it was necessary to improve on the technical facilitator. Also, discussant profiles and summaries were deemed necessary to increase trust levels, transparency and efficiency of collaboration. This should also lead to a better insight in the rules and the roles in the discussions underlying the reports.

22.5.3.3 Design Recommendations

Taking into account the points of attention identified in the SCM analysis, we focused on three main intranet design recommendations.

- Convergence in the discussions was improved by offering an online discussion environment with features for creating documents to reflect decisions made or consensus reached, and to work on progress reports and work-plans. The role of the author of the reports (filled by the campaign coordinator) needed to be better supported.
- Personal profiles should enhance trust and reputation, to make participants in discussions "real people", and allow others to see what roles a person plays in the network as a whole or in other campaigns.
- Designated document libraries for the formal documents of campaigns provide a single point of reference to monitor progress and accountability in campaign activities, accessible by all members in the network in a simple and direct way.

The intranet should allow participants to keep a focus on the agenda of upcoming decisions, and to help prepare these decisions. As an online tool, e-mail depends upon each campaigner and member organizing their documents and discussions into folders and deleting e-mails not considered relevant due to the constraints of storage space. The intranet, however, has the tools to streamline and organize this information and make management of a campaign or network entity like the Bi-annual General Meeting more collective and coordinated. With the proposed setup, campaigners are better prepared with the documents that are needed for the meetings and can also easily view pre-discussions in relation to a policy development.

22.5.4 Results and Observations

The proposed intranet has gradually been taken in use. The full impact of the development process and the available tools is still hard to assess, but some interesting observations can be made [184].

In the application of our model, we placed most emphasis on the collaboration layer in our social context model. The focus was on the implementation of the campaign coordinator role within a campaign, as the person playing this role is responsible for managing the work-plan and progress reports, as well as the discussions in the campaign.

The Web-based discussion tool provided far more support to keep the discussion and documents focused. One of the users, however, remarked that he saw how the work of the coordinator became easier, yet the ease of participation as campaigner and document reviewer had diminished. This mainly relates to aspects of the support layer in our model: the current software no longer offers "notification" of new contributions, and does not point users to such contributions on the intranet. Since this was an inherent (and useful) feature in the old e-mail system, it also became the most requested feature for the new system. We should have used the SCM to more precisely identify such successful communication features of the mailing list system in advance, to make sure there is a similar support in the new system.

Our focus on the upper layers of the SCM resulted in an intranet in terms of the collaboration and production processes in the FoEI network. The tools are tailored to these processes and not presented as generic functionality. As a result, the intranet addresses a fundamental challenge facing FoEI and other networks in the social and environmental realm: the evaluation of the network in terms of its collaboration and production processes. The network as an entity can now see a model and representation of itself in a "cyber-mirror". This should not be seen as an end in itself, but a means to enhance the network's understanding of itself.

As a result of this "gaze", the network refined certain norms, for instance around access by members to specific discussions in the "Executive Committee". Also, our impression is that the intranet has worked as catalyst for the introduction of new communication processes such as regional discussions, allowing them to be more visible as such within the network. The focus in the SCM model on communication processes and ways to support these, rather than tools and features in isolation, has definitely contributed to this more useful view.

22.6 Conclusions

In this chapter, we presented our Social Context Model, extended with a Community Activity Cycle. The model is intended to enable an efficient analysis of communication processes in collaborative communities, primarily in the field of policy-making for sustainable development (both intra- and inter-organizational). It should help with the diagnosis and prioritization of communication problems and potential improvements.

We illustrated the application of the model with a brief case analysis. The introduction of a new technology for communication in a network as diverse as Friends of the Earth (FoEI) is a political process, and takes time. Within the FoEI community, members attach different values to the social aspects and to the process aspects of their relations. The now fine-tuned practice of decision-making and accountability is the result of a long history of discussions and experimentation ith people, technologies, and procedures. "Just" introducing a new technology xactly this realm would not have worked, however.

Doing an analysis with the extended Social Context Model, it is possible to "grow" rather than "impose" a technological infrastructure that is embedded in the social context and governance structure of a grassroots organization. Essential, yet hard-to-capture organizational requirements such as trust, effectiveness, outreach, accountability, rewards, and so on can be more easily woven into the socio-technical fabric of the organization with this kind of context analysis.

Of course, this chapter could only show some of the complexity of a full analysis. Furthermore, a single application of the model is not very useful. However, as an index to real-world communication process problems and solutions in their complicated context, the SCM can allow for the acquisition, classification, and navigation of best practices and benchmarks and reference models. More case analysis, resulting in more sophisticated reference models will eventually lead to more widely useful applications.

Chapter 23
Online Tools for a Sustainable Collaborative Economy

Gary Alexander

Abstract. *This chapter investigates the theory and practice of online systems aimed at building an economy based upon trust. It looks at practical ways in which we can create an economy driven directly by social and economic need rather than commercial pressures. It describes current projects which are developing some of the needed tools. It gives an extended fictional example in which these tools are integrated and extended, and provides a summary of the theoretical principles upon which such a sustainable collaborative economy would need to be based.*

23.1 Introduction

An image of a radically different future has been forming in response to the various environmental, social and economic problems facing the world. It is exemplified by such documents as Agenda 21 and the Earth Charter and the many submissions by non-governmental organizations (NGOs) to the World Summit on Sustainable Development in Johannesburg in 2002 [432; 464]. Global governance is a theme of some work [161], with some using an explicit systems perspective as in this chapter [348]. Previous work has attempted to flesh out what such a future might look like, particularly in the area of exchange and economy [8; 10], adding to a considerable existing literature [233; 337; 593].

We have gone through many economic crises over the past century or so, but now that we are also up against environmental limits, crises beyond all our past experience loom. Earlier civilizations collapsed when they over-reached environmental constraints. With our present huge population of six billion set to rise at least another 50% before it stabilizes, our impact is sure to increase unless there is major change. We have created a period of massive global extinction of species that could reach a point at which humanity is swept away or at best decimated. There is really no alternative to radical change.

The core element of such a future would be a changed relationship between people, moving towards a collaborative world based upon trust and away from competition. Much of this chapter will look at online tools to help build and maintain such relationships. This new relationship would be the basis for the growth of new forms of economic activity. These would begin to be driven directly by environmental and social need rather than determined by money flows (for example: a farmer who, through connection with local consumers, could aim to farm his land to best maintain its fertility and beauty, while providing local organic food that suited the customer tastes, rather than simply farming to maximize income from the food distribution industry).

The pressures created by a need to maximize money flows are a dominant reason for the continuing environmental destruction and advancing poverty and inequality [139]. The nature of an economy with this correct driving force – environmental and social need – is the other key issue this chapter will develop, giving principles, existing examples, and future possibilities.

The change to such a future is likely to come about more through grassroots and civil society initiatives linking together, than through governmental or international regulation and initiatives. Governments are more likely to follow civil society than to take the lead.

Central to this vision is collaboration and the linking of groups. This means that it will need a communications infrastructure, and moreover, one that is capable of matching the diversity and changing needs of the participating population. The few-to-many broadcast media of the past do not have the required flexibility, so that such a change will require the use of the kinds of interactions the modern Internet and Web are providing.

The purpose of this chapter is to further develop the economic aspects of this image of the future, showing how online systems could enable and support it. It will give practical examples and develop theoretical principles. The chapter will start by taking a brief look at a selection of current projects that are consistent with this approach. It will then present an extended fictional example, set a few years in the future, which combines the features of current examples, and adds extra dimensions to both relationship and trading aspects. This is to provide a context for more practical details than a purely theoretical approach would allow. The final section will give theoretical principles of a sustainable, collaborative economy drawing on both the current and the fictional examples.

23.2 Current Developments

There are a huge number of projects and Web sites that provide or are developing the social structures and tools required for a sustainable, collaborative economy. This section describes just a few which illustrate the principles needed.

23.2.1 Trust-based Information Systems

Early human societies were economic cooperatives, in which people gave to and received from their communities without explicit balancing returns on each transaction [465]. This worked because the close, ongoing relationship between people created trust. As larger-scale agricultural and then city-based cultures developed, this trust was lost and so barter and then money systems were developed to enable exchange between strangers with very limited trust.

Our modern information systems are bringing back the possibility of trust-based exchange, and a considerable literature is developing showing the importance of trust, and the significance of reputation in maintaining it [376; 403]. Alexander conjectures that indirect reciprocity requires a continual assessment and re-assessment of the group members' reputation and status [12].

23.2.1.1 Reputation and Rating in eBay and Slashdot

A large part of the success of the eBay online auction/sales sites [692] is their rating system. At the completion of every transaction, buyer and seller are invited to rate and comment upon each other. The set of these ratings and comments forms a public reputation for all members, accessible to anyone considering either buying from or selling to them. It is widely used by members, and creates an atmosphere in which people seek to be seen as trustworthy. This mechanism provides internal, natural self-regulation that does not require external policing and enforcement of standards.

Similarly, there are Web sites giving comparative product information and ratings on sellers by previous customers, on almost anything one can imagine buying. Through this means, public reputation based on customer ratings is becoming an increasingly powerful means of self-regulation in the economy.

Another form of reputation that is becoming widespread is illustrated by the Weblog "Slashdot, News for Nerds" [770], a heavily used news and discussion site for people interested in open source software and related issues. People active in the discussions get opportunities to rate other people's contributions. Thus every message has a score, with those that are disliked given negative scores and those that are liked given positive scores. Users only see messages that score above a threshold that they can adjust. Thus rubbish gets hidden (but not actually censored) while good messages get highlighted. The total of anyone's scores is called their "karma" and can be seen as a measure of the quality of their contributions. Again, this provides self-regulation by users.

23.2.1.2 Local Currencies and the Open Money Project

Local currencies are trading systems using various forms of money that are restricted to use by a defined group, usually quite small. They may have their own notes and coins or use checks, slips of paper or other records [337]. Some of the best-known systems are Time Dollars [782], Time Banks UK [780], and the Local Exchange Trading Systems [737]. Within the group to which the currency is restricted, there is an on-going relationship and sense of reputation which means that trust features more highly than in conventional financial transactions. Thus local currencies are a big step towards a trust-based economy.

The Open Money Project [754] is a recent development by the originators of the Local Exchange Trading Systems, emphasizing the role of local businesses in promoting their community. It is developing core software for rich trading systems using local currencies. It will allow individuals to be members of many overlapping groups, which might have their own currencies or which might share one.

23.2.1.3 Living Directory, an Extended Trust-based Community

The Living Directory Network [736], formerly known as "Friendly Favors" is a Web-based community that has been online since August 1999 and claims over 36,000 participants in over 150 countries. It is explicitly trust-based, with membership by invitation from existing members, but with mechanisms to enable angers to meet members and build trust with them. Its function is to enable bers to support each other's projects by sharing know-how, connections and resources. It has an internal currency called Thankyou's or T's, which functions like a discount on conventional monetary transactions. It enables members to find others with shared interests by enabling them to create a personal profile with key words that is open to other members designated as "friends".

23.2.2 Linking up Existing Groups

The Internet is generally thought of as the great linking technology. It has created new social forms, such as chat rooms and Weblogs. It has facilitated an unprecedented growth of non-governmental organizations (NGOs) concerned with every conceivable environmental and social cause. However, the next level of linking, where separate groups join forces or share resources, is not well supported on the Internet. This social linking was not part of its original purpose or architecture, and there are various social and technical barriers to it. Each organization has its own online database of members that are not easily linked. Individuals have multiple, incompatible online identities with each organization to which they belong. However, there are now many initiatives aimed at overcoming these barriers.

23.2.2.1 One World – Linking the Major NGOs

One World [753] is a network of over 1,500 organizations in 90 countries that "is dedicated to harnessing the democratic potential of the Internet to promote human rights and sustainable development" and has "a vision of equitable and sustainable distribution of wealth amongst the world's population, underpinned by global attainment and protection of human rights and by governance structures which permit local communities control over their own affairs".

23.2.2.2 ManyOne – the Ethical Alternative to AOL

ManyOne Networks [753] is designing a platform that will enable the linking of groups and individuals with social and environmental concerns. It is developing a slick multi-media browser that provides educational features and enables trade and communication.

23.2.2.3 Persistent Identity and the Augmented Social Network Project

The Augmented Social Network (ASN) project [278] was announced at a recent conference on Networking a Sustainable Future [759] and was acclaimed there as the technical solution to linking groups and individuals that everyone had been looking for. Its purpose is to extend the functionality of the Internet through open standards and software development to provide "persistent identity" online. Persistent identity provides an alternative to an individual having a separate, incompatible identity, password, and personal profile with every online organization of which they are a part. Instead, you are registered with an "identity broker", who controls your online profile. When you log-on to a group, the id and password are automatically referred back to your identity broker to be checked.

The ASN project is developing a form of persistent identity that is suitable for non-commercial civil society organizations; there are already well-advanc commercial versions, such as Microsoft's "Passport". It will enable individu control their own profile and who is allowed to see it. The project also includes "relationship brokers" who have access to personal profiles and can suggest useful connections and alliances.

The ASN project intends the software needed to implement its systems to be open source and freely available for non-commercial uses. If successful, it will spawn a new industry, a set of businesses offering identity and relationship brokering services. They will be the guardians and facilitators of online trust-based relationships.

23.3 An Extended Example

The preceding examples of current projects show many of the needed ingredients for a sustainable, collaborative economy: groups with an intention of linking for social and environmental purposes, reputation systems, local currencies within a group with a trust relationship, ways of linking people with common interests, of matching people's wants and offers, new software tailored to support these functions and systems of persistent identity and relationship brokering.

The fictional example that follows is meant to incorporate most of these ingredients in a single system, while adding extra features to enhance the maintenance of a good relationship. It adds an environmental dimension to exchange by adding environmental information as part of an extended "cost". It promotes community service rather than monetary gain as a motivation for exchange. The purpose of the example is to help people to envision a sustainable, collaborative world and to provide a context for the theory in the next section.

23.3.1 The Setting

23.3.1.1 Keith K and Friends in 2010

Let us imagine a group of people centered on 'Keith K', in the year 2010. The year is chosen only a few years into the future so that we can assume that life is largely as it is now, but that there is the beginning of a movement, which we could call "Planetary Citizenship", to build a sustainable, collaborative economy. It has grown to perhaps 1% of the population. Keith is a member of several groups who know each other quite well, and with overlapping memberships, one of many variants on a theme that have formed part of this worldwide movement.

The result is that in the small groups there is an initial basis of trust. Similarly, in the larger movement, many groups are linked to other groups through shared members, while others are linked indirectly, as through "friends of friends of friends". Thus the movement as a whole starts with a set of personal links, some direct and some indirect which are the basis of trust and reputation [376]. It is a cooperative of cooperatives. This explicit, known connection between people has become a central part of Planetary Citizenship.

23.3.1.2 Online Infrastructure – the Community Support Providers

ne group of Keith's friends, with radical IT and media interests, have set up a l business, "PlaNet" which is a Community Support Provider (CSP). It is part lobally linked set of similar businesses that collectively develop, supply, and support open source software for use by their customers (for example, of the sort being developed by ManyOne and the Augmented Social Network project). They have developed friendships around the world through their links with other CSPs. This helps with trading and other activities.

The CSPs provide the usual services of an Internet service provider such as electronic mail or Web space, but with better support on a more personal basis. They also act as an Identity Broker. Beyond this, they offer a local homepage which provides local news, events, wants and offers, and generally provides the basis for the growing new collaborative economy, as will be described below. This builds on their connection with the Planetary Citizenship movement, which enables them to offer local and global trade on an ethical and sustainable basis.

Although linked globally, each CSP provides services and support customized to the needs and culture of their members, so Third World CSPs might be based upon one connection point in a village, while in the West access is widespread.

23.3.1.3 An Example of the Collaborative Economy – the Food Cooperative

Another group of Keith's friends, with interests in food, gardening and cooking, have set up another local business, the "Funky Foods" cooperative. Their basic premise is to find ways of providing an alternative to the industrial food system for at least some of their members' foods. This is a key step in creating a more environmentally-friendly lifestyle. They do this by organizing local gardeners and sympathetic local farmers, first of all, but also more distant sources of food on a Fair Trade basis through links regionally and globally with similar businesses in the Planetary Citizenship movement. They also supply non-members through local shops and cafés.

As a cooperative, they encourage their members to spend small amounts of time either helping out in associated gardens or farms, or with food distribution or with administering the co-op. The CSP software helps them to collect orders, arrange deliveries, organize voluntary labor and provides low-effort accounts.

The food cooperative is just one of a number of cooperatives and local businesses that have become associated with the Planetary Citizenship movement in Keith's area.

23.3.2 Support for Connection and Relationship

If the basis for Planetary Citizenship is a changed relationship between people, then considerable support will be needed for that. Cooperation and mutual support require skills of communication and organization [376], which need to be learned. Moreover, e-mail and online discussions are notorious for their bad communication. That is a reflection of the general lack of awareness of what good communication means, compounded by an absence of body language and tone of voice [9]. The software being developed now, and assumed to exist in this example, can function as social templates, creating expectations and encouraging good communication. Thus it can provide natural, self-regulating mechanisms for collaborative behavior.

23.3.2.1 The Address Book

If connection and relationship are central to Planetary Citizenship, then the personal address book is likely to be central to its software. It provides a natural place to indicate connection and reputation. Figure 23.1 shows what such an entry in Keith K's address book might look like (it is based on a modern address book, courtesy of Apple Computers, with extra features added graphically).

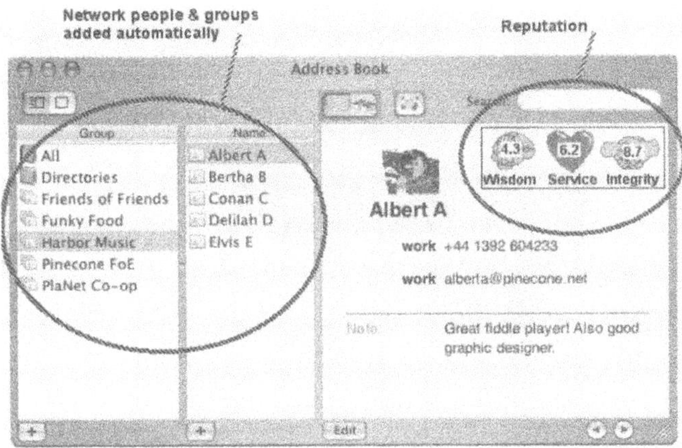

Figure 23.1: A sample address book entry

As with current address books, it includes lists of individuals and groups. However, through the connections with other CSPs and the use of persistent identity, it can automatically add entries from people and groups with whom there is a known relationship of trust and can keep them up to date. It also displays reputation. The details of this are likely to vary from group to group, as will the degree to which it is publicly displayed. Keith K's groups use three notional dimensions of reputation, as shown in Figure 23.1, all of which are based upon polls.

"*Wisdom*" is based on ratings given to the person's contributions to group discussions, as is in done on Slashdot. High scores mean that people like what you have to say in group discussions. "*Service*" comes out of the exchange system, and indicates hours freely given to the community. The point of putting it so prominently is to encourage that. "*Integrity*" is the kind of rating you get on eBay and many eCommerce sites on the Web today. High scores mean people think you have done a good job when you have sold something or served someone.

23.3.2.2 Encouraging Good Communication

One of the deeper issues underlying the social and environmental problems the world faces is the poor communication skills endemic in modern societies. In a competitive and hierarchical society, the skills of understanding another's point of view, of conflict resolution and consensus building get left behind. People talk across each other rather than check that they have been understood. Communications engineers have learned what is necessary to get two machines to communicate effectively. There is a basic communication cycle of statement followed by a check that the statement has been correctly understood.

The enhanced mail system in this fictional example attempts to bring this issue to people's awareness. As shown in Figure 23.2, it distinguishes three types of message (notionally), normal informal messages, somewhat more formal ones in which people need to be clear about what they are agreeing on, and messages about delicate and sensitive issues, where feelings are on the line and conflict could arise or has arisen.

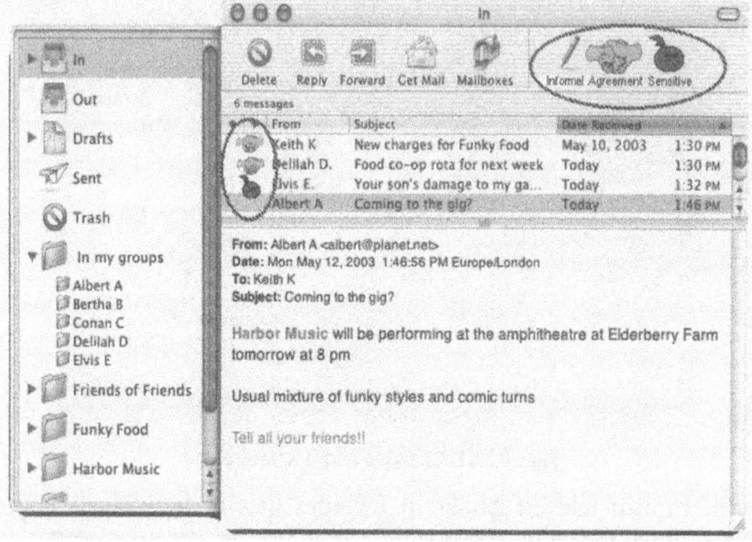

Figure 23.2: A mailbox with sensitivity of message types distinguished

The message displayed in Figure 23.2 is casual and informal. To complete the communication cycle, all it needs is a simple acknowledgement, such as is shown in Figure 23.3. It is automatically generated so that all the user has to do is click on "Send". A similar form can be generated for the next type of message, which makes agreements explicit.

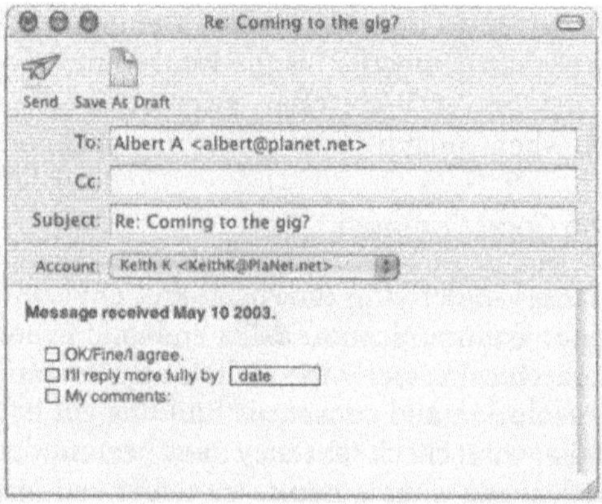

Re: Coming to the gig?

Send Save As Draft

To: Albert A <albert@planet.net>

Cc:

Subject: Re: Coming to the gig?

Account: Keith K <KeithK@PlaNet.net>

Message received May 10 2003.

☐ OK/Fine/I agree.
☐ I'll reply more fully by [date]
☐ My comments:

Figure 23.3: An automatic reply to an informal message

Figure 23.4 shows a template for sensitive messages. It tries to stop people simply arguing across one another, by encouraging them each to express and acknowledge the other's point of view.

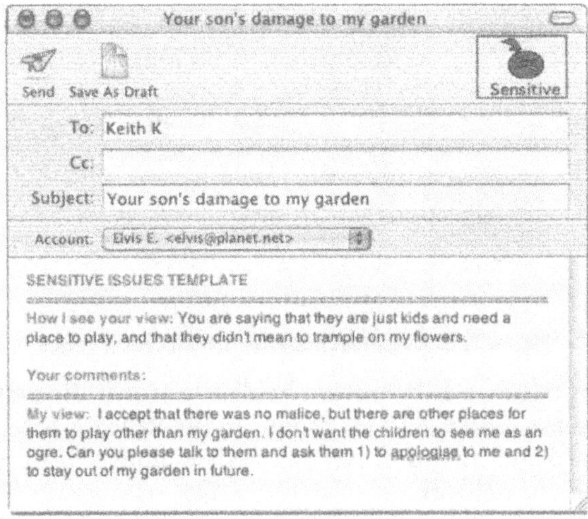

Figure 23.4: An automatic reply for a sensitive message

It is a social template that people can choose to use or not. At the very least, its existence points out that seeing someone else's point of view is important if you are hoping to come to some agreement with them or to resolve a conflict.

The facilities for supporting collaborative relationships are all about improving communication by helping people to realize that it is an issue and creating expectations and suggesting constructive solutions.

23.3.3 Support for Cooperation and Exchange

A key issue in the Planetary Citizenship movement we are imagining is an economy that is driven by social need and is sensitive to environmental pressures. This is difficult in a competitive economy where financial pressures often oppose it. In the more collaborative economy we are envisaging it becomes more possible. It can be greatly assisted by online tools.

23.3.3.1 Information on Quality and Environmental Impact

One of the pleasures of the Web now is that you can look up anything you want to buy and find out what is available, what is good and bad, and compare prices. You can often find ratings of products and suppliers too. Ratings, reviews, reputations are a great help, but these sites almost never take the environment into account, or the social conditions of manufacture (for example, trainers made by poor children in the Third World). Services such as EcoTorch (www.ecotorch.com) are a first step, but the concept could be taken much further. In Keith K's world, de-'led information for buyers is thoroughly integrated into the software provided e PlaNet CSP, as shown in Figure 23.5.

	Real Costs		Money Costs
Toby's Tot Trader 23 Apple Ave. Quality new & used goods 23 Reviews Organic cotton baby carrier 18 Reviews	Footprint: 2.5 Energy: 6.3	Hours: 1.8 Conditions:	€ 45 20% €
Bertha's Baby Boutique 750 Pinecone Place Everything for your baby 16 Reviews Best buy baby carrier 7 Reviews	Footprint: 3.6 Energy: 8.4	Hours: 1.6 Conditions:	€ 28 0% €

Figure 23.5: An extract from the "Green Yellow Pages"

If Keith wants to buy a baby carrier, he does a search. He gets a list that puts those companies associated with his groups in the network first, thus promoting firms that are local and trusted. It includes reviews both of the product and of the firm, by people in the network.

Cost is not given simply in national currency. "Real costs", which are the impact of consuming that item, are given too. They are a complex concept that can be expressed in many ways. As shown here they are divided into an environmental and a social section. The environmental side is then further broken down into ecological footprint, which is a fairly broad measuring tool, plus the energy costs. The social costs are broken down into the number of hours it took to make the object, plus a rating of the social conditions of the workers.

On the money side, there is not only the conventional cost in national currency (Euros for no good reason), but also a local currency (the Green Euro?). Support for local currencies must come built into the CSP software. Comparing these two products, it is clear that the one with the greater cost in money is also better in real cost.

23.3.3.2 A Low-effort Account System

If a collaborative economy is to be driven by information flows rather than money flows, the information systems will need to be really easy to use, in the same way that e-mail is easier than paper-based mail. In this example, we assume that all members potentially have online accounts with all other individuals or businesses. Transactions and records are an extension of the e-mail system.

When Keith K gets his food from Funky Foods, his purchases are recorded, perhaps using a swipe card, and put on his account. Once a month he receives a bill through his e-mail, summarizing his account. This generates an automatic reply, which is in the form of a check as in Figure 23.6.

To pay this bill, Keith simply sends the e-mail, after checking that it is correct. He has the option of filling in the rating and review. The rating contributes to Funky Food's integrity reputation, while the review goes on the Green Yello Pages. So here we have several parts of the software working together to make easy, and also to encourage collaborative behavior. The strength of this c system is that it provides an opportunity for and encourages ratings and reviews on each transaction. Such feedback could not easily be provided with conventional checks or with cash.

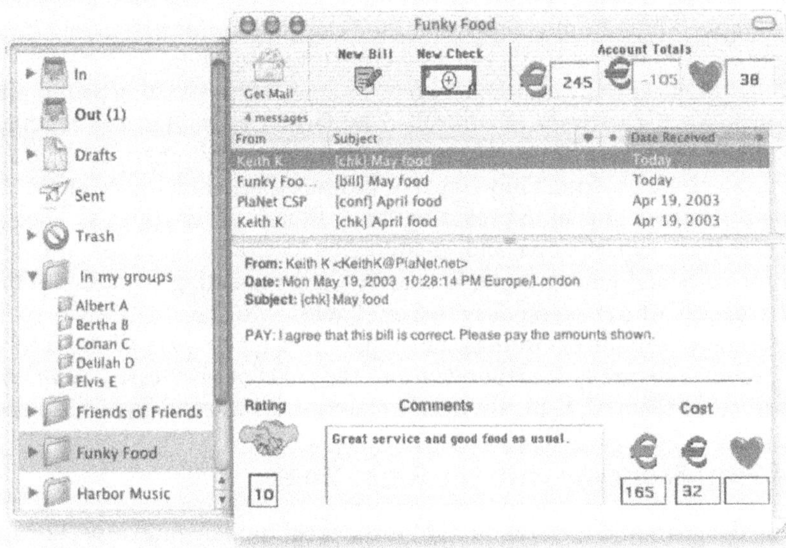

Figure 23.6: An automatically generated check

Notice that this account system divides cost into three categories, conventional currency (Euros in this case), local currency (the Green Euro) and Service (the heart symbol). If Keith does some work for Funky Foods he may get paid either in a currency or may choose to designate some of that work as community service, denominated in hours. That will form part of his public rating for service (see Figure 23.1). By highlighting this, the software encourages public service.

23.4 Principles of a Sustainable, Collaborative Economy

23.4.1 Biological Models of Economic Organization

One of the big lessons from events of the twentieth century is that centralized, command and control economies are not suitable organizing approaches as an alternative to the competitive market. They cannot obtain the information needed to match the variety needed and are prone to corruption. Repression is almost inevitable as a side effect of their structure.

We are in new territory here. The best models appear to be based upon biological and cybernetic principles [348]. The decentralized, distributed, self-regulating, feedback control of an organism is a good metaphorical starting point. This is well illustrated by the examples. Rather than central command and control, they have structures that are networks of networks (or cooperatives of cooperatives). This permits decisions to be taken at a level where information is available, by the people who are affected.

Businesses like the Planet CSP and the Funky Food cooperative have escaped from commercial pressures through their close relationship with their customers and suppliers. They have a stable niche, a territory in which they can operate constrained by feedback not competition. They are thus free to cooperate with similar groups serving different customers, as do all the CSPs, and food cooperatives in the example. Freed from competition they can share information and best practice and can support each other in a synergistic relationship.

Rather than centrally imposed targets, or goals of maximizing profit, the economy in Keith K's network is controlled by feedback, ratings and reputation by users. Businesses have direct information about their customer's reactions, rather than just a crude buy/don't buy signal or indirect marketing information.

The software is crucial to providing this feedback. It functions as a social template, creating expectations and encouraging desirable behavior, reducing the need for rules and enforcement. It creates a self-regulating economy. What happens to people who cheat on eBay? No one trades with them.

Information about environmental impact is a major part of the feedback. It thus becomes naturally incorporated into the goals of businesses. With competitive pressures reduced, firms are freed to take environmental goals seriously.

23.4.2 An Economy with the Right Goals

Getting the goals and motivation right is the most fundamental aspect of applying systems principles to a sustainable, collaborative economy. From a systems perspective, the direct motivation and all corrective feedback on the economic behavior of individuals, businesses, and other organizations should be determined by:

- The health of the natural world (including wilderness areas as well as fisheries, farmland, waterways, etc.), its diversity and ecological stability.
- The health and well-being of all of humanity: adequate food, a healthy lifestyle, emotional health and support, security, community, etc.

This is clearly radically different from the present. In a competitive economy, where economic survival means maintaining adequate money flows, these goals are a luxury which gets very low priority [337]. It is very difficult for a business to give social and environmental considerations a high priority if it means being unable to survive in the market.

The most radical part of the picture given in the examples is the move away from an economy driven by money and towards an economy driven by goals of service towards the environment and the community. It remains to be seen how far in that direction we can go.

In ManyOne and other systems, local currencies are often limited to a discount on conventional monetary prices. In Keith K's world, for local transactions there was a choice of national or local currency or community service. The point of that was to encourage a group to move towards service as it became able to. More distant exchange was based upon Fair Trade, which puts social and environmental concerns before lowest price or highest profit. These are all useful but limited steps, still a long way from a complete change of goals for the economy.

With businesses constrained by information and feedback rather than competitive pressures, and in niches where they serve customers based upon relationship and reputation, lowest price or highest profit no longer determine what is purchased or how it is produced. Much of the function served by money now is removed.

The final step is to move towards an economy in which information and feedback fully constrain activities so that money can be phased out. As Keith K and friends do more for their local co-ops as service, they gradually receive more as service, with no payment required. The extent to which this can be extended between groups and globally is an open question which is probably unanswerable from where we are at present.

23.4.3 Community and Relationship as the Basis for an Economy

In ancient times, when people lived in small, largely self-sufficient communities, the "economy" was much more closely aligned with these goals (although "humanity" was often restricted to one's own cultural group, and detrimental environmental effects were quite local). The disconnection from what seems like obvious goals has occurred gradually over the past few thousand years reaching an extreme as economic globalization reached its full extent in the latter part of the twentieth century [10].

For an economy to be based upon these global goals, the goals of the individuals and organizations that make it up need to be aligned rather than opposed, collaborative rather than competitive. That is the reason for the emphasis on community and the need for good relationships and communication to enable it

The key social skills needed to keep goals aligned are those of understanding another's perspective and of conflict resolution. These are skills that need to be developed and learned. The software used by Keith K gave some support for such skills, but what is most important is an understanding of their importance and a desire to develop them.

23.5 Moving to a Sustainable Collaborative Economy

The kind of radical transformation described in this chapter could be called revolutionary, but if it is to succeed it will need to be more like the industrial revolution than like a political revolution: starting small and within the existing system, learning what works, gradually becoming more influential and continuing to learn as it grows.

The present starting points are very small, but are numerous. They include experiments with new forms of cooperatives, eco-villages, local currencies, fair trade and much more. They involve that small minority of people who can get beyond cynicism to envision a positive future.

They are now at the stage where they are beginning to link together. They will have a lot to learn from each other, and a lot more to invent. They are beginning to form a network of networks, a cooperative of cooperatives. This will form the framework for the nervous system of a metaphorical global organism. It is only as stage matures that a larger-scale acceptance will have a good chance of success; the fictional example of Keith K and friends assumed a 1% global take-up by 2010. As this co-op of co-ops matures, it will become very attractive to the many individuals, businesses and organizations that are seeking a more environmentally sound and stable way of living. They will find that it offers them a way out of the competitive traps in which they have been caught and so the movement will grow rapidly. Participating businesses will be able to offer discounts in local currencies and to pay their employees partly in local currencies. Their contributions to the community as service will form part of their public reputation and so will be encouraged.

As the growing network becomes a significant market to be served by the business community, larger organizations will begin to take it seriously. They too will begin to participate in local currency systems and to contribute more community service. However, for them to do so successfully, they will also need to participate in the feedback mechanisms it has built: the ratings, reviews and information on real social and environmental costs. Through these feedback mechanisms, the mainstream economy will become ensnared and transformed. It is probably only at this late stage that it is likely to become acceptable to the political mainstream as well, and from that point on the transition will be very rapid.

Chapter 24
CBNRM Net: From Managing Natural Resources to Managing Ecosystems, Knowledge, and People

Lars T. Soeftestad • Prakash Kashwan

Abstract. *Environmental online communication epitomizes the opportunities offered by information and communication technologies in managing the changes brought about because of globalization of environmental and developmental concerns, and the increased integration of environmental issues with socio-political concerns of development. Taking the Community-Based Natural Resource Management Network (CBNRM Net) as a case study, this chapter argues how environmental online communication could look beyond an emphasis on information to focusing on knowledge, and to management of such knowledge that can be instrumental in achieving the goals of more equitable and sustainable natural resource management. An assessment of the present and future role and scope of environmental online communication in development cooperation, and in community-based natural resource management in particular, is provided.*

24.1 Introduction

Our interest in Environmental Online Communication (EOC) is connected with our work in development cooperation. We are optimistic about the potential role that EOC can play in addressing the issues and remedying the problems we are seeing, specifically as related to poverty reduction and to sustainable management of natural resources. The chapter aims to assess the usefulness and limitations of EOC. This term as such may be new, but it comprises established concepts. Initially, we understand EOC to be communication about the environment using the Internet, specifically the World Wide Web (Web). We conclude by suggesting a broader understanding of the term as it applies to the area of Natural Resource Management (NRM) and Community-Based Natural Resource Management (CBNRM), as these terms are used within development cooperation. EOC is discussed in the context of networks.

The structure of the chapter is: Section 24.2 presents a broad context for understanding and assessing EOC. Section 24.3 addresses development cooperation. Section 24.4 is a case study of an EOC application, namely the Community-Based Natural Resource Management Network (CBNRM Net). In Section 24.5, evidence from related networks is presented. Section 24.6 analyzes EOC as found within CBNRM Net and other networks. Finally, Section 24.7 contains our conclusions.

24.2 The Context

EOC represents the bringing together of four lines of development: (1) globalization, which makes possible the lines of development and social change given below, and also integrates them, (2) environmental issues brought on to the center stage, and the integration of social and ecological issues, (3) a growing focus on communication, in both its intra- and inter-cultural aspects and (4) the seemingly endless possibilities represented by information and communication technology (ICT).

24.2.1 Globalization

Globalization is understood to represent the many unfolding processes that, taken together, cause an accelerating pace of social transformation and innovation. In this sense it represents the context for the three other lines of development.

Globalization is neither positive or negative [92]. It has negative implications while also representing possibilities for change, for innovation and for betterment in livelihoods. One important positive effect is the increasing focus at the local level, that is, *localization* [629]. Taken together, localization and globalization are responsible for the growing interconnectedness and interdependence in the contemporary world. ICTs are the most visible expression of these seemingly converging processes, as well as the vehicle that makes it possible [497]. Another positive outcome is the increasing focus on global commons that has led to a concern with public policy and a call for global public policy. This is of particular interest as viewed from the vantage point of EOC.

24.2.2 The Environment

The environment began to be addressed as early as the 1970s. At the 1992 Earth Summit emphasis was placed on understanding the environment in *context*. Governments committed to an idea with profound implications for sustainable development: every person should have access to information about the environment, opportunities to participate in decision-making processes affecting the environment, and access to redress and remedy. Articulated in Principle 10 of the Rio Declaration, these "access principles" represent fundamental global norms of equitable and environmentally sound decision-making [432].

EOC, concerned with facilitating access to information and knowledge and with participation, can provide the pivot for reaching these goals. The environment is no longer considered for its own sake, but is mainstreamed in poverty reduction strategies (see Section 24.2.5). This adds to the significance of EOC in multi-stakeholder dialogues, and to the communicative aspect of such consultations.

24.2.3 Communication

Communication is here understood as "the creation of shared understanding through interaction among two or more actors" [402]. Communication is esse tially a relationship *between people*. Traditionally communication was direct took place between individuals who knew each other and were similar in most respects. The medium of communication was mostly oral, and the content was complex. Modern communication is also complex, but in different ways. It takes place between many more stakeholders, often located at different societal levels [342]. The medium of communication is largely written and in electronic form. Modern-day communication is often asymmetrical, the content is often instrumental, and it increasingly contains information without a contextual frame of reference. The interpretation and use of information – as knowledge – is becoming a separate and less prioritized exercise [517].

Communication in the context of development cooperation is understood as a discourse between stakeholders over what development is, with the – in some cases – realized outcome of arriving at an agreed definition [393]. It is useful to remind us that communication is a potential and not a solution. Thus it becomes important to address how communication is changing, why this is problematic, and what can be done about it [136].

24.2.4 Information and Communication Technologies

Information and communication technologies in the context of supporting development cooperation comprise the following:

- *Broadcasting and publishing.* Including newspapers, radio (analog and digital) and television.
- *Computing capacity.*
- *The Internet.* Including chat, e-mail, mailing lists, newsgroups, FTP, video conferencing, the Web, Web conferencing and Web-to-mail.
- *Telecommunications.* Including mobile phones, satellite communication and telephones.

The Internet is the backbone of global computer-mediated communication [93]. Here EOC is taken to be partly a subset of computer-mediated communication, and partly a broader concept and idea, as it focuses mostly on what flows between users, rather than the actual hardware. ICTs hold great potential for helping developing countries and countries in transition. The key reasons for this are because they can provide low cost and accessible means for people to communicate with each other. Exactly how to do this, however, remains an open question. While ICTs hold great potential they also represent countless possibilities for mistakes [122].

The cultural, historical and technological contexts for the invention and development of ICTs are complex. Some are still under development while others appear to have found their niche or are lagging behind. This is, perhaps, less a question of lack of innovative potentials than of innovative incentives. Technological innovation today is mostly a question of economics and profit. The emphasis is on the "new" ICTs, which in many cases do not fit well with the situation at the local level in developing countries and countries in transition.

24.2.5 Implications and Visions

The differences between the haves and the have-nots are widening and poverty is growing. Various analyses refer this situation back to different factors, but many seem to agree on differences in power – whether cultural, economic or political – between different population segments in any one country, and between countries. The old adage that "knowledge is power" is brought to bear on the analysis, and it is argued that ICTs can help leverage the situation for disadvantaged poor by delivering the right knowledge at their doorsteps at the right time [517].

The fundamental question that arises in applying ICTs to developmental cooperation – specifically to poverty reduction strategies – is how to understand and assess ICTs within the overlapping contexts of globalization, the environment, and communication [201].

24.3 Development Cooperation

Development cooperation is a reflection of the above lines of development. In this section, two of them are given special attention.

24.3.1 Knowledge Management and Communication

To understand the concept of knowledge management (KM), it is important to realize the meaning ascribed to the term "knowledge". "Information" is data arranged in meaningful patterns, whereas "knowledge" is something that is believed and is true [517]. In a similar vein Lloyd-Laney [340, p4] argues that "knowledge is the sense that people make of information". Barring the implications of recent discussions over the concepts of truth, reliability and cultural relativism, the distinction holds, perhaps with the added provision that knowledge is contextual and broad, and often needs translation and interpretation. Information and knowledge is shared or transmitted through the process of communication. It is important to keep in mind that while it is unproblematic to communicate information, it is often difficult to communicate knowledge. KM is here understood as a broad and applied context for development cooperation, inter-cultural communication and ICTs [203; 456; 516; 518; 630]. Furthermore, we understand "knowledge" to cover Traditional Knowledge (TK), Indigenous Knowledge (IK), and modern knowledge systems, as occurring within their respective cultural and scientific paradigms.

KM is often misunderstood, in fact, so much so that a list of the "deadliest sins" of KM have been compiled [171], including the following: (1) emphasizing knowledge stock over knowledge flow, (2) failing to see that managing knowledge must also be about creating contexts for sharing, (3) not heeding the role and importance of tacit knowledge and (4) separating knowledge from its uses.

There are three dimensions to KM: (1) sharing knowledge, (2) the reach of ICTs, which gives a new dimension to sharing knowledge and (3) explicating knowledge, i.e. capturing, organizing, and disseminating it. Likewise, key dimensions of KM programs involve decisions about: (1) with whom to share, (2) what to share, (3) how to share and (4) deciding to share. The decision to share is especially important. It involves four further "how to" questions [630]:

- *Connecting versus collecting.* Connecting refers to linking people who need to know with those who know. Collecting refers to capturing and disseminating knowledge through ICTs, aimed at codification, storage and retrieval.
- *Creating social processes within which knowledge sharing occurs.* Networks that connect relevant stakeholders are a key building block.
- *The use of alliances and partnerships.* Alliances and partnerships are fundamental to achieving agreed-upon goals.
- *Choice of ICTs.* The Internet, specifically the Web, is a key factor in catalyzing KM. But there are problems with relying too heavily on the Web.

Closely connected with dissemination of knowledge is the concept of *knowledge networking.* Here, knowledge is not passively disseminated but actively shared, typically with likeminded organizations, networks and individuals.

Communication as understood here is identical with *development communication* as this term emerged in the 1950s onwards. It refers to the application of communication strategies and principles in development cooperation. As such, development communication has, from the beginning, been a reflection of the dominant theoretical paradigms in development cooperation. The dominant communication paradigm has been behavior change models with their attendant emphasis on, for example, modernization and diffusion of innovations. The other broad approach is participatory models and approaches [178; 597]. We find that the latter most closely agrees with the reality as we see it.

24.3.2 ICTs, Social Change and Equity

There is today an intense focus on the potential role of ICTs to bring about sustainable livelihoods and to reduce poverty. This potential has, so far, not been realized, partly because of little access to knowledge, and partly because access to the technologies themselves is limited for many people. Much attention is given to the digital divide, which is only part of a broader and more problematic *development divide* [237]. The way to realize this potential and understand ICTs' societal and developmental role is to locate ICTs within: (1) the broad contextual variability of socio-cultural and economic-political realities that exist in developing countries and countries in transition and (2) the context of communication and KM.

Intellectual roots for a relevant conceptualization of the relationship between ICT and development can be found, among others, in the work of "appropriate technology". Stretching back to Schumacher's credo "small is beautiful" [489], appropriate technology supports the development and use of sustainable approaches to meeting human and ecological needs through the appropriate use of technology. Today's complex problems cannot be solved by using technology independent of its context [224]. To be appropriate, technology must be connected to the place, resources, economics, culture and impact of its use. This necessitates a strong human and culture-centered approach to applying ICTs in a development context. According to appropriate technology, the impact of ICTs is emergent and dependent upon its social context [Soeftestad, 2004 #949].

ICTs as applied to development cooperation represents efforts to scale up tra-`nal means of communication, in terms of numbers of stakeholders involved, e volume, content and speed of communication. The *networked society* has been touted as the outcome [93]. At the same time there are the extremely complex and heterogeneous situations that ICTs are being applied to, and questions as to the rationale behind this work, and its short- and long-term implications. Access to ICTs is, perhaps, a *necessary* step in improving the climate for development in developing countries and countries in transition, but it is never a *sufficient* one.

24.4 The Community-Based Natural Resource Management Network

ICTs have led to a surge in networking activities in development cooperation. CBNRM Net [668] is one such network. To achieve its goal of contributing to social change, leading to more equitable and sustainable NRM, CBNRM Net reaches beyond EOC, and moves from an emphasis on information to knowledge, and to management of such knowledge. CBNRM Net aims to provide relevant KM services for the emerging global network of CBNRM stakeholders [516; 517; 518].

24.4.1 History

CBNRM Net grew out of several intellectual lines of thought and operational activities in the World Bank in the 1990s in which one of us was actively involved. Two activities proved to be significant. The first was CPRNet, the World Bank Common Property Resource Management Network [677]. Established in 1995, it was the first in-house effort to network between staff and outside practitioners and experts, and aimed at serving the needs of World Bank investment operations. The second was the international CBNRM workshop, Washington D.C., May 1998 [516]. The workshop's applied context was training and capacity building in CBNRM. As the members of CPRNet and the participants in the May 1998 workshop realized, means of codification, storage, retrieval and dissemination of CBNRM knowledge were not available. Establishing a network of these stakeholders, supported by ICTs, seemed the correct way to proceed.

Both these activities were success stories, but neither was mainstreamed in World Bank operations. This paved the way for organizing them within a network of CBNRM stakeholders (in the case of the May 1998 workshop, a key recommendation presented to the World Bank was to establish such a network). Thus, CBNRM Net was established in 2001, and CPRNet was incorporated into CBNRM Net at the same time. CBNRM Net in this way represents continuity in networking stretching back to 1995.

24.4.2 Organization

CBNRM Net consists of a mailing list and a Web site, and is formally organized as a project of the Norwegian non-profit "CBNRM Networking" [669]. The management structure is horizontal and devolved, and consists of a coordinator (currently one of us), supported by input of members. While CBNRM Networking covers all expenses, those involved in managing CBNRM Net so far work for free. As the present coordinator runs an own consulting firm, there are no conflicti allegiances, which is important with regard to network vulnerability and t parency. Membership is free, and carries the right to submit knowledge for posting in the Newsletter and on the Web site. The only obligation attached to membership is that members are requested to be actively involved in the running of the network, through producing knowledge and sharing it with fellow CBNRM stakeholders. There are two types of membership: individual and institutional.

Although there is increased interest among members and others in submitting knowledge, be it to the Newsletter or the Web site, the resources required to manage the network are manageable within the present administrative set-up. Requests for help and advice from members are increasing, and this, together with management of the membership database, is gradually becoming a time-consuming task. As CBNRM Net grows, a better-adapted management structure will have to evolve.

24.4.3 Knowledge Management and Knowledge Production

The rationale behind CBNRM Net is that, as individual CBNRM stakeholders and members of CBNRM Net – whether located in the public sector, the private sector or civil society – we all experiment and learn from our work. CBNRM Net provides the opportunity as well as the means for members to share their experiences with others. The key organizing principles for CBNRM Net's activities are a structured and detailed approach to KM, together with knowledge production. The organization of the Web site, including the CBNRM knowledge architecture that underlies it, is a testimony to this.

The KM aspect of CBNRM Net is currently being implemented. Regarding knowledge production the situation is different. Given the situation as regards, for example, member involvement, it has not yet been possible to work on this. One scenario is to network the knowledge already existing with members through intra-group sharing, leading to discussions and brainstorming that, in turn, would result in members reaching new understanding and insight of the issues concerned, located at a higher level of applicability and validity. This has implications for CBNRM Net's communication model.

24.4.4 ICTs and Evolution of a Communication Model

Initially, the emphasis was on the Web site. From the very beginning a cautious strategy was adopted, including using simple HTML without any embellishments. This is done in order to minimize download time and increase access for users with low hardware and software configurations and/or slow connections.

Gradually it became clear that members requested knowledge dissemination also through the Newsletter. The Newsletter is increasingly becoming the preferred means of communication between the members and the CBNRM Net management, as well as between the members (see Section 24.6.2). It is in the nature of CBNRM Net as an ongoing experiment in EOC that the communication model itself is changing over time.

To complement this emphasis on communication through use of ICTs, there is evidence that members who attend conferences and training at local, regional and bal levels meet informally. This face-to-face and more personal contact and munication is crucial, and is expected to increase as the network evolves.

24.4.5 Membership: Thematic and Geographic Coverage

Regarding sectors and themes, members have a very broad expertise. When it comes to geographic coverage, members currently live and/or work in almost 100 countries. The majority of members in the North are working in developing countries and countries in transition. The relatively high membership in Africa is the result of a conscious effort to target this region (Table 24.1).

Table 24.1: CBNRM Net, membership in regions of the world (October 2003)

Region	Number of members
Africa (Sub-Sahara)	209
East Asia (incl. parts of peninsular South-East Asia)	30
Eastern Europe, West Asia and Central Asia (incl. Russia and countries in transition)	3
Latin America and the Caribbean (incl. Central America)	21
North Africa and the Middle East	4
North America	100
The Pacific (incl. parts of peninsular South-East Asia)	15
South Asia	42
Western Europe	79
Total membership	**503**

24.5 Related Networks

In order to understand and assess CBNRM Net, this section presents a few related networks and networking activities. There are a number of related networks, most of which are best understood not as networks but as Web sites, projects or programs where networking is but one of several activities. The following are chosen for comparison (geographic references are included where necessary):

- CBNRM Asia Virtual Resource Centre [667].
- CBNRM Support Programme in Botswana [670].
- Community-Based Coastal Resources Management (CBCRM) Resource Center, Philippines [666].
- Decentralized Natural Resource Management, India [757].
- FRAME, Africa [707].
- IDRC CBNRM Asia [718].
- Livelihoods Connect, global [734].
- MekongInfo [740].
- South Asia Human Development Forum Net [775].
- Tracker, Africa [784].

These networks and networking activities have several things in common. Their approaches to KM reflect the specific circumstances under which they were established, and because of this it is not straightforward to compare them. Some differences can nonetheless be pointed out. A set of criteria has been selected present these differences in Table 24.2 (see Table 24.3 for the complete set).

Table 24.2: CBNRM Net and other networks, a comparison of selected criteria

Criterion	CBNRM Net	Other networks
Web site architecture	Very simple	Often very complex
Membership-based or not	Yes	As a rule not
Thematic focus	Very broad	Often specific issues
Geographic focus	Global	Mostly regional and country levels
Emphasis on political-economic dimensions	Substantial	Varies
Emphasis on knowledge, as opposed to information	Substantial	Varies
Flexibility	High	Low or non-existent
Directional flow	Both ways, but mostly to the target group	Almost completely to the target group
Hard/software and connectivity used	Very low	Often high, sometimes substantial
Technical insight expected of members/users	Very low	Often high
Concerned with using a broad suite of ICTs	Partly the Web and partly e-mail	A majority use only or primarily the Web
Openness	Substantial	Varies, mostly low

One of these networks warrants more scrutiny, namely Decentralized Natural Resource Management (DNRM) in India. It consists of a mailing list and a Web site. DNRM advocates democracy, equity, human rights and justice. The 325 members include, among others, government officials, implementers, academics, NGOs and activists. Discussions tend to be empirical rather than theoretical [449]. Some discussions have led to advocacy and action in favor of communities.

24.6 Analysis and Discussion

Keeping in mind that the focus is on EOC and on KM in CBNRM, this section is framed in terms of networking and networks.

24.6.1 On Networks

At its most fundamental, a network is a set of interconnected nodes. It may have a hierarchy but has no centre. Relationships between nodes are more or less asymmetrical, but all are necessary for the functioning of the network [92]. Networks exist at several levels, from global to local levels [93]. The global economy is networked, and globalization itself is an expression of the level, size and extent of networked interaction between its constituent elements.

The new more complex and inclusive approach to communication between stakeholders and to sharing of knowledge leads to networking between these stakeholders. Such networking becomes formalized in one way or another into networks.

Networks can be *providers* and *generators* of information and knowledge. They also be *intermediaries*, that is, information and knowledge service providers th responsibility for information and knowledge brokerage and delivery.

The special type of network under focus here is understood as more or less formalized communication between numbers of like-minded stakeholders that share knowledge in various ways on a more or less regular basis. Whitten and Wolfe [620] understand a network as relevant series of linkages between individuals that, under specific conditions, may form a basis for the mobilization of people for specific purposes. Networks can be informal and formal. These *social networks* are constituted by members that interact face-to-face. One type of social network is the *community of interest*, consisting of people who share a common interest in a particular topic, and that come together informally to share knowledge. Another type is the *community of practice*, a group of people involved in similar activities or in similar disciplines, and that come together to develop and share knowledge.

The advent of ICTs sets networks in contemporary society apart from earlier networks in some respects. The difference is fundamentally one of quantity and not of quality, in that communication goes faster and that one can reach many people at the same time. The availability of ICTs has led to increase not just in the number of networks, but also in tasks, size and scale. The use of ICTs is often a *sine qua non* for such networks, as ICTs make it possible for members to be located physically apart. Because of this, such networks are often referred to as *virtual communities* or *virtual networks*. Based on the weaker sense of community that virtual networks constitute, they are sometimes called *discourse communities* [499; 512].

The members of a discourse community participate in a genre, that is, they have shared goals, they communicate with one another, and they use participatory mechanisms to provide information and feedback [156]. Because of, among others, the physical separation of participants, the relationship between virtual networks and communities of practice is not clear. In the case of CBNRM Net, we prefer to understand it as the latter. This is, perhaps, based less on present realities than on the direction in which CBNRM Net is likely to move.

Virtual networks are a visible effect of the growing clout of civil society and NGOs. They are also key avenues to further this growing importance. The effect of globalization on the global NGO sector is the realization that other people are grappling with the same problems. The promise of scaling up networks should be approached with care. Depending upon the circumstances, such scaling up may be difficult to achieve [516], and may be better approached by networking like-minded networks, that is, knowledge networking (see Section 24.3.1). Networks have a number of characteristics as shown in Table 24.3.

The criterion "flexibility" is important. Flexibility is here understood as the unused potential for change [40]. Understanding the essence of flexibility follows from the recognition and understanding that complexity and flexibility are closely connected. Flexibility is an essential aspect of functioning ecological and social systems, a key factor behind their high degree of adaptability. The same applies to technological systems, and to integrated social and technological systems. In CBNRM Net, flexibility is the result of an ongoing assessment of methods of working and priorities, resulting from members' input and participation. EOC can, as an integrated social and technological system, and within the context of networks, maintain flexibility in several ways:

- Cover the extent of cultures and types of social organization.
- Aim to be available to all stakeholders, across sectors (public, private, and civil society) and levels (from local to global).
- Balance the emphasis on information and knowledge.
- Maintain a broad approach to issues considered as useful or relevant.
- Be sensitive to changes in the situation for stakeholders and how this affects the means-goals set-up.
- Emphasize the whole array of ICTs at disposal.
- Focus on the ultimate goal(s) and use ICTs as means to reach those goals.

Some of the criteria listed in Table 24.3, especially under the characteristics "structure" and "interactions", are primarily aimed at social networks and not at virtual networks. In the case of CBNRM Net some characteristics are less applicable at the present time. However, as CBNRM Net evolves, it is a matter of time before they apply equally well.

Table 24.3: Network characteristics and criteria

Characteristic	
	1. Organization – owners and managers 2. Management structure – horizontal or vertical 3. Funding – source 4. Membership-based or not – if membership, what are the members' rights and obligations 5. Web site architecture
ICTs	
ther	

24.6.2 CBNRM Net's Communication Model

As an integrated unity consisting of a Newsletter, a Web site, online databases and organizational aspects, CBNRM Net amounts to a *Web information system.* CBNRM Net is a *provider* and *generator* of information and knowledge, and it is an *intermediary.* In performing these functions and capacities it operates under a set of constraints and incentives that are partly identical with those that other networks experience. CBNRM Net is, however, situated apart from other networks in some respects (see Table 24.2):

- It is very much a bottom-up approach.
- It has no formal institutional affiliation.
- It is situated in between the accepted ways of organizing and networking, that is, it crosses, among others, administrative areas, cultures, languages, national borders, projects and sectors.
- It has developed a unique niche as a provider of services for the global CBNRM community of practice. In fact, it has contributed in a major way to developing this network, which is now becoming synonymous with CBNRM Net.
- It is integrated horizontally (connecting members in, for example, projects and sectors) and vertically (connecting members across organizational hierarchies in the public and private sectors as well as internationally).

The evolving communication model is influenced by these conditions. The key issue that comes out of discussions with members, as well as a member survey in 2002, is a concern with the Web site. While a few have suggestions for improving it, most members in developing countries and countries in transition have problems in accessing it [460]. Several members in the North report that they do not access the Web at all because of lack of time and a general feeling of information overload. This speaks to a potentially serious problem with communicating knowledge via the Internet, specifically the Web, and has led to increasing the emphasis on e-mail and the Newsletter (see Section 24.4.4).

However, also in the case of the Newsletter there are some hurdles. To give an example, recently some members, staff at a development project in a remote location in Mali, requested that they be taken off the CBNRM Net distribution list. They are hooked up via mobile phones and satellite, and their Internet service provider has set very low levels for the size of attachments. The Newsletters are now e-mailed to the project's head office in Bamako from where they are sent on via regular mail service.

This points to a limitation in the use of ICTs to network a constituency that is as varied and as physically separated as CBNRM stakeholders. At the same time, the present understanding of CBNRM Net's communication model leaves some things to be desired. The aspect of personal contact is by and large missing. It would be necessary to extend the communication model from the macro- to the micro-level to broaden the type and kind of contact between members. A gradual extension of the model is evident in that, as the network grows, people increasingly become aware of each other (see Section 24.4.4).

Regarding retrieving and presenting knowledge, be it in the Newsletter or on the Web site, it is difficult to please all members. And, in the final analysis, whatever knowledge is made available is largely up to the members themselves. To succeed in its mission, CBNRM Net will have to make the case that it is a network *by* its members, *with* its members and *for* its members. Building this sense of ownership is a key task that runs parallel with the above-mentioned functions.

24.6.3 Networks and Environmental Online Communication

So far, we have understood EOC to be communication about the environment using the Internet, specifically e-mail and the Web. We believe, however, that EOC has a greater potential. Based on the earlier discussion (see Sections 24.2, 24.6.1, 24.6.2) we present the following arguments:

- *Adaptability and flexibility.* Provide an emphasis on the goals with development cooperation and not just the means, that is, ICTs. Both networks and EOC would seem to be easily adaptable to each other and to the changing contexts and circumstances on the ground.
- *Move from information to knowledge as defining element.* Emphasizing the contextual dimension of information will enable better modeling of KM flows and needs in a given setting.
- *Networks as discourse communities or communities of practice.* Networks understood in this way (see Section 24.6.1) have important potentials in communicating environmental knowledge.
- *Knowledge networks* (see Sections 24.3.1, 24.6.1). There are major advantages to be gained in networking existing networks [183].
- *Focus on incentive structures.* The incentives in developing countries and countries in transition to use objective information and knowledge are not strong. Furthermore, incentives are often found in the wrong areas, for example, as connected with funding and as focused on technical solutions. Incentives should take note of organizational processes and focus on institutional incentives [456]. It is important to think the long term. The situation at the micro-level, including the community and the household, needs to be addressed. Motivational behavior change using, for example, the techniques of social marketing should be considered [22; 597; 608].
- *Ways and means of communicating to be broadened.* It is necessary to search for new ways of communicating. The role of broadcasting is being looked at anew, especially digital radio [152]. Web-to-mail or distributing material on CD-ROMs may be alternative options. Open source solutions would contribute to decreasing costs and to development of local and specific-purpose Web applications.
- *The array of useful ICTs to be expanded.* This is supported in part by the growth in the convergence of ICTs through digitalization technology, which facilitates the integration of computers, telecommunications, broadcasting and consumer electronics, and in part by a concern with adapting ICTs to the cul-tural, social, socio-economic and technical realities at the point of use [103; [illegible]].

Such a broad understanding of EOC would, we argue, lead to an emphasis on civil society and its relationship with the wider society. Just as the overarching goals of decentralization, governance, participation, stakeholder involvement and transparency are key elements in development cooperation, a broad conceptualization of EOC will benefit from this while, at the same time, contributing to further it. It follows that EOC operates on two levels: (1) an abstract, ideal and theoretical level and (2) a concrete and practical level.

We propose elements of a process for arriving at relevant EOC approaches, consisting of data collection and analysis. Depending on the circumstances, this can be an iterative process. Two different but connected types of analyses will have to be performed to assess relevant EOC-based networking activities:

- *Stakeholder analysis.* Determines the interests of the stakeholders in relation to the overall interest of the project or program [631; 632]. This is best done through regular stakeholder analysis.
- *Content analysis.* Determines what types of knowledge and information flows between which stakeholders, when and how (see Table 24.3). This is best done through detailed participatory observation and interviews, coupled with more general Poverty and Social Impact Analysis (PSIA).

24.6.4 The Network Analytical Model

The analyses used to arrive at relevant EOC approaches can be used as input to construct a formal Network Analytical Model, which has been introduced previously [519]. This model can, in turn, be used to analyze the networks discussed in this chapter. The model consists of three interacting modes:

- *Modes of coverage.* Identical with the stakeholder categories commonly recognized in development cooperation.
- *Modes of organization.* The way stakeholders organize themselves. The Modes are: (1) *sector* (including agriculture, infrastructure and education), (2) *project*, (3) *issue* (including biodiversity, CBNRM, common property, conservation and research; see "issue" under modes of integration) and (4) *donor* (activities funded and implemented by donors).
- *Modes of integration.* To maintain the cohesiveness and integration as networks scale up, what is lost in direct communication is compensated for by a new form of integration at higher levels. The modes are *space* and *issue* (understood in an inter-disciplinary way; see "issue" under modes of organization).

Networks can be established and integrated along the modes of integration in two ways: (1) horizontally and (2) vertically (see Section 24.6.2). Horizontal integration means integration of stakeholders within the same level (in the sense of society or social organization). At the lower levels, horizontal integration will mostly be around issues. As one moves up towards the macro-levels the integration will increasingly take place also in space. Whereas horizontal integration takes place *within* levels, vertical integration takes places *between* levels. Horizontal and vertical integration often coexist, especially at the higher levels. Vertic integration can be identical with co-management or collaborative managemen

24.7 Conclusions

The societal and developmental role of ICTs is to apply them in support of the major agendas of this day and age, including furthering democracy, decentralization and transparency, to ensure the wellbeing of the disadvantaged masses in developing countries and countries in transition. Social development is dependent upon our ability to establish interaction and synergetic relationships between human values and technological innovations. Thus, the challenge of ICTs is how to sensitize and humanize them to play a key role in this crucial agenda. Scaling ICTs to serve various needs within development cooperation, that is, aligning the needs of various stakeholders with the means at disposal, becomes a humanizing project, addressing both inter-personal and inter-cultural relations [467].

ICTs must strive to emulate key elements and values of the countries and cultures in which they are being used. At the same time, as ICTs are being applied to the area of inter-cultural communication, they must contribute to aligning the diversity and heterogeneity of cultures. In this way ICTs may contribute to engender a cultural pluralism and a plurality of knowledge systems [635]. Several of these arguments come together in a call for EOC to emphasize flexibility (see Section 24.6.1, Table 24.3).

There are short-term and long-term obstacles to realizing relevant applications built around this agenda. In the area of knowledge we need to understand more about specific ICTs and their point of interaction with people and cultures. In the area of communication there is much to be desired in connection with understanding inter- and intra-cultural aspects of communication and their interfaces with ICTs. We need to establish networks of stakeholders to bridge various disconnects, disparities and fault lines. For this to happen we need to talk together, and for this ICTs and EOC are a suitable means and tool, bearing in mind that – depending upon the choice of ICTs – EOC can be both inclusive and exclusive.

As an instrument in achieving sustainable development the emphasis on public consultation in environmental decision-making has, together with the focus on communication (but somewhat contrary to the rationale behind the use of ICTs), been understood as a bottom-up approach. It follows that public involvement is being used and understood in conjunction with emphasis on a number of related approaches and tools, including participation and stakeholder consultation. It also follows from a reorientation of governments away from "command-and-control" and market mechanisms towards what has been referred to as "societal instruments" in order to enable widespread public and civil society involvement. There is a close connection between public consultation and EOC. The latter can contribute decisively and in a meaningful way to public consultation and thus in public involvement, at several societal levels, in generating discussions and agreed-upon outcomes. Good examples at the global level include issues in the domain of global commons and global public policy, including the Clean Development Mechanism from the Kyoto Protocol, the management of straddling and highly migratory fish stocks, the conservation of the living resources of the high seas, and the sustainable use of seabed resources, as foreseen in the United Nations Convention on the Law of the Sea. Other relevant topics concerning environment and development are included in several multilateral treaties and agreements.

CBNRM Net's EOC model concentrates on fostering an open and inclusive dialogue involving all stakeholders, aimed at participation in defining and implementing sustainable NRM strategies, and ensuring ownership of these strategies by all involved stakeholders. The strategic use of available communication tools, as defined by the emerging EOC process and model, will help ensure such inclusion.

EOC is a promising tool for addressing key issues in development cooperation connected with communication, knowledge and meaning. But first we might do well to think through what EOC is, namely a mode of speeding up human communication across time and space. EOC can repackage human communication and it can increase its reach – but it cannot improve upon the essence of human communication. On the other hand, it stands the risk of reducing the content and usefulness of human communication. Herein lies the key to harnessing the promises of EOC: exploring the potentials while keeping clear of the dangers.

Acknowledgements

We are indebted to the following CBNRM Net members for their feedback on this chapter: Daniela Diz, Pontificia Universidade Catolica do Rio de Janeiro, Brazil and Hamish G. Rennie, University of Waikato, New Zealand. The points of view expressed in this chapter are our own and are not necessarily shared by our employers.

Chapter 25
Patterns of Internet Use by Coastal Managers: Results of a Survey

Emma Woodward

Abstract. *Patterns of Internet use for coastal management purposes were investigated for the period June/July 2000. Results were obtained through paper-based and online questionnaires. A separate screening survey gave an indication of the likely demographics of the online questionnaire respondents. Eighty-seven responses were obtained from 21 countries, with the majority from the United States, Australia and the United Kingdom. The results suggest that online participants were most likely to be male, aged 25–39, live in the city, and be employed by the government at a State or Provincial level. The research questionnaire administered to around 200 people at a conference returned 99 responses, whilst the questionnaire available on the Web returned 66 responses. Coastal managers communicate online most frequently with practitioners from their own region. They are most likely to only access information from Web sites originating within their own region or from Web sites originating in North America.*

25.1 Introduction

This research focuses on use of the Internet to facilitate communication and obtain information by coastal management practitioners. A study was undertaken in June/July 2000 to determine patterns of Internet use with the aim of producing baseline information against which future usage could be examined. The research identified gaps in the current provision and availability of online coastal management information as well as demand for existing and future information. The research also determined what technologies coastal managers use to communicate online. An analysis of the results includes suggestions as to possible reasons for the explained limitations of Internet use and some developments that have occurred in the period of time since.

25.1.1 Context

Recent attention has been paid to developing online information management systems to assist in the sustainable development and management of natural systems worldwide. This includes the development of Web portals and complex databases of information. In contrast, information and resources that managers access, and wish to access, via the Internet has received little attention. Likewise, the structure of communication networks that managers use to access and exchange information via the Internet, which is important in determining patterns of information use, has been under-examined. This information is invaluable in ensuring that both the structure and content of new information systems is user driven and the end product is both useful and viable. The aims of the research were to:

- Identify current patterns of Internet use in the field of coastal management, including place and frequency of Internet access and the demographics of users;
- Identify current use of Internet technologies by coastal managers (including e-mail, bulletin boards, newsgroups, and listservs);
- Compare the patterns of Internet use of two geographically different groups of coastal managers and explain any similarities and differences;
- Determine what information is currently accessed by coastal managers and identify what the coastal management community desires from the Internet; and
- Obtain an indication of current use of the Internet for coastal management in a developing region and limitations of its use.

25.1.2 The Internet and the Coastal Manager

Successful management of the coastal zone often requires coordination among several levels of government with overlapping jurisdiction and involves numerous user and interest groups [290; 571]. Effective communication between stakeholders within the coastal zone is paramount for successful implementation of coastal zone management plans and policies.

Coastal management is undertaken in various forms all over the world [106; 108; 109; 229] with many management experiences duplicated. Since environmental management is largely about trial and error and learning by mistake, sharing of experiences can reduce the error to success ratio. As communicating and sharing information is a large part of what environmental professionals do [463], the Internet has created unprecedented networking opportunities for the coastal management community. This research aims to determine the degree to which this networking potential is being realized today.

The Internet has shrunk the world of communications making interaction possible at a distance in real time [570]. It has contributed greatly to the exchange of ideas and information, and creates a medium for collaboration and interaction between individuals and their computers without regard for geographic location [328].

The Internet allows environmental professionals to use e-mail to exchange ideas and information on new technologies and to come together at a low cost to work together successfully on proposals and projects [463]. The Internet has been described as a technology that has created global integration and has been referred to as the "information superhighway". However uptake of the Internet has progressed unevenly, creating a "digital divide". The digital divide refers to the growing gap between different communities in terms of access to technology and the Internet. However, the main issue of the digital divide is not about access to the technology. It is about access to information, knowledge, and communica-

opportunities. The digital divide occurs both within and between countries [280] and affects Internet use by coastal managers in many countries. However while Internet uptake has been slower in some regions, access issues have been overcome by combining the Internet with other communication technologies, which will be discussed further later.

25.2 Methodology

Research information was collected in four ways. An automated "screening" survey was constructed and posted on the Web site CoastalManagement.com [674] and potential participants in the research were notified through listservs, newsgroups and other e-mail distribution lists. The online screening survey identified the demographics of respondents including; age, gender, country of residence, town/city of residence, classification of the area of residence (city, town or rural; town population 10,000- 50,000), place of employment, and whether they were in a paid or volunteer position.

An automated questionnaire was constructed and posted on the CoastalManagement.com Web site. Each of the participants in the screening survey was e-mailed the link to the questionnaire. The online questionnaire was also promoted via listservs, newsgroups, and e-mail distribution lists. The introductory questions focused on the respondent's frequency of Internet use, place of access and use for coastal management purposes. The coastal manager was then asked to select the Internet technologies they use for both communicating and obtaining information for coastal management and also for other purposes. The respondent was also asked to list any coastal management related Internet newsgroups to which they subscribed. Respondents were then given a sliding scale, from "never" through to "very frequently" to answer a set of questions. The questions included the frequency of obtaining information from various sources for research purposes and frequency of "difficulty" in obtaining information or communicating via the Internet. Respondents were also asked how often they communicate with coastal mangers from different world regions, and obtain information from Web sites originating in those same regions.

The same questionnaire in word-format was distributed in hardcopy to delegates of the Coastal Society's 17th International Coastal Management Conference in Portland, USA in June 2000. The purpose of the hardcopy questionnaire was to make the questionnaire accessible to coastal managers regardless of their level of computer or Internet literacy.

Interviews were conducted with members of the South Pacific Regional Environmental Programme (SPREP) in Western Samoa. These interviews did not require any level of Internet or computer literacy and barriers to both the Internet and computer use in general were highlighted for the Pacific Region.

25.3 Results

The results of the four data sets; the screening survey, the online questionnaire, the hardcopy questionnaire and the interview with SPREP are presented in the following sections.

25.3.1 Screening Survey

Of the 87 screening survey respondents, 64 (74%) were male and 23 (26%) were female. The majority were in the 25–39 years age group (63%) followed by 30% in the 40–55 years age group. Respondents in the 18–24 years and 55+ age groups each constituted 3.5% of the population.

Respondents were questioned about the population size of the settlement in which they live. In many countries the size of the settlement dictates the degree of available infrastructure and demand for telecommunications. Larger settlements such as cities and towns are also the usual site for educational and academic institutions and state and federal government departments. The majority of respondents (70%) live in the city (population > 50,000) followed by 22% in towns and the remainder (8%) in rural settings (population < 10,000).

The majority of screening survey participants is employed by state/provincial level government, educational institutions and research institutions. Respondents were also asked to provide a brief job description. The range of work undertaken by coastal managers utilizing the Internet is wide and varied. It includes work on integrated coastal management projects; environmental consultancy; public education, appreciation and participation; numerical modeling, coordination of grants programs, hazard assessment and coastal engineering. Responses were received from 21 countries. The countries and number of responses from each are shown in Table 25.1.

Table 25.1: Country of origin of screening survey respondents

United States	26	Italy	2	Mexico	1
Australia	23	Netherlands	2	New Zealand	1
United Kingdom	12	Spain	2	Portugal	1
Canada	4	Colombia	2	Saint Kitts and Nevis	1
Brazil	2	Indonesia	1	Scotland	1
Germany	2	Ireland	1	Thailand	1
India	2	Japan	1	Turkey	1

25.3.2 Research Questionnaire

25.3.2.1 Patterns of Internet Use

The vast majority of both online and conference respondents access the Internet more than once daily (online 66%, conference 75%) with all respondents accessing the Internet at least once per week. Nearly all respondents use the Internet for coastal management purposes, with conference and online questionnaire respondents showing similar frequencies of Internet access to obtain data and information for coastal management use. Twenty percent of the online respondents do not use English as their native language, and 12.3% "occasionally" find that language prevents them from communicating efficiently or understanding information on the Internet. Seventy-eight percent of Internet respondents and 51% of conference respondents subscribe to coastal management related newsgroups.

Respondents were asked to list the coastal management related newsgroups and listservs they subscribe to. A selection from over 100 responses is given in Table 25.2 to indicate the range of topics discussed by coastal managers online.

Table 25.2: Selection of newsgroups and listservs to which respondents subscribe

e-kenyacoast	Cturtle (sea turtle discussion)	Alaska water issues
SEA-GIS@listserv.heanet.ie	Waterforum	conservation newsgroup
Marinelife	Marine Protected Area	Marine & coastal community network
CoastNet	Coastal Guide News	Mangrove Discussion List
Pacific Ballast Water Group	Oceanspace	Coasthuggers
Ramsar Forum	Marine Ecotourism Group	World Watch Institute
Delta Colorado	Water recycling	Eco-compass
Bight of the Californias	Fish Ecology	Arctic Info
Tsunami Bulletin Board	Global futures	Yarra Mouth Friends

25.3.2.2 The Internet and Geographic Information Systems (GIS)

The Internet has assisted coastal managers in their GIS work in several ways. Sixty-seven percent of Internet respondents and 58% of conference respondents use GIS in coastal management. Responses to "How has the Internet helped coastal management-related GIS?" included those related to:

- *Data access:* Improved access to spatial and non-spatial data-sets, both historical and real time.
- *Data availability:* Able to publicize what information is out there.
- *Transmitting and receiving data:* Internal transfer of data from Head Office to regional offices and transfer of information from other agencies.
- *Technical support:* E-mail discussion with more able GIS colleagues and Arc Info training over the Internet.
- *Sharing of data and ideas, networking:* Obtain examples of how others have applied GIS to coastal management, pool data sources and allow specific area research.

Other GIS users commented that "at this point it has not helped due to difficulty in accessing the right information" and "sometimes I can download it, but often I can't due to connections, cost, or not posted".

25.3.2.3 Information Accessed Online

Coastal managers access a wide variety of information online. A selection of responses is listed below under the major categories identified.

- *Environmental issues:* Coral bleaching, coastal erosion and sea level rise.
- *Education and training:* Educational materials, tools and training opportunities.
- *Community involvement:* Capacity building, community participation successes, co-management information, local knowledge and citizen group information.

- *Events and news:* Meetings, conferences, press releases.
- *Organizations, institutions and agencies:* University research Web sites, NGOs, and industry information.
- *Data sets:* Water quality, coastal landslide and erosion, non-spatial, topography, bathymetry, oceanographic, population, satellite imagery, aerial photos.
- *Law and legislation:* Coastal zone plans and regulations.
- *Policies, plans, and strategies:* Planning news and policy options.
- *Research information:* Fellowships, grants, techniques and models.
- *Projects and programs:* Program reports and summaries, global ICM success and failures, other states' coastal zone management programs.
- *Publications:* Reports, studies, journal articles, newspaper articles, conference papers.
- *Planning and management:* Restoration, species management, dune management, practice examples, collaborative efforts.

Coastal managers also wish to have greater online access to data sets, particularly GIS-specific data, employment and grant information, and in particular "practical coastal management" experiences and lessons.

25.3.2.4 Patterns of Communication

A comparison of results from online and conference respondents was undertaken to compare patterns of communication between two geographically different populations. The vast majority of conference respondents were from North America. It is assumed that the countries of origin of online respondents are similar to those of the screening survey, with about 90% of respondents evenly distributed between the regions of North American, Europe and Scandinavia, and Australia and New Zealand.

Respondents were asked how often they communicate with coastal managers and obtain information from Internet sites developed in different world "regions".

The most interesting result was that coastal managers from the regions of Europe and Scandinavia and Australia and New Zealand are likely to obtain online information from Web sites developed in their own regions as well as from Web sites developed in North America, while coastal managers from North America tend only to obtain information from Web sites from North America.

Coastal managers also have a strong tendency to communicate via the Internet with coastal managers living only in their own region. This was a somewhat surprising result given the number of listservs and newsgroups that coastal managers subscribe to, however many of these are locally focused.

25.3.2.5 Interview with SPREP Representative

The South Pacific Regional Environment Program (SPREP) is an organization established by the governments and administrations of the Pacific region with vision to work towards achieving a community of countries and territories

the capacity and commitment to implement programs for environmental management and conservation [524]. SPREP was selected because the organization represents a wide range of small island communities, many of which face coastal management problems likely to occur in response to predicted sea level rise due to the Greenhouse Effect [281; 603].

The coordinator of community coastal management programs at SPREP is in contact with all Pacific member nations to organize amongst other things workshops and competitions for communities in an endeavor to create opportunities for interaction and environmental awareness. A reliable Internet service provider in Samoa was established in 1998 and SPREP's Web site [776] aims to act as a clearing house for environmental information in the region. Whilst many of the SPREP member countries have some degree of Internet access, this is not the preferred method for communication between the island nations. Members from different nations often send requests for information such as educational tools by fax to SPREP, after which the coordinator will search the Web for suitable information, print it out and fax it back to the requestor.

As a newsgroup coordinator, members will send the coordinator information, such as stories of successful initiatives or programs, either by e-mail or fax. After assembling and editing the information, all members will receive the newsletter by the same medium: fax. The use of the Internet for communication purposes is therefore proving beneficial for some Pacific islands, but its use is not widespread.

25.4 Discussion and Conclusions

Information providers from both government agencies and the commercial sector showed interest in the results of this research throughout the study. Such information providers included a database software developer in the process of developing Global Eco Database [709] and a British Colombia government agency developing a comprehensive coastal resource inventory [673]. Both wanted to know what information coastal managers wish to access on the Internet, and other researchers have since stressed the need for such information.

Together with the construction of comprehensive databases of environmental information, for example the virtual coastal and marine data warehouse Coastbase [675], is the underlying and essential requirement of knowledge about what potential information the users want from the system. As clearly stated by Eleveld et al. [150, p504] in reference to the Coastbase project: "Similar systems that aim to serve users of environmental data and information are being developed, and will probably continue to be developed. However, very little has been published about the real needs of these users, and about how these needs could actually be used for the design and evaluation of such systems".

Kay and Christie [291], in their paper addressing possible futures of the Internet in coastal management, identify one of the key research issues as the ability of coastal managers and information technology specialists to combine to actually shape its future. And this is required at all levels of information provision.

This study showed that coastal managers want to know about lessons learned, successes and failures, best practice and what is working, with more case studies working solutions and publication of successful coastal management pro-s. Although there is a desire for such information to be made available on the Web, this type of information is usually not even generated in the "grey literature" of the government agency. The reporting process itself needs to be designed and tailored and included in the coastal management plan or project budget to assist future project managers or participants who wish to avoid the negative lessons learned from previous coastal managers. Certainly the Web is an opportune medium for such information, however informal, to be made available in the public domain once it is generated. Often a constraint in the availability of Web-based information is the low priority assigned to it by the information holder. With continued budget constraints squeezing many government departments and environmental agencies, the priority of finding enough money to undertake projects and "inessential" extension activities such as making information available on the Web, is the first to suffer.

The Internet has however been used progressively for coastal management purposes by some organizations and government departments. The public participation process in coastal management planning has benefited in some areas from the Internet and the ease with which information can be disseminated. This has assisted in notifying the public that the submission process has begun and making draft management plans available, to interactive online discussion groups and surveys. One example is the Town Meeting on America's Coastal Future [391] where the Internet was used as the primary medium to communicate coastal management issues at the grassroots level. While the Internet was seen as holding indisputable power for making technological information more readily available and accessible to the general public, again the importance of knowing your audience and designing the Web site for that audience was stated [64].

Attempts are continually being made by a wide range of groups to realize the Internet's potential in the provision of information. Whether information is being provided at a cost or as an extension service by government departments, one limiting factor in the success of the venture and which appears to be receiving less research and attention than the design of the database or Web site, is the audience itself, the people who the information is targeting. While distinguishing a target audience, accessing them and then determining their needs and desires for information is certainly not a simple task, it is one that is essential in producing and providing a product that will succeed.

On the other hand, while coastal managers jumped at the chance to list all the information they wished to access online, some were obviously frustrated that it was not available. The mentality that seemed to prevail was "why isn't the information on the Web, because it obviously exists elsewhere". There are high expectations of the Internet as a panacea for all information requirements, which is often unrealistic given that the information has not been documented and doesn't exist for it to be made available on the Web. To some degree perhaps Internet users should reassess their expectations of what the Internet can deliver.

In conclusion, the importance of research identifying users and their requirements for online information cannot be understated in designing an information-providing product. Greater interaction must be forged between information users and information technology specialists to ensure that the real needs of users are determined and that these needs are used in the design and evaluation of such systems.

Bibliography

1. Abbott, R.J. (1983). "Program Design by Informal English Descriptions", *Communications of the ACM*, 26(11): 882-894.
2. ACCA (1997). *Guide to Environment and Energy Reporting and Accounting 1997*. London: Association of Chartered Certified Accountants.
3. ACCA (2003). *The Big Picture: How the Environment Influences Corporate Profit*. London: Association of Chartered Certified Accountants.
4. Ackerman, M.S. (2000). "The Intellectual Challenge of CSCW: the Gap Between Social Requirements and Technical Feasibility", *Human-Computer Interaction*, 15(2): 179-203.
5. Adhikarya, R. (2000). "Mainstreaming the Environment through Environmental Education, Training and Communication", *Communicating the Environment*. M. Oepen and W. Hamacher. Frankfurt: Peter Lang. 82-92.
6. Adriaanse, A., Bringezu, S., et al. (1998). *Stoffströme: Die materielle Basis von Industriegesellschaften*. Berlin: Birkhäuser.
7. Ahrens, A., Böhm, E., Heitmann, K. and Hillenbrand, T. (2003). *Leitfaden zur Anwendung umweltverträglicher Stoffe*. Berlin: Umweltbundesamt.
8. Alexander, G. (2000). "Information-Based Tools for Building Community and Sustainability", *Futures*, 32(3-4): 317-337.
9. Alexander, G. (2000). "Netiquette, or the Social Conventions of Computer Conferencing", *The Open University*. http://sustainability.open.ac.uk/gary/onlinelearn/netiquette.html.
10. Alexander, G. (2002). *eGaia - Growing a Peaceful, Sustainable Earth Through Communications*: Lighthouse Books.
11. Alexander, J.E. and Tate, M.A. (1999). *Web Wisdom: How to Evaluate and Create Information Quality on the Web*. Mahwah: Lawrence Erlbaum Associates.
12. Alexander, R. (1987). *The Biology of Moral Systems*. New York: Aldine de Gruyter.
13. Allen, M., Kettleborough, J. and Stainforth, D. (2003). "Model Error in Weather and Climate Forecasting", *ECMWF Predictability Seminar*. Reading, UK. 275-294.
14. Allen, M. and Stainforth, D. (2002). "Towards Objective Probabilistic Climate Forecasting", *Nature*, 419(6903): 228.
15. Allen, M.R. (1999). "Do-it-yourself Climate Prediction", *Nature*, 401(6754): 642.
16. Altman, P. (2001). E-Mail to the Author (Sept 4, 2001).
17. Altman, P. (2001). Personal Interview (July 9, 2001).
18. Altman, P. (2001). Personal Interview (May 11, 2001).
19. Anderson, R.P., Laverde, M. and Peterson, A.T. (2002). "Geographical Distributions of Spiny Pocket Mice in South America: Insights from Predictive Models", *Global Ecology and Biogeography*, 11(2): 131-141.
20. Anderson, R.P., Laverde, M. and Peterson, A.T. (2002). "Using Niche-based GIS Modeling to Test Geographic Predictions of Competitive Exclusion and Competitive Release in South American Pocket Mice", *Oikos*, 98(2): 3-16.
21. Anderson, R.P., Lew, D. and Peterson, A.T. (2003). "Evaluating Predictive Models of Species' Distributions: Criteria for Selecting Optimal Models", *Ecological Modelling*, 162(3): 211-232.
22. Andreasen, A.R. (1995). *Marketing Social Change: Changing Behavior to Promote Health, Social Development, and the Environment*. San Francisco: Jossey-Bass.
23. Armstrong, E., Bodoff, S., et al. (2003). "The Java Web Services Tutorial", *Sun Microsystems*. http://java.sun.com/webservices/docs/1.1/tutorial/doc/.
24. Arnberger, E. (1993). *Thematische Kartographie* (3rd ed). Braunschweig: Westermann.
25. Arndt, H.-K., Christ, M. and Günther, O. (2001). "Environmental Reporting in the Public Interest", *Environmental Information Systems in Industry and Public Administration*. C. Rautenstrauch. Hershey: Idea Group Publishing. 347-354.

26. Arndt, H.-K. and Günther, O., Eds. (2000). *Environmental Markup Language (EML)*. Marburg: Metropolis.

27. Arthurs, L., Maurice, P., et al. (2003). "Agent-based Simulation of Biocomplexity: Effects of Adsorption on Natural Organic Mobility Through Soils", *226th American Chemical Society National Meeting*. New York, USA: American Chemical Society. http://pubs.acs.org/meetings/.

28. AusCERT, The University of Queensland and CERT/CC (2002). "CERT UNIX Security Checklist v2.0". http://www.auscert.org.au/render.html?it=1935&cid=1920.

29. Australian Bureau of Statistics (2002). Australian Social Trends (Cat. No. 4102.0). Canberra, Australian Bureau of Statistics. http://www.abs.gov.au/.

30. Australian Human Rights and Equal Opportunity Commission (2000). "Emerging Themes: National Inquiry into Rural and Remote Education", *Australian Human Rights and Equal Opportunity Commission*.
http://www.hreoc.gov.au/humanrights/ruraleducation/briefing/report/index.html.

31. Azzone, G., Brophy, M., et al. (1997). "A Stakeholders' View of Environmental Reporting", *Long Range Planning*, 30(5): 699-709.

32. Baard, M. (2003). "Hydrogen Is No Gas, Yet", *Wired News (June 23, 2003)*.
http://www.wired.com/news/technology/0,1282,59322,00.html.

33. Baedeker, C., Kalff, M. and Welfens, M.J. (2001). *Clever leben: MIPS für KIDS. Zukunftsfähige Konsum- und Lebensstile*. Munich: Ökom.

34. Baeza-Yates, R. and Ribeiro-Neto, B. (1999). *Modern Information Retrieval*. Harlow: ACM Press Books.

35. Baier, A. (1986). "Trust and Antitrust", *Ethics*, 96(2): 231-260.

36. Bakkenes, M., Alkemade, J.R.M., et al. (2002). "Assessing Effects of Forecasted Climate Change on the Diversity and Distribution of European Higher Plants for 2050", *Global Change Biology*, 8(4): 390-407.

37. Ball-Rokeach, S.J. and Gutierrez Hoyt, E. (2001). "Communication Technology and Community", *Communication Research*, 28(4): 355-357.

38. Baltensweiler, A. (2003). "Using the Java Connector to Publish Heterogeneous Data", *23nd Annual ESRI International User Conference*. San Diego, USA: ESRI.
http://gis.esri.com/library/userconf/proc03/p0500.pdf.

39. Barnes, J.A. (1972). *Social Networks* (Addison-Wesley Module in Anthropology, 26). Reading: Addison-Wesley.

40. Bateson, G. (2000). *Steps to an Ecology of Mind*. Chicago: University of Chicago Press.

41. Bauer, C. and Scharl, A. (1999). "Acquisition and Symbolic Visualization of Aggregated Customer Information for Analyzing Web Information Systems", *32nd Hawaii International Conference on System Sciences (HICSS-32)*. Maui, USA: IEEE Computer Society Press.

42. Bauer, C. and Scharl, A. (2000). "Quantitative Evaluation of Web Site Content and Structure", *Internet Research: Networking Applications and Policy*, 10(1): 31-43.

43. Baumberger, N. and Haegeli, M. (2000). "Using Metadata in Multistep Preprocessing and Long-term Monitoring", *GIS Work Session UN/ECE, Conference of European Statisticians*. Neuchâtel, Switzerland: UN/ECE. http://www.unece.org/stats/documents/2000.04.gis.htm.

44. BBC (2003). "Boost for Offshore Wind Power", *BBC News World Edition (July 14, 2003)*.
http://news.bbc.co.uk/2/hi/uk_news/3063433.stm.

45. Behrensmeier, R. and Bringezu, S. (1995). Zur Methodik der volkswirtschaftlichen Material-Intensitäts-Analyse: Der bundesdeutsche Umweltverbrauch nach Bedarfsfeldern. Wuppertal, Wuppertal Institut für Klima, Umwelt und Energie (Wuppertal Papers No. 46).
http://www.wupperinst.org/.

46. Bell, D. (1973). *The Coming of Post-Industrial Society: A Venture in Social Forecasting*. New York: Basic Books.

47. Bell, D. (2001). *An Introduction to Cybercultures*. London: Routledge.

48. Ben-Ner, A. and Putterman, L. (2001). "Trusting and Trustworthiness", *Boston University Law Review*, 81(3): 523-550.

49. Berkel, A.v. and Jong, M.d. (1999). "Coherence Phenomena in Hypertextual Environments", *Textproduktion: HyperText, Text, KonText*. E.-M. Jakobs et al. Frankfurt: Peter Lang. 29-40.

50. Berkeley, S. (2000). "Corporate Reputation and the Internet", *Environmental Data Services*. http://www.endsdirectory.com/articles/index.cfm?action=dir12.

51. Berlecon Research and ProActive International (2002). "Free/Libre and Open Source Software (FLOSS) Survey and Study", *International Institute of Infonomics*. http://www.infonomics.nl/FLOSS/.

52. Berry, P.M., Dawson, T.P., Harrison, P.A. and Pearson, R.G. (2002). "Modelling Potential Impacts of Climate Change on the Bioclimatic Envelope of Species in Britain and Ireland", *Global Ecology and Biogeography*, 11(6): 453-462.

53. Beulshausen, T. (2003). "Die Umsetzung des Weißbuchs – Wo ist das Problem?", *Zeitschrift für Umweltchemie und Ökotoxikologie*, 15(1): 45-47.

54. Bhagat, V. (2001). "The Internet: A Powerful Relationship Management Tool for Fundraisers", *The Nonprofit Handbook: Fund Raising*. J.M. Greenfield. New York: John Wiley and Sons. 16-48.

55. Biber, D., Conrad, S. and Reppen, R. (1998). *Corpus Linguistics – Investigating Language Structure and Use*. Cambridge: Cambridge University Press.

56. Bijker, W.E. (1995). *Of Bicycles, Bakelites, and Bulbs: Toward a Theory of Sociotechnical Change*. Cambridge: MIT Press.

57. Blankenbach, K. (1999). "Multimedia-Displays: Von der Physik zur Technik", *Physikalische Blätter*, 5(55): 33-38.

58. Blasius, J. (1994). "Correspondence Analysis in Social Science Research", *Correspondence Analysis in the Social Sciences*. M. Greenacre and J. Blasius. London: Academic Press. 23-52.

59. Blumberg, J., Korsvold, Å. and Blum, G. (2000). "Environmental Performance and Shareholder Value", *World Business Council for Sustainable Development*. http://www.wbscd.org/.

60. Blumenthal, R.G. (2003). "Good Vibes – Socially Responsible Investing is Gaining Fans ... and Clout", *Barron's*, 83(27): L2-L6.

61. BMWA (2002). "Betriebliches Mitarbeitervorsorgegesetz, Bundesgesetzblatt für die Republik Österreich (10. Juli 2002)", *Bundesministerium für Wirtschaft und Arbeit*. http://www.bmwa.gv.at/BMWA/Themen/Arbeitsrecht/Arbeitsrecht/BMVG/.

62. Boeder, P. (2002). "Nonprofits on E", *First Monday*, 7(7): http://firstmonday.org/issues/issue7_7/boeder/index.html.

63. Boeker, S. (2002). *Fundraising im Internet. Nutzungsmöglichkeiten des Internet als Kommunikationsmedium zwischen Nonprofit-Organisationen und potenziellen und existierenden Unterstützern*. Konstanz: Nonprofit Verlag.

64. Bookman, C.A. (2000). "Town Meeting on America's Coastal Future: Using the Internet to Promote Coastal Stewardship", *Ocean & Coastal Management*, 43: 937-951.

65. Bowker, G.C. (2000). "Biodiversity Datadiversity", *Social Studies of Science*, 30(5): 643-683.

66. Bowker, G.C. and Star, S.L. (1999). *Sorting Things Out: Classification and its Consequences*. Cambridge: MIT Press.

67. Boyce, J.K. (2002). *The Political Economy of the Environment*. Cheltenham: Edward Elgar.

68. Braun, R., Russ, M. and Krcmar, H. (2002). "Web-based Tools for Environmental Performance Indicators", *EnviroInfo Vienna 2002 – Environmental Communication in the Information Society – 16th International Symposium Environmental Informatics 2002*. Vienna, Austria. 225-232.

69. Brew, C. and Moens, M. (2000). "Data-Intensive Linguistics", *University of Edinburgh, HCRC Language Technology Group*. http://www.ltg.ed.ac.uk/~chrisbr/dilbook/dilbook.html.

70. Brill, H., Brill, J.A. and Feigenbaum, C. (1999). *Investing with Your Values – Making Money & Making a Difference*. Princeton: Bloomberg Press.

71. Bringezu, S. (2000). *Ressourcennutzung in Wirtschaftsräumen – Stoffstromanalysen für eine nachhaltige Entwicklung*. Berlin: Springer.

72. Browning, G. (1997). "Updating Electronic Democracy", *Database*, 20(3): 47-50.

73. Brunner, K. and Neudeck, S. (2002). Graphische und kartographische Aspekte der Bildanzeige. *Telekartographie und Location Based Services, Geowissenschaftliche Mitteilungen Nr. 58*. Vienna, Technical University Vienna.

74. Buchanan, R.W. and Lukaszewski, C. (1997). *Measuring the Impact of Your Web Site: Proven Yardsticks for Evaluating Customer Impact, Site Evolution, Revenue Growth, and Return on Investment.* New York: John Wiley & Sons.

75. Buckingham Shum, S., De Roure, D., et al. (2002). "CoAKTinG: Collaborative Advanced Knowledge Technologies in the Grid", *Eleventh IEEE International Symposium on High Performance Distributed Computing (HPDC-11), Workshop on Advanced Collaborative Environments.* Edinburgh, Scotland: IEEE. http://www-unix.mcs.anl.gov/fl/flevents/wace/wace2002/.

76. Bucy, E.P., Lang, A., Potter, R.F. and Grabe, M.E. (1999). "Formal Features of Cyberspace: Relationships Between Web Page Complexity and Site Traffic", *49th Annual Conference of the International Communication Association (ICA-99).* San Francisco, USA.

77. Buehler, K. and McKee, L. (1996). The OpenGIS Guide. Wayland, USA, OpenGIS Consortium. http://www.opengis.org/techno/guide.htm.

78. Bullinger, H.-J., Baumann, T., et al. (2002). *Business Communities – Professionelles Beziehungsmanagement von Kunden, Mitarbeitern und B2B-Partnern im Internet.* Bonn: Galileo Press.

79. Burbules, N.C. and Callister, T.A.J. (2000). *Watch IT: The Risks and Promises of Information Technology for Education.* Boulder: Westview Press.

80. Bush, G.W. (2003). "State of the Union Speech", *Los Angeles Times (Feb 28, 2003).* http://www.latimes.com/news/nationworld/nation/la-sotu-text,0,5127144.story.

81. Bush, V. (1945). "As We May Think", *The Atlantic Monthly,* 176(1): 101-108.

82. Cabaniss, S., Leenheer, J., et al. (2003). "Stochastic Synthesis: Simulating the Environmental Transformations of Natural Organic Matter", *National Science Foundation.* http://www.nd.edu/~nom/.

83. Cabaniss, S., Madey, G., et al. (2003). "Stochastic Synthesis Model for the Evolution of Natural Organic Matter", *225th American Chemical Society National Meeting.* New Orleans, USA: American Chemical Society. http://pubs.acs.org/meetings/.

84. Callow, K. (1998). *Man and Message. A Guide to Meaning-Based Text Analysis.* Boston: University Press of America.

85. Cantrill, J.G. (1992). "Understanding Environmental Advocacy: Interdisciplinary Research and the Role of Cognition", *Journal of Environmental Education,* 24(1): 35-42.

86. Caraveo, S. (2002). "Introduction to SOAP for PHP", *PHPCON 2002.* Millbrae, USA. http://www.php-con.com/2002/view/session.php?s=1001.

87. Carey, P.D. and Brown, N.J. (1994). "The Use of GIS to Identify Sites that will Become Suitable for a Rare Orchid, *Himantoglossum hircinum* L., in a Future Changed Climate", *Biodiversity Letters,* 2(4): 117-123.

88. Carter, B. (2000). "A New Playing Field for all Non-profits", *Fund Raising Management,* 31(8): 40.

89. Carter, D.S.G. (1998). "Implementing New Generation Instructional Information Management Systems: A Western Australia Example", *Australian Educational Computing,* 13: 22-28.

90. Cash, D.W., Clark, W.C., et al. (2003). "Knowledge Systems for Sustainable Development", *Proceedings of the National Academy of Sciences of the United States of America,* 100(14): 8086-8091.

91. Castells, M. (1993). "The Informational Economy and the New International Division of Labor", *The New Global Economy in the Information Age – Reflections on Our Changing World.* M. Carnoy et al. University Park: Pennsylvania State University Press. 15-44.

92. Castells, M. (1999). Information Technology, Globalization and Social Development. *UNRISD Discussion Papers 114.* Geneva, United Nations Research Institute for Social Development. http://www.unrisd.org/.

93. Castells, M. (2000). *The Rise of the Network Society* (2nd ed). Oxford: Blackwell.

94. Castells, M. (2001). *The Internet-Galaxy: Reflections on the Internet, Business, and Society.* Oxford: Oxford University Press.

95. Cavaness, C. (2003). *Programming Jakarta Struts.* Sebastopol: O'Reilly & Associates.

96. CBC (2003). "Fuel Cell Products Ready for Sale Soon", *CBC News (June 10, 2003).* http://www.cbc.ca/storyview/CBC/2003/06/10/fuel_cells030610.

97. CBS (2003). "Fill 'Er Up ... With Hydrogen?" *CBS News (April 24, 2003)*. http://www.cbsnews.com/stories/2003/04/24/tech/main550987.shtml.

98. CBS (2003). "Hydro-Fuel Faces Bumpy Road", *CBS News (June 30, 2003)*. http://www.cbsnews.com/stories/2003/01/31/tech/main538846.shtml.

99. CBS (2003). "Nuclear Power Gets A Boost", *CBS News (June 10, 2003)*. http://www.cbsnews.com/stories/2003/06/10/politics/main557969.shtml.

100. CEC (2003). "Glossary of Energy Terms", *California Energy Commission*. http://www.energy.ca.gov/glossary/.

101. Chamblee, R., Gilmore, R., Thomas, G. and Soldow, G. (1993). "When Copy Complexity Can Help Ad Readership", *Journal of Advertising Research*, 33(3): 23-28.

102. Chang, G., Healey, M.J., McHugh, J.A.M. and Wang, J.T.L. (2001). *Mining the World Wide Web - An Information Search Approach*. Norwell: Kluwer Academic Publishers.

103. Chapman, R. and Slaymaker, T. (2002). ICTs and Rural Development: Review of the Literature, Current Interventions and Opportunities for Action. *Working Paper 192*. London, Overseas Development Institute. http://www.odi.org.uk/publications/working_papers/wp192.pdf.

104. Chervenak, A., Foster, I., et al. (2000). "The Data Grid: Towards an Architecture for the Distributed Management and Analysis of Large Scientific Datasets", *Journal of Network and Computer Applications*, 23(3): 187-200.

105. CICA (1994). Reporting on Environmental Reporting (Carried Out in association with the Canadian Standards Association, the Financial Executives Institute Canada and the International Institute for Sustainable Development). Toronto, Canadian Institute of Chartered Accountants.

106. Cicin-Sain, B. and Knecht, R.W. (1998). *Integrated Coastal and Ocean Management: Concepts and Practices*. Washington: Island Press.

107. CIPR (2002). "Integrating Intellectual Property Rights and Development Policy Report", *Commission on Intellectual Property Rights,*. http://www.iprcommission.org/graphic/documents/final_report.htm.

108. Clark, J.R. (1996). *Coastal Zone Management Handbook*. New York: CRC/Lewis Publishers.

109. Clark, J.R. (1998). *Coastal Seas: The Conservation Challenge*. Oxford: Blackwell Science.

110. Clarke, A.E. and Fujimura, J.H., Eds. (1992). *The Right Tools for the Job: At Work in Twentieth-century Life Sciences*. Princeton: Princeton University Press.

111. Clarke, D.G. and Clark, D.M. (1995). "Lineage", *Elements of Spatial Data Quality*. S.C. Guptill and J.L. Morrison. Oxford: Elsevier. 13-30.

112. Clausen, J., Loew, T., et al. (2001). The INEM Sustainability Reporting Guide. Hamburg, International Network for Environmental Management. http://www.inem.org/free_downloads/.

113. Cohen, E. (2002). *Challenges of Information Technology Education in the 21st Century*. Hershey: Idea Group Publishing.

114. Cohn, L. and Brady, D. (2000). "B2B: The Hottest Net Bet Yet?", *Business Week*. January 17, 2000: 36.

115. Collins, C., Kenway, J. and McLeod, J. (2000). Factors Influencing the Educational Performance of Males and Females in School and Their Initial Destinations After Leaving School. Canberra, Commonwealth Department of Education, Training and Youth Affairs. http://www.dest.gov.au/.

116. Colwell, R.R. (2003). "Obstinate Issues, Sophisticated Solutions: Environmental Science and Education for a New Age (John H. Chafee Memorial Lecture on Science and the Environment)", *Third National Conference on Science, Policy, and the Environment*. Washington, USA: National Council for Science and the Environment.

117. Corning, P.A. (2002). "Thermoeconomics: Beyond the Second Law", *Journal of Bioeconomics*, 4(1): 57-88.

118. Corson-Finnerty, A. and Blanchard, L. (1998). *Fundraising and Friend-raising on the Web*. Chicago: American Library Association.

119. Cox, P.M., Betts, R.A., et al. (2000). "Acceleration of Global Warming Due to Carbon-Cycle Feedbacks in a Coupled Climate Model", *Nature*, 408(6809): 184-187.

Cullen, M.J.P. (1993). "The Unified Forecast/Climate Model", *Meteorological Magazine*, 122(1449): `1-94.

121. Cuptill, S.C. and Morrison, J.L., Eds. (1995). *Elements of Spatial Data Quality*. Amsterdam: Elsevier Science.

122. Curtain, R. (2003). Information and Communications Technologies and Development: Help or Hindrance? Melbourne, AusAID. http://www.developmentgateway.com.au/pdf/ict/CurtainICT4Development040903.pdf.

123. Daniels, P.L. and Moore, S. (2002). "Approaches for Quantifying the Metabolism of Physical Economies – Part I: Methodological Overview", *Journal of Industrial Ecology*, 5(4): 69-93.

124. Davenport, E. and Cronin, B. (2000). "The Citation Network as a Prototype for Representing Trust in Virtual Environments", *The Web of Knowledge: a Festschrift in Honor of Eugene Garfield*. B. Cronin and H.B. Atkins. Medford: Information Today, American Society for Information Science. 517-534.

125. Davies, E.K., Glick, M., Harrison, K.N. and Richards, W.G. (2002). "Pattern Recognition and Massively Distributed Computing", *Journal of Computational Chemistry*, 23(16): 1544-1550.

126. Davis, R. (1999). *The Web of Politics: The Internet's Impact on the American Political System*. New York: Oxford University Press.

127. de Moor, A. (1998). "Information Tools for Sustainable Development: Enabling Distributed Human Intelligence", *Journal of Failure & Lessons Learned in Information Technology Management*, 2(1): 21-31.

128. de Moor, A. and Kleef, R. (2001). "Authoring Tools for Effective Societal Discourse", *15th International Informatics for Environmental Protection Symposium: Sustainability in the Information Society*. Zurich, Switzerland. 751-756.

129. de Moor, A. and Kleef, R. (2003). "A Social Context Model for Discussion Process Analysis", *Information Systems for Sustainability*. Hershey: Idea Group Publishing. In print.

130. Deeter-Schmelz, D.R. and Kennedy, K.N. (2002). "An Exploratory Study of the Internet as an Industrial Communication Tool: Examining Buyers' Perceptions", *Industrial Marketing Management*, 31: 145-154.

131. Denisov, N. and Christoffersen, L.E. (2000). Impact of Environmental Information on Decision-Making Processes and the Environment. Arendal (Norway), GRID Arendal. http://www.grida.no/impact/papers/fullimpact.pdf.

132. DeSimone, L.D. and Popoff, F. (1997). *Eco-Efficiency: The Business Link to Sustainable Development*. Cambridge: MIT Press.

133. Dill, S., Kumar, R., et al. (2002). "Self-Similarity in the Web", *ACM Transactions on Internet Technology*, 2(3): 205-223.

134. DOE (2003). "Glossary of Renewable Energy Terms", *U.S. Department of Energy*. http://www.eere.energy.gov/power/consumer/energyglossary.html.

135. Doering, D.S., Cassara, A., et al. (2002). *Tomorrow's Markets – Global Trends and Their Implications for Business*. Washington: World Resources Institute.

136. Domatob, J.K., Ausmus, K.W. and Butler, J.M. (1996). "New Communication Technologies in Tropical African Development", *Development in Practice*, 6(3): 228-239.

137. Dore, J., Woodhill, J., Andrews, K. and Keating, C. (2003). "Sustainable Regional Development: Lessons from Australian Efforts", *Managing Australia's Environment*. S. Dovers and S. Wild River. Leichhardt: The Federation Press. 154-180.

138. Dormann, J. and Holliday, C. (2002). *Innovation, Technology, Sustainability & Society*. Geneva: World Business Council for Sustainable Development.

139. Douthwaite, R. (1996). *Short Circuit, Strengthening Local Economies for Security in an Unstable World*. Totnes: Green Books.

140. Dumanski, J. and Pieri, C. (1996). "Application of the Pressure-State-Response Framework for the Land Quality Indicators (LQI) Programme", *Land Quality Indicators and Their Use in Sustainable Agriculture and Rural Development*. Rome, Italy: FAO. http://www.fao.org/docrep/W4745E/w4745e08.htm.

141. Dumas, J.S. and Redish, J.C. (1999). *A Practical Guide to Usability Testing* (Revised Edition). Exeter: Intellect.

142. Dunning, T. (1993). "Accurate Methods for the Statistics of Surprise and Coincidence", *Computational Linguistics*, 19(1): 61-74.

143. Dupmeier, C. and Werner, G. (2002). "Theme Park Soil: A Case Study for Using Template Technology in Web-based Environmental Information Systems", *16th International Conference Informatics for Environmental Protection, Environmental Communication in the Information Society.* Vienna, Austria: IGU/ISEP. 196-203.

144. Dzbor, M., Domingue, J.B. and Motta, E. (2003). "Magpie – Towards a Semantic Web Browser", *2nd International Semantic Web Conference.* Sanibel Island, USA. 47-62.

145. Eastwood, D. (2002). "Education for Sustainable Development by Web-based Distance Learning: Postgraduate Teaching of Virtual PgDip/MSc Courses", *Planet,* December 2002 (Special Edition 4): 18-19.

146. Eddy, J.A. (1993). "Environmental Research: What We Must Do", *Environmental Modeling with GIS.* M.F. Goodchild et al. New York: Oxford University Press. 3-7.

147. EIA (2003). "Energy Glossary", *Energy Information Association.* http://www.eia.doe.gov/glossary/glossary_main_page.htm.

148. Eilon, R. (2002). "Building Interactive Web Services with WSIA & WSRP", *Web Services Journal,* 2(12): 2-6.

149. Eisenstadt, M., Komzak, J. and Dzbor, M. (2003). "Instant Messaging + Maps = Powerful Collaboration Tools for Distance Learning", *TelEduc'03.* Havana, Cuba. http://www.gestaenlinea.com/teleduc03/.

150. Eleveld, M.A., Schrimpf, W.B.H. and Siegert, A.G. (2003). "User Requirements and Information Definition for a Virtual Coastal and Marine Data Warehouse", *Ocean & Coastal Management,* 46: 487-505.

151. Elmadfa, I. and Burger, P. (1998). "Österreichischer Ernährungsbericht 1998", *Institut für Ernährungswissenschaften der Universität Wien.* http://www.univie.ac.at/Ernaehrungswissenschaften/download.htm.

152. Eltzroth, C. and Kenny, C. (2003). Broadcasting and Development: Options for the World Bank. Washington, World Bank.

153. Engelbart, D.C. (1962). Augmenting Human Intellect: A Conceptual Framework. Stanford, Stanford Research Institute. http://www.bootstrap.org/institute/bibliography.html.

154. Environmental Protection Authority (1998). State of the Environment Report. Perth, Government of Western Australia. http://www.environ.wa.gov.au/downloads/1064_SOE98.pdf.

155. Erasmus, B.F.N., Van Jaarsveld, A.S., et al. (2002). "Vulnerability of South African Animal Taxa to Climate Change", *Global Change Biology,* 8(7): 679-693.

156. Erickson, T. (1997). "Social Interaction on the Net: Virtual Community as Participatory Genre", *30th Annual Hawaii International Conference on System Sciences (HICSS-30).* Hawaii, USA: IEEE. 13-21.

157. Erlingsson, U. and Krishnamoorthy, M. (1997). "Interactive Graph Drawing on the World Wide Web", *6th International World Wide Web Conference.* Santa Clara, USA. http://decweb.ethz.ch/WWW6/Posters/703/igdraw.html.

158. Ernst & Young, KPMG, PricewaterhouseCoopers and House of Mandag Morgen (1999). *The Copenhagen Charter: A Management Guide to Stakeholder Reporting.* Copenhagen: Mandag Morgen.

159. ESRI (2000). "Multiuser Geographic Information Systems with ArcInfo 8", *Environmental Systems Research Institute.* http://www.esri.com/library/whitepapers/pdfs/multiuser_ai8.pdf.

160. ESRI (2002). "ArcIMS 4 Architecture and Functionality", *Environmental Systems Research Institute.* http://downloads.esri.com/support/whitepapers/ims_/arcims4_architecture.pdf.

161. Esty, D. and Ivanovna, M. (2002). *Global Environmental Governance.* New Haven: Yale School of Forestry and Environmental Studies.

162. European Commission (2000). "Report from the Commission to the Council and the European Parliament on the Experience Gained in the Application of Council Directive 90/313/EEC of 7 June 1990, on Freedom of Access to Information on the Environment", *Commission of the European Communities.* http://europa.eu.int/comm/environment/docum/00400_en.htm.

163. European Commission (2001). "White Paper – Strategy for a Future Chemicals Policy", *Commission of the European Communities.* http://europa.eu.int/comm/environment/chemicals/0188_en.pdf.

164. European Commission (2003). "Directive 2003/4/EC of the European Parliament and of the Council of 28 January 2003 on Public Access to Environmental Information and Repealing Council Directive 90/313/EEC", *Commission of the European Communities.* http://europa.eu.int/comm/environment/aarhus/.

165. European Commission (2003). "Linking up Europe: The Importance of Interoperability for E-Government Services", *Commission of the European Communities.* http://europa.eu.int/ISPO/ida/export/files/en/1523.pdf.

166. European Committee for Standardization (1999). *EN ISO 14031: Environmental Performance Evaluation Guidelines.* Brussels: European Committee for Standardization.

167. European Parliament (2001). "Regulation (EC) No. 761/2001 of the European Parliament and of the Council of 19 March 2001 Allowing Voluntary Participation by Organisations in a Community Eco-Management and Audit Scheme", *Official Journal of the European Communities,* L: 114.

168. European Working Group on Libre Software (2001). "Free Software/Open Source: Information Society Opportunities for Europe?" http://eu.conecta.it/paper.pdf.

169. EUROSTAT (2001). *Economy-wide Material Flow Accounts and Derived Indicators. A Methodological Guide.* Luxembourg: Statistical Office of the European Union.

170. Evans, J.R. and King, V.E. (1999). "Business-to-Business Marketing and the World Wide Web: Planning, Managing, and Assessing Websites", *Industrial Marketing Management,* 28: 343-358.

171. Fahey, L. and Prusak, L. (1998). "The Eleven Deadliest Sins of Knowledge Management", *California Management Review,* 40(3): 265-276.

172. Fairclough, N. (1989). *Language and Power.* London: Longman.

173. Fairclough, N. (1992). *Discourse and Social Change.* Cambridge: Polity Press.

174. Fairclough, N. (2001). *New Labour, New Language?* London: Routledge.

175. Farrell, N. (2002). "Silicon Chips the Environment", *Vnunet.com.* http://www.vnunet.com/News/1136800.

176. FEA (2002). *Global Voluntary Corporate Environmental Reporting – The Corporate Register Directory.* Berlin: Federal Environmental Agency.

177. Federico, A., Hinterberger, F. and Musmeci, F. (2002). "Application of the MIPS Concept in Italy: Two Case Studies", *Sustainable Europe Research Institute (SERI).* http://www.seri.at/Data/personendaten/fh/mips_italy.pdf.

178. Feek, W. and Morry, C. (2003). Communication and Natural Resources Management. Rome, Food and Agriculture Organization of the United Nations. http://www.comminit.com/pdf/communication-and-natural-resource-management.pdf.

179. Fellbaum, C. (1998). *WordNet – An Electronic Lexical Database.* Boston: MIT Press.

180. Fichter, D. (2003). "Why Web Projects Fail", *Online,* 27(4): 43-46.

181. Fichter, K. (1998). *Umweltkommunikation und Wettbewerbsfähigkeit. Wettbewerbstheorien im Lichte empirischer Ergebnisse zur Umweltberichterstattung von Unternehmen.* Marburg: Metropolis.

182. Fisher, J. (2000). "Poison PCs", *Salon Media Group.* http://dir.salon.com/tech/feature/2000/09/18/toxic_pc/index.html.

183. Flor, A.G. (2001). "ICT and Poverty: The Indisputable Link", *Third Asia Development Forum on 'Regional Economic Cooperation in Asia and the Pacific'.* Bangkok, Thailand: Asian Development Bank. http://www.worldbank.org/html/extdr/offrep/eap/eapprem/infoalexan.pdf.

184. FoEI (2003). Grant Report to Ford Foundation: Strengthening the Message, Friends of the Earth International.

185. Fogg, B.J., Soohoo, C., et al. (2002). How Do People Evaluate a Web Site's Credibility? Results from a Large Study. New York, Consumer Web Watch. http://www.consumerwebwatch.org/news/report3_credibilityresearch/stanfordPTL.pdf.

186. Forest, C.E., Stone, P.H., et al. (2002). "Quantifying Uncertainties in Climate System Properties with the Use of Recent Climate Observations", *Science,* 295(5552): 113-117.

187. Foster, I. and Kesselman, C. (1997). "Globus: A Metacomputing Infrastructure Toolkit", *International Journal of Supercomputer Applications,* 11(2): 115-128.

188. Foster, I. and Kesselman, C., Eds. (1999). *The Grid: Blueprint for a New Computing Infrastructure.* San Francisco: Morgan Kaufman.

189. Fox, R. (1999). "The Social Identity of Management Ergolect", *English for Specific Purposes,* 3: 261-278.

190. Freeman, L. (1999). "Agencies See 37% Growth Amid B2B Revolution", *Business Marketing,* 84(7): 1-18.

191. Freitag, U. (2001). "WAPLuis - Umweltdaten für mobile Endgeräte", *Neue Methoden für das Wissensmanagement im Umweltschutz: 4. Workshop des GI-Arbeitskreises Hypermedia im Umweltschutz und Workshop 3 der GI-Initiative Environmental Markup Language.* K. Tochtermann and W.F. Riekert. Ulm, Germany: Metropolis. http://www.metropolis-verlag.de/.

192. FSF (2003). "The Free Software Definition", *Free Software Foundation.* http://www.gnu.org/philosophy/free-sw.html.

193. Fukuyama, F. (1995). *Trust: the Social Virtues and the Creation of Prosperity.* New York: Free Press Paperbacks.

194. Future and Institute for Ecological Economy Research (1994). *Environmental Reports - Environmental Statements: Guidelines on Preparation and Distribution.* Osnabrück: Future.

195. Gahan, M. (2003). "Changing the Face of Energy", *BBC News World Edition (Feb 24, 2003).* http://news.bbc.co.uk/2/hi/business/2793587.stm.

196. Galison, P. (1997). *Image and Logic: a Material Culture of Microphysics.* Chicago: University of Chicago Press.

197. Garcia, L. (2001). Personal Interview (July 9, 2001).

198. Garnett, A. and Lewis, P. (2000). "Population and Labour Movements in Rural Australia", *Australasian Journal of Regional Studies,* 6(2): 157-171.

199. Gattuso, G. (1996). "Fundraisers and Cyberspace", *Fund Raising Management,* 27(8): 29.

200. German Institute for Standardization (1997). *DIN 33922: Environmental Reports for the Public.* Berlin: Beuth.

201. Gerster, R. and Zimmermann, S. (2003). Information and Communication Technologies (ICTs) for Poverty Reduction? *Discussion Paper.* Bern, Swiss Agency for Development and Cooperation, Gerster Consulting. http://www.gersterconsulting.ch/docs/ICT_for_Poverty_Reduction.pdf.

202. Ghiselin, M.T. (2000). "A Bibliography for Bioeconomics", *Journal of Bioeconomics,* 2(3): 233-270.

203. Gibbons, M., Limoges, C., et al. (1994). *The New Production of Knowledge: The Dynamics of Science and Research in Contemporary Societies.* London: Sage.

204. Giljum, S. and Hinterberger, F. (2000). "Wie misst man ökologische Nachhaltigkeit? Ein Vergleich ausgewählter Methoden des "Physical Accounting"", *Natur und Kultur,* 1(2): 18.

205. Glass, S. (1999). *The Changing World of Foundation Fundraising: New Challenges and Opportunities.* San Francisco: Jossey-Bass.

206. Godehardt, E. (1990). *Graphs as Structural Models: The Application of Graphs and Multigraphs in Cluster Analysis* (2nd ed). Braunschweig: Vieweg.

207. Goldfarb, B. (2002). *Visual Pedagogy: Media Culture in and Beyond the Classroom.* Durham: Duke Press.

208. Goldman, A.I. (1999). *Knowledge in a Social World.* Oxford: Clarendon Press.

209. Gongla, P. and Rizzuto, C.R. (2001). "Evolving Communities of Practice: IBM Global Services Experience", *IBM Systems Journal,* 40(4): 842-862.

210. Gordon, C., Cooper, C., et al. (2000). "The Simulation of SST, Sea Ice Extents and Ocean Heat Transports in a Version of the Hadley Centre Coupled Model without Flux Adjustments", *Climate Dynamics,* 16(2-3): 147-168.

211. Graham, P. (2003). "Hackers and Painters". http://www.paulgraham.com/hp.html.

212. Graves, G.R. (2000). "Costs and Benefits of Web Access to Museum Data", *Trends in Ecology and Evolution,* 15(9): 374.

213. Gray, J. and Szalay, A. (2002). "The World-wide Telescope", *Communication of the ACM,* 45(11): 51-55.

214. Grefenstette, G. (1995). "Comparing Two Language Identification Schemes", *3rd International Conference on Statistical Analysis of Textual Data*. Rome, Italy. 263-268.

215. GRI (2000). "Sustainability Reporting Guidelines on Economic, Environmental, and Social Performance", *Global Reporting Initiative*. http://www.globalreporting.org/.

216. Grossmann, W.D. (2000). "Realising Sustainable Development with the Information Society – The Holistic Double Gain-Link Approach", *Landscape and Urban Planning*, 50(3): 179-193.

217. Gurak, L.J. and Bobendrier, L. (1999). A Meta-analysis of Guidelines for Web-based Research. Twin Cities, University of Minnesota. http://www.isc.umn.edu/research/papers/Bowen-web.htm.

218. Haddad, W.D. and Draxler, A. (2002). "The Dynamics of Technologies for Education", *Technologies for Education – Potential, Parameters and Prospects*. W.D. Haddad and A. Draxler. Paris: United Nations Educational, Scientific and Cultural Organization (UNESCO).

219. Hadden, K. (2001). Personal Interview (Aug 13, 2001).

220. Hadden, K. (2001). Personal Interview (May 11, 2001).

221. Hair, J.F., Anderson, R.E., Tatham, R.L. and Black, W.C. (1998). *Multivariate Data Analysis* (5th ed). Upper Saddle River: Prentice Hall.

222. Halliday, M.A.K. and Hasan, R. (1976). *Cohesion in English*. London: Longman.

223. Halme, M. and Huse, M. (1997). "The Influence of Corporate Governance, Industry and Country Factors on Environmental Reporting", *Scandinavian Journal of Management*, 13(2): 137-157.

224. Hamelink, C. (1999). ICTs and Social Development: The Global Policy Context. *UNRISD Discussion Papers 116*. Geneva, United Nations Research Institute for Social Development. http://www.unrisd.org/.

225. Hammerton, J., Osborne, M., Armstrong, S. and Daelemans, W. (2002). "Introduction to Special Issue on Machine Learning Approaches to Shallow Parsing", *Journal of Machine Learning Research*, 2: 551-558.

226. Han, J. and Kamber, M. (2001). *Data Mining: Concepts and Techniques*. San Francisco: Morgan Kaufmann.

227. Hanson, W. (2000). *Principles of Internet Marketing*. Cincinnati: South-Western College Publishing.

228. Harrison, H. (1997). Trends in the Delivery of Rural Health, Education and Banking Services (Discussion Paper Vol 11). Canberra, National Farmers' Federation. http://www.nff.org.au/.

229. Harvey, N. and Caton, B. (2003). *Coastal Management in Australia*. Oxford: Oxford University Press.

230. Heemskerk, B., Pistorio, P. and Scicluna, M. (2003). *Sustainable Development Reporting – Striking the Balance*. Geneva: World Business Council for Sustainable Development.

231. Hein, G., Eisfeller, B., Winkel, J.O. and Öhler, V. (2001). "Determining Location Using Wireless Networks", *GPS World*, 12(3): 26-37.

232. Helms, R.C. (1996). "Can Computer Chip Makers Reduce Environmental Impact?" *Stanford University News Service*. http://www.stanford.edu/dept/news/relaged/960605chipsenvir.html.

233. Henderson, H. (1996). *Building a Win-Win World*. San Francisco: Berrett-Koehler.

234. Henseler, J., Isenmann, R. and Müller-Merbach, H. (2004). *Internetbasierte Geschäftsberichterstattung. Stand und Entwicklungstendenzen*. Berlin: Schmidt.

235. Hermann, B. (2002). Software Engineering – Developing a Web Application for the Project Ecoradar. Stuttgart, ITM Beratungsgesellschaft mbH: http://www.itm-consulting.de/.

236. Hess, D.J. (1997). *Science Studies: An Advanced Introduction*. New York: New York University Press.

237. Hewitt de Alcántara, C. (2001). The Development Divide in a Digital Age. An Issues Paper. *Technology, Business and Society Programme, Paper No. 4*. Geneva, United Nations Research Institute for Social Development. http://www.unrisd.org/.

238. Hilty, L.M., Page, B., Radermacher, F.J. and Riekert, W.F. (1995). "Environmental Informatics as a New Discipline of Applied Computer Science", *Environmental Informatics – Methodology and Applications of Environmental Information Processing*. N. Avouris. Dordrecht: Kluwer Academic Publishers. 1-11.

239. Hinterberger, F., Luks, F. and Stewen, M. (1996). *Ökologische Wirtschaftspolitik: Zwischen Ökodiktatur und Umweltkatastrophe*. Berlin: Birkhäuser.

240. Hirsch, T. (2003). "Staggering Scale of Development", *BBC News World Edition (July 14, 2003)*. http://news.bbc.co.uk/2/hi/science/nature/3065419.stm.

241. Hock-Hai, T., Oh, L.B., Liu, C. and Wei, K.K. (2003). "An Empirical Study of the Effects of Interactivity on Web User Attitude", *International Journal of Human-Computer Studies*, 58(3): 281-306.

242. Hohn, B. (2001). *Internet-Marketing und Fundraising für Nonprofit-Organisationen*. Wiesbaden: Deutscher Universitäts-Verlag.

243. Holmes, D.I. (1998). "The Evolution of Stylometry in Humanities Scholarship", *Literary and Linguistic Computing*, 13(3): 111-117.

244. Honig, M.A., Cowling, R.M. and Richardson, D.M. (1992). "The Invasive Potential of Australian Banksias in South-African Fynbos – A Comparison of the Reproductive Potential of *Banksia Ericifolia* and *Leucadendron Laureolum*", *Australian Journal of Ecology*, 17(3): 305-314.

245. Hontelez, J. (2003). Towards a Green EU Constitution – Greening the European Convention Proposal. Bruxelles, European Environmental Bureau. http://www.eeb.org/publication/G8-IGC-EN-final-8-8-03.pdf.

246. Houghton, J.T., Ding, Y., et al., Eds. (2001). *Climate Change 2001: The Science of Climate Change*: Cambridge University Press.

247. Hrebicek, J., Pitner, T. and Benko, V. (2003). "Standardization of Environmental Data and Information Management in the Czech and Slovak Republic", *17th International Conference Informatics for Environmental Protection, The Information Society and Enlargement of the European Union*. Cottbus, Germany: Metropolis-Verlag. 117-124.

248. Huang, Y., Madey, G., Xiang, X. and Chanowich, E. (2003). "Web-based Molecular Modeling Using Java/Swarm, J2EE, and RDBMS Techologies", *Seventh Annual Swarm Researchers Meeting*. Notre Dame, USA. http://www.nd.edu/~swarm03/.

249. Hull, D.A. and Grefenstette, G. (1996). "Querying Across Languages: A Dictionary-based Approach to Multilingual Information Retrieval", *19th Annual International ACM SIGIR Conference on Research and Development in Information Retrieval*. Zurich, Switzerland: ACM Press. 49-57.

250. Huntley, B., Bartlein, P.J. and Prentice, I.C. (1989). "Climatic Control of the Distribution and Abundance of Beech (*Fagus* L.) in Europe and North America", *Journal of Biogeography*, 16(6): 551-560.

251. Hussen, A. (2000). *Principles of Environmental Economics: Economics, Ecology and Public Policy*. London and New York: Routledge.

252. Iacono, S. and Weisband, S. (1997). "Developing Trust in Virtual Teams", *30th Annual Hawaii International Conference on System Sciences*. Los Alamitos, USA: IEEE Computer Society Press. 412-420.

253. ICCA (2002). "Responsible Care Status Report 2002", *International Council of Chemical Associations*. http://www.icca-chem.org/pdf/icca004.pdf.

254. Ide, N. and Véronis, J. (1998). "Word Sense Disambiguation: The State of the Art", *Computational Linguistics*, 24(1): 1-40.

255. IEEE (1984). *Software Engineering Standards*. New York: IEEE Press.

256. Iguchi, K., Matsuura, K., et al. (2003). "Predicting Invasions of North American Basses in Japan Using Native Range Data and a Genetic Algorithm", *Transactions of the American Fisheries Society*, Submitted: http://afs.allenpress.com/.

257. IISD (1993). *Coming Clean: Corporate Environmental Reporting – Opening up for Sustainable Development*. London: International Institute for Sustainable Development, Deloitte Touche Tohmatsu International and SustainAbility.

258. Innes, J.E. (1998). "Information in Communicative Planning", *Journal of the American Planning Association*, 64(1): 52-63.

259. Innes, J.E. and Booher, D.E. (2000). "Indicators for Sustainable Communities: A Strategy Building on Complexity Theory and Distributed Intelligence", *Planning Theory and Practice*, 1(2): 173-186.

260. Isenmann, R. (2004). “Internet-based Sustainability Reporting”, *International Journal of Environment and Sustainable Development,* 3(2): Forthcoming.

261. Isenmann, R. and Lenz, C. (2002). “Internet Use for Corporate Environmental Reporting: Current Challenges – Technical Benefits – Practical Guidance”, *Business Strategy and the Environment,* 11(3): 181-202.

262. Isenmann, R., Lenz, C., et al. (2003). "Standardisierung XML-basierter DTDs zur betrieblichen Umweltberichterstattung. Integration von Umweltinformationen in betriebliche Informationssysteme", *11. Tagung der Fachgruppe Betriebliche Umweltinformationssysteme der Gesellschaft für Informatik e.V. 2003.* Stuttgart, Germany: Shaker. 69-82.

263. Isenmann, R., Lenz, C. and Müller-Merbach, H. (2002). "The Future of Corporate Environmental Reporting. Environmental Communication in the Information Society", *16th Conference Informatics for Environmental Protection 2002.* Vienna, Austria: International Society for Environmental Protection. 234-241.

264. Isenmann, R., Lenz, C., Reitz, C. and Müller-Merbach, H. (2001). "Nachhaltigkeitsberichterstattung. Internet als Medium zur Integration der Geschäfts- und Umweltberichterstattung von Unternehmen. Information Age Economy. Innovations, Methods and Applications for Electronic Commerce", *5. Internationale Tagung Wirtschaftinformatik 2001.* Augsburg, Germany: Physica. 817-830.

265. ISO19139 (2002). Geographic Information – Metadata, Implementation Model v0.7, Digital Geographic Information Working Group. http://metadata.dgiwg.org/metadata/public/iso19115/impl/model/webver/.

266. ISO (2002). ISO 19115:2003 Geographic Information – Metadata. Geneva, International Organization for Standardization. http://www.iso.ch/iso/en/ISOOnline.frontpage.

267. Ivory, M.Y. and Chevalier, A. (2002). A Study of Automated Web Site Evaluation Tools (Technical Report UW-CSE-02-10-01), University of Washington, Department of Computer Science and Enginnering. ftp://ftp.cs.washington.edu/tr/2002/10/UW-CSE-02-10-01.pdf.

268. Ivory, M.Y. and Hearst, M.A. (2002). "Statistical Profiles of Highly-rated Web sites", *SIGCHI Conference on Human Factors in Computing Systems.* Minneapolis, USA. 367-374.

269. Janesick, V.J. (1994). "The Dance of Qualitative Research Design: Metaphor, Methodolatry, and Meaning", *Handbook of Qualitative Research.* N. Denzin and Y. Lincoln. Thousand Oaks: Sage. 209-219.

270. Jarvis, C. and Kupiec, J. (2002). "Accessing Public Sector Environmental Data & Information: Development & Demonstration of National Portal", *16th International Conference Informatics for Environmental Protection, Environmental Communication in the Information Society.* Vienna, Austria: IGU/ISEP. 343-351.

271. Jasanoff, S., Markle, G.E. and Pinch, T., Eds. (1995). *Handbook of Science and Technology Studies.* Thousand Oaks: Sage Publications.

272. Johansen, P.H., Karatzas, K., et al. (2001). "Citizen-Centred Information Dissemination with Multimodal Information Channels", *Environmental Informatics.* Zurich, Switzerland. http://www.empa.ch/uio1/iepo1/.

273. Johnston, M. (1999). *The Nonprofit Guide to the Internet: How to Survive and Thrive* (2nd ed). New York: John Wiley and Sons.

274. Johnston, M. (2002). "Evaluating Online Fundraising Success", *Fundraising on the Internet: The ePhilanthropyFoundation.org's Guide to Success Online.* M. Warwick et al. San Francisco: Jossey-Bass. 39-46.

275. Jones, C.D. and Palmer, J.R. (1998). Spinup Methods for HadCM3L. *Climate Research Technical Note 84.* Exeter, Hadley Centre for Climate Prediction and Research. http://www.met-office.gov.uk/research/hadleycentre/.

276. Jones, K. and Walton, J. (1999). "Internet-based Environmental Reporting. Key Components", *Sustainable Measure, Evaluation and Reporting of Environmental and Social Performance.* M. Bennett and P. James. Sheffield: Greenleaf Publishing. 412-425.

277. Jones, R. and Tonts, M. (1995). “Rural Restructuring and Social Sustainability: Some Reflections on the Western Australian Wheatbelt”, *Australian Geographer,* 26(2): 133-140.

278. Jordan, K., Hauser, J. and Foster, S. (2003). "The Augmented Social Network: Building Identity and Trust into the Next-Generation Internet", *PlanetWork*. http://collaboratory.planetwork.net/linktank_whitepaper/.

279. Jorgenson, A.K. (2003). "Globalization and the Environment", *Journal of World-Systems Research*, 9(2): 195-203.

280. Jurich, S. (2000). "Under Observation: The Information Revolution and the Digital Divide: A Review of Literature", *TechKnowLogia*, 2(2): http://www.techknowlogia.org/.

281. Kaluwin, C. (1993). "Vulnerability Assessment of Western Samoa", *IPCC/WCC '93 Eastern Hemisphere Workshop*. Tsukuba, Japan: Department of Environment, Sport and Territories.

282. Kamminga, H. (1995). "Interpreting the World, Changing the World, and Living in the World: Is a 'Science for the People' Possible?" *Science for the Earth: Can Science Make the World a Better Place?* T. Wakeford and M. Walters. Chichester: John Wiley. 321-346.

283. Kaplan, A.W. (1999). "From Passive to Active about Solar Electricity: Innovation Decision Process and Photovoltaic Interest Generation", *Technovation*, 19: 467-481.

284. Karatzas, K. (2003). "Environmental Informatics: Concepts and Definitions", *17th International Conference on Informatics for Environmental Protection*. Cottbus, Germany: Metropolis Verlag. 146-151.

285. Karatzas, K., Bassoukos, A., et al. (2003). "Environmental Informatics and Urban Air Quality Management and Information Systems", *5th International Symposium on Environmental Software Systems (ISESS 2003)*. Semmering, Austria: International Federation for Information Processing. 134-146.

286. Karatzas, K., Moussiopoulos, N. and Papadopoulos, A. (2001). "Web-Based Tools for Environmental Management", *Environmental Management and Health*, 12: 356-363.

287. Kassas, M. (2002). "Environmental Education: Biodiversity", *The Environmentalist*, 22(4): 345-351.

288. Kassem, N. and Enterprise Team (2000). *Designing Enterprise Applications with the Java 2 Platform, Enterprise Edition*. Reading: Addison-Wesley.

289. Katerattanakul, P. and Siau, K. (1999). "Measuring Information Quality of Web Sites: Development of an Instrument", *20th International Conference on Information Systems (ICIS-99)*. Charlotte, USA: International Conference on Information Systems. 279-285.

290. Kay, R. and Alder, J. (1999). *Coastal Planning and Management*. London: Routledge.

291. Kay, R. and Christie, P. (2001). "An Analysis of the Impact of the Internet on Coastal Management", *Coastal Management*, 29: 157-181.

292. Keniry, J., Blums, A., et al. (2003). Regional Business: A Plan for Action. Canberra, Department of Transport and Regional Services. http://www.rbda.gov.au/.

293. Kenwood, C.A. (2001). A Business Case Study of Open Source Software. Bedford, MITRE Corporation. http://www.mitre.org.

294. Kenyon, P., Sercombe, H., Black, A. and Lhuede, D. (2001). Creating Better Educational and Employment Opportunities for Rural Young People. Hobart, National Youth Affairs Research Scheme. http://www.acys.utas.edu.au/ncys/nyars/N20.htm.

295. Keohane, R.O. and Nye, J.S. (1998). "Power and Interdependence in the Information Age", *Foreign Affairs*, 77(5): 81-94.

296. Killingsworth, M.J. and Jacobsen, M. (1999). "The Rhetorical Construction of Environmental Risk Narratives in Government and Activists Web Sites: A Critique", *Narrative and Professional Communication*. J.M. Perkins and N. Blyler. Stamford: Ablex. 167-177.

297. Kim, G.M. and Lee, G.S. (2003). "E-Catalog Evaluation Criteria and Their Relative Importance", *Journal of Computer Information Systems*, 43(4): 55-62.

298. Kleinberg, J., Papadimitriou, C. and Raghavan, P. (1998). "A Microeconomic View of Data Mining", *Data Mining and Knowledge Discovery*, 2(4): 311-324.

299. Klotz, R.J. (1998). *Surfing Through the Eyes of Aristotle: Political Advocacy on the World Wide Web*. Dissertation, Washington: Washington University.

300. Knorr Cetina, K. (1999). *Epistemic Cultures: How the Sciences Make Knowledge*. Cambridge: Harvard University Press.

301. Knuth, D.E. (1992). *Literate Programming* (CSLI Lecture Notes No 27). Stanford: CSLI Publications.

302. Knutti, R., Stocker, T.F., Joos, F. and Plattner, G.-K. (2002). "Constraints on Radiative Forcing and Future Climate Change from Observations and Climate Model Ensembles", *Nature,* 416(6882): 719-723.

303. Köhl, M., Traub, B. and Päivinen, R. (2000). "Harmonisation and Standardisation in Multinational Environmental Statistics – Mission Impossible?", *Environmental Monitoring and Assessment,* 63(2): 361-380.

304. Kolk, A. (1999). "The Internet as a Green Management Tool", *Harvard Business Review,* 6(3): 307-316.

305. Kolk, A. (2003). "Trends in Sustainability Reporting by the Fortune Global 250", *Business Strategy and the Environment,* 12(5): 279-291.

306. Kolk, A. and Van der Veen, M. (2002). "KPMG International Survey of Corporate Sustainability Reporting 2002", *KPMG, University of Amsterdam.* http://www.wimm.nl/publicaties/KPMG2002.pdf.

307. Kolk, A., Van der Veen, M., et al. (1999). "KPMG International Survey of Environmental Reporting 1999", *KPMG, University of Amsterdam.* http://www.wimm.nl/publicaties/kpmg1999.pdf.

308. Kolk, A., Walhain, S. and Van de Wateringen, S. (2001). "Environmental Reporting by the Fortune Global 250: Exploring the Influence of Nationality and Sector", *Business Strategy and the Environment,* 10(1): 15-28.

309. Korpela, E., Werthimer, D., et al. (2001). "SETI@home – Massively Distributed Computing for SETI", *Computing in Science and Engineering,* 3(1): 78-83.

310. Koslow, S.H. and Huerta, M.F. (2000). *Electronic Collaboration in Science* (Progress in Neuroinformatics Research, Volume 2). Mahwah: Lawrence Erlbaum Associates.

311. Kraak, M.J. and Brown, A., Eds. (2000). *Web Cartography: Developments and Prospects.* Andover, UK: Taylor & Francis.

312. Krcmar, H. (2003). *Informationsmanagement.* Berlin: Springer.

313. Kreeb, M. and Schulz, W.F. (2002). "Implementing an Environmental Management Web-Portal", *EnviroInfo Vienna 2002 – Environmental Communication in the Information Society – 16th International Symposium Environmental Informatics 2002.* Vienna, Austria. 254-261.

314. Kress, G. (1998). "Visual and Verbal Modes of Representation in Electronically Mediated Communication: The Potential of New Forms of Text", *Page to Screen: Taking Literacy into the Electronic Era.* I. Snyder. London: Routledge. 53-79.

315. Krishtalka, L. and Humphrey, P.S. (2000). "Can Natural History Museums Capture the Future?", *BioScience,* 50(7): 611-617.

316. Krüger, M. (2001). *Konzeption und Entwicklung einer automatisierten Umweltberichterstattung am Beispiel der Hasseröder Brauerei GmbH.* Master's Thesis, Magdeburg: Otto-von-Guericke-Universität.

317. Krut, R. and Moretz, A. (2000). "The State of Global Environmental Reporting: Lessons from the Global 100", *Corporate Environmental Strategy,* 7(1): 85-91.

318. Kubasek, M. (2002). "Enviweb – Enter Gate to Environment Protection", *Tvorba Softwaru 2002.* Ostrava, Czech Republic: Tanger. 99-108.

319. Kubasek, M. (2002). "Enviweb and its Role in Public Environmental Communication in the Czech Republic", *16th International Conference Informatics for Environmental Protection, Environmental Communication in the Information Society.* Vienna, Austria: IGU/ISEP. 329-331.

320. Kubasek, M. (2003). "Semantic Web Technology – Ontology Extraction from Environmental Web", *17th International Conference Informatics for Environmental Protection, The Information Society and Enlargement of the European Union.* Cottbus, Germany: Metropolis. 905-909.

321. Kuper, A. and Kuper, J., Eds. (1985). *The Social Science Encyclopedia.* London: Routledge.

322. Landesman, C. (1995). "Nonprofits and the World-Wide Web", *Internet Nonprofit Center.* http://www.nonprofits.org/lib/website.html.

323. Larsen, L.B. (2000). "Strategic Implication of Environmental Reporting", *Corporate Environmental Strategy,* 7(3): 276-287.

324. Lave, J. and Wenger, E. (1991). *Situated Learning: Legitimate Peripheral Participation.* Cambridge: Cambridge University Press.

325. Law, J. (2001). "Notes on the Theory of the Actor Network: Ordering, Strategy, and Heterogeneity", *Department of Sociology, Lancaster University.* http://www.comp.lancs.ac.uk/sociology/soc054jl.html.

326. Lebart, L., Salem, A. and Berry, L. (1998). *Exploring Textual Data.* Dordrecht: Kluwer Academic Publishers.

327. Lee, Y., Strong, D., Kahn, D. and Wang, R. (2002). "AIMQ: A Methodology for Information Quality Assessment", *Information & Management,* 40(2): 133-146.

328. Leiner, B.M., Cerf, V.G., et al. (2000). "A Brief History of the Internet", *Internet Society.* http://www.isoc.org/internet-history/brief.html.

329. Leitzmann, C. (1996). *Vegetarische Ernährung.* Stuttgart: UTB.

330. Lenz, C. (2003). *Empfängerorientierte Unternehmenskommunikation. Einsatz der Internet-Technologie am Beispiel der Umweltberichterstattung.* Köln: Eul.

331. Lenz, C., Isenmann, R. and Reitz, C. (2001). "Zielgruppenorientierte Umweltberichterstattung von Unternehmen mit XML", *Neue Methoden für das Wissensmanagement im Umweltschutz: 4. Workshop des GI-Arbeitskreises Hypermedia im Umweltschutz und Workshop 3 der GI-Initiative Environmental Markup Language.* K. Tochtermann and W.F. Riekert. Ulm, Germany: Metropolis. 57-69.

332. Lerner, J. and Tirole, J. (2000). "The Simple Economics of Open Source". http://ideas.repec.org/p/nbr/nberwo/7600.html.

333. Lerner, R.M. (2003). "At the Forge: Content Management", *Linux Journal,* 2003(118): http://portal.acm.org/dl.cfm.

334. Lévy, P. (2001). *Cyberculture.* Minneapolis: University of Minnesota Press.

335. Lewis, A. (2002). *Morals, Markets and Money: Ethical, Green and Socially Responsible Finance.* Harlow: Financial Times Prentice Hall.

336. Liedtke, C. and Rohn, H. (1997). Zukunftsfähiges Unternehmen (1): Öko-Audit und Ressourcenmanagement bei der Kambium Möbelwerkstätte GmbH. Wuppertal, Wuppertal Institut für Klima, Umwelt und Energie (Wuppertal Papers No. 69). http://www.wupperinst.org/.

337. Lietaer, B. (2001). *The Future of Money.* London: Century.

338. Line, M., Hawley, H. and Krut, R. (2003). "The Development of Global Environmental and Social Reporting", *Corporate Environmental Strategy,* 9(1): 69-78.

339. Lippman, S. (2001). "Supply Chain Environmental Management", *Environmental Quality Management,* 11(2): 11-14.

340. Lloyd-Laney, M. (2003). Making Knowledge Networks Work for the Poor, Intermediate Technology Consultants. http://www.itcltd.com/final_reports.htm.

341. Lo, C.P. and Yeung, A.K.W. (2002). *Concepts and Techniques of Geographic Information Systems.* New Jersey: Prentice Hall.

342. Long, N. (2000). "Exploring Local/Global Transformations. A View from Anthropology", *Anthropology, Development and Modernities. Exploring Discourses, Counter-Tendencies and Violence.* A. Arc and N. Long. London: Routledge. 184-201.

343. Lorek, S. and Spangenberg, J.H. (2001). Environmentally Sustainable Household Consumption. From Aggregate Environmental Pressures to Indicators for Priority Fields of Action. Wuppertal, Wuppertal Institut für Klima, Umwelt und Energie (Wuppertal Papers No. 117). http://www.wupperinst.org/.

344. Love, J. (2002). "Managing Donor Data Online", *Fundraising on the Internet: The ePhilanthropy-Foundation.org's Guide to Success Online.* M. Warwick et al. San Francisco: Jossey-Bass. 155-169.

345. MacLean, R. and Gottfrid, R. (2000). "Corporate Environmental Reports: Stuck Management Processes Hold Back Real Progress", *Corporate Environmental Strategy,* 7(3): 244-255.

346. Madej, P. (2002). "Map-based Internet Information Services", *16th International Conference Informatics for Environmental Protection, Environmental Communication in the Information Society.* Vienna, Austria: IGU/ISEP. 325-328.

347. Madey, G., Huang, Y., et al. (2003). "Simulation of Biocomplexity: Interactions of NOM, Mineral Surfaces, and Microorganism in Soils", *ASLO 2003 Aquatic Science Meeting*. Salt Lake City, USA. http://aslo.org/meetings/slc2003/.

348. Madron, R. and Jopling, J. (2003). *Gaian Democracies*. Totnes: Green Books.

349. Makower, J., Pernick, R. and Wilder, C. (2003). Clean Energy Trends 2003, Clean Edge. http://www.cleanedge.com/reports/index.php?report=trends2003.

350. Manderfeld, C. (2002). "Umweltinformationen im Internet – Ein 'State of the Art' Bericht", *16th International Conference on Informatics for Environmental Protection (EnviroInfo-2002)*. Vienna, Austria. 2-5.

351. Manstein, C. (2003). "Material Intensity Calculator (MIC)", *MIPS Academy*. http://www.faktor10.at/mipsacademy/.

352. Marshall, S.R. and Brown, D. (2003). "Corporate Environmental Reporting: What's in a Metric?", *Business Strategy and the Environment*, 12(2): 87-106.

353. Marthoz, J.-P. (1999). "Freedom of the Media", *World Information and Communication Report 1999-2000*. M. Tawfik et al. Paris: United Nations Educational, Scientific and Cultural Organization (UNESCO). 72-82.

354. Marx-Gómez, J. and Rautenstrauch, C., Eds. (2001). *Von der Ökobilanzierung bis zur automatisierten Umweltberichterstattung mit Stoffstrommanagementsystemen – eine Fallstudie*. Aachen: Shaker.

355. Maute, S. (2000). Evaluating the Content Management System Idea. Stuttgart, ITM Beratungsgesellschaft mbH: http://www.itm-consulting.de/.

356. Maute, S. (2001). Corporate Design and Corporate Identity of the Project Ecoradar. Stuttgart, ITM Beratungsgesellschaft mbH: http://www.itm-consulting.de/.

357. Maute, S. (2001). Operational Concept of the Technical Infrastructure Behind Ecoradar. Stuttgart, ITM Beratungsgesellschaft mbH: http://www.itm-consulting.de/.

358. Maute, S. (2002). Web Development Kit for Ecoradar. Stuttgart, ITM Beratungsgesellschaft mbH: http://www.itm-consulting.de/.

359. Mayer-Föll, R., Zetzmann, K., et al. (2002). "Untersuchung zum Einsatz von WAP-basierten Diensten im UIS Baden-Württemberg und im Hessischen Landesamt für Umwelt und Geologie", *Ministerium für Umwelt und Verkehr Baden-Württemberg*. http://www.lfu.baden-wuerttemberg.de/.

360. McAndrews, M.A. (2000). *Presidential Campaigns, Political Messages and the Internet: An Analysis of Communication and Design Strategies on Web Sites with Political Content*. Thesis, Washington: Georgetown University.

361. McDonough, W. and Braungart, M. (2002). *Cradle to Cradle: Remaking the Way We Make Things*. New York: North Point Press.

362. McEnery, T. and Wilson, A. (1996). *Corpus Linguistics*. Edinburgh: Edinburgh University Press.

363. McLaughlin, B. (2001). *Java & XML* (2nd ed). Sebastopol: O'Reilly & Associates.

364. McLuhan, M. and Quentin, F. (1967). *The Medium is the Message, An Inventory of Effects*. New York: Random House.

365. Meadows, D. (1998). Indicators and Information Systems for Sustainable Development – A Report to the Balaton Group. Hartland Four Corners, Sustainability Institute. http://www.nssd.net/pdf/Donella.pdf.

366. Meppem, T. and Bourke, S. (1999). "Different Ways of Knowing: A Communicative Turn Toward Sustainability", *Ecological Economics*, 30(3): 389-404.

367. Merrick, J. and Crookshanks, C. (2001). Report on a Survey of Environmental Reporting Costs and Benefits. Tempe, UK Department of Environment, Food, and Rural Affairs (DEFRA). http://www.defra.gov.uk/environment/envrp/environ/environ.pdf.

368. Mertz, T. (1998). *Zur Materialintensität des Personenverkehrs – Eine Analyse nach dem MIPS-Konzept*, Trier: Universität Trier.

369. Mesterharm, M. (2001). *Integrierte Umweltkommunikation von Unternehmen*. Marburg: Metropolis.

370. Midgely, G.E., Hannah, L., et al. (2002). "Assessing the Vulnerability of Species Richness to Anthropogenic Climate Change in a Biodiversity Hotspot", *Global Ecology and Biogeography*, 11(6): 445-451.

371. Millen, D.R., Fontaine, M.A. and Miller, M.J. (2002). "Understanding the Benefit and Costs of Communities of Practice", *Communications of the ACM*, 45(4): 69-73.

372. Mitra, A. (1999). "Characteristics of the WWW Text: Tracing Discursive Strategies", *Journal of Computer-Mediated Communication*, 5(1): http://www.ascusc.org/jcmc/.

373. Mobasher, B., Dai, H., Nakagawa, M. and Luo, T. (2002). "Discovery and Evaluation of Aggregate Usage Profiles for Web Personalization", *Data Mining and Knowledge Discovery*, 6(1): 61-82.

374. Mohammadian, M. (2000). "Bioeconomics: A New Vision for Sustainability", *Sustainable Development*. Elohimjl et al. Munich: Rainer Hampp. 55-63.

375. Mohler, J.L. and Duff, J.M. (2000). *Designing Interactive Web Sites*. New York: Delmar Thomson Learning.

376. Mohtashemi, M. and Mui, L. (2003). "Evolution of Indirect Reciprocity by Social Information: The Role of Trust and Reputation in Evolution of Altruism", *Journal of Theoretical Biology*, 223(4): 523-531.

377. Momirlan, M. and Veziroglu, T.N. (2002). "Current Status of Hydrogen Energy", *Renewable and Sustainable Energy Reviews*, 6: 141-179.

378. Morhardt, E.J. (2002). *Clean, Green, and Read All Over. Ten Rules for Effective Corporate Environmental and Sustainability Reporting*. Milwaukee: ASQ Quality Press.

379. Murray, G. and Costanzo, T. (1999). "Usability and the Web: An Overview", *National Library of Canada*. http://www.nlc-bnc.ca/9/1/p1-260-e.html.

380. Nakicenovic, N. and Swart, R. (2000). IPCC Special Report on Emission Scenarios (SRES), Intergovernmental Panel on Climate Change. http://www.grida.no/climate/ipcc/emission/.

381. National Research Council (2003). Government Data Centers – Meeting Increasing Demands. Washington D.C., The National Academy Press: 30. www.nap.edu.

382. Navarro-Sigüenza, A.G., Peterson, A.T. and Gordillo-Martinez, A. (2002). "A Mexican Case Study on a Centralised Database from World Natural History Museums", *Data Science Journal*, 1(1): 45-53.

383. Navarro-Sigüenza, A.G., Peterson, A.T. and Gordillo-Martínez, A. (2003). "Museums Working Together: The Atlas of the Birds of Mexico", *Why Museums Matter: Avian Archives in an Age of Extinction*. N.J. Collar et al: Bulletin of the British Ornithologists' Club. 207-225.

384. NCSE (2003). "Education, Sustainability and Security Conference Report to be Released on Capitol Hill June 26", *National Council for Science and the Environment*. http://csf.colorado.edu/forums/ncse/current/msg00265.html.

385. NCSE (2003). "Recommendations for Education for a Sustainable and Secure Future", *Third National Conference on Science, Policy, and the Environment*. Washington, USA: National Council for Science and the Environment. http://www.NCSEonline.org/EFS.

386. Nelson, T.H. (1993). *Literary Machines 93.1*. Sausalito: Mindful Press.

387. Nickols, F. (2002). "Communities of Practice, Definition, Indicators & Identifying Characteristics". http://home.att.net/~discon/KM/CoPCharacteristics.htm.

388. Nielsen, J. (2000). *Designing Web Usability: The Practice of Simplicity*. Indianapolis: New Riders Publishing.

389. Nissenbaum, H. (2001). "Securing Trust Online: Wisdom or Oxymoron?", *Boston Law Review*, 81(3): 635-664.

390. NOAA (1999). "World Ocean Atlas 1998", *National Oceanographic Data Center*. http://www.nodc.noaa.gov/OC5/data_woa.html.

391. NOAA (2000). Summary Report: Town Meeting on America's Coastal Future: Using the Internet to Promote Coastal Stewardship. Washington, National Atmospheric and Oceanographic Administration: National Oceans Service. http://oceanservice.noaa.gov/.

392. Nonaka, I., Reinmoeller, P. and Senoo, D. (1998). "The 'ART' of Knowledge: Systems to Capitalize on Market Knowledge", *European Management Journal*, 16(6): 673-684.

393. Nustad, K.G. (2000). "Formell Utvikling og Uformelle Strategier: Et Post-apartheid Utviklingsprosjekt", *Norsk Antropologisk Tidsskrift*, 11(3): 212-226.

394. O'Dell, J. (2003). "Initial Praise for Bush's Hydrogen Fuel Cell Plan", *Los Angeles Times (Jan 29, 2003)*. http://www.latimes.com/news/nationworld/nation/la-na-hydrogen29jan29,0,6067556.story.

395. OECD (2000). *Environmental Performace Reviews (1st Cycle): Conclusions and Recommendations for 32 Countries*. Paris: Organisation for Economic Co-operation and Development.

396. OEGUT (2003). "Nachhaltiges Investment in den Mitarbeitervorsorgekassen", *Austrian Society for Environment and Technology*. http://www.gruenesgeld.at/service/mvk.htm.

397. Oepen, M. (2000). "Environmental Communication in a Context", *Communicating the Environment*. M. Oepen and W. Hamacher. Frankfurt: Peter Lang. 41-61.

398. Olsina, L., Godoy, D., Lafuente, G.J. and Rossi, G. (1999). "Specifying Quality Characteristics and Attributes for Websites", *First ICSE Workshop on Web Engineering (WebE-99)*. Los Angeles, USA.

399. Olson, G.M. and Olson, J.S. (2000). "Distance Matters", *Human-Computer Interaction*, 15(2-3): 139-178.

400. Olympia-Verlag (2001). "Der Markt der Telekommunikation: Marktentwicklung, Zielgruppen, Medien", *Kicker Sportmagazin*. http://www.olympia-verlag.de/kicker/Download/Telekom2001.pdf.

401. Orthofer, R., Humer, H., Kutschera, P. and Loibl, W. (2003). "EnviroMap Austria – a Novel System for an Environmental on the Web", *5th International Symposium on Environmental Software Systems (ISESS'03)*. Semmering, Austria: International Federation of Information Processing, IFIP Conference Series. 122-133.

402. Ostwald, J. (1995). "Knowledge Construction Glossary", *University of Colorado*. http://www.cs.colorado.edu/~ostwald/glossaries/kc-glossary.html.

403. Oxendine, A., Borgida, E., Sullivan, J. and Jackson, M. (2003). "The Importance of Trust and Community in Developing and Maintaining a Community Electronic Network", *International Journal of Human-Computer Studies*, 58(6): 671-696.

404. Page, B. and Rautenstrauch, C. (2001). "Environmental Informatics – Methods, Tools and Applications in Environmental Information Processing", *Environmental Information Systems in Industry and Public Administration*. C. Rautenstrauch and S. Patig. Hershey: Idea Group Publishing. 2-11.

405. Palmer, S. (1998). "Current Issues and Limitations in Using the Internet for Teaching and Learning", *Australian Educational Computing*, 13(1): 12-17.

406. Panetta, F.D. and Dodd, J. (1987). "Bioclimatic Prediction of the Potential Distribution of Skeleton Weed *Chondrilla Juncea* L. in Western Australia", *Journal of the Australian Institute of Agricultural Science*, 53: 11-16.

407. Papes, M. and Peterson, A.T. (2003). "Predicting the Potential Invasive Distribution for *Eupatorium Adenophorum* Spreng. in China", *Journal of Wuhan Botanical Research*, 21(2): 137-142.

408. Parush, A. (2001). "Usability Design and Testing", *Interactions*, 8(5): 13-17.

409. Pattison, F. (2003). Personal Interview (January 15, 2003).

410. Pavlik, J. (1997). "The Future of Online Journalism: Bonanza or Black Hole?", *Columbia Journalism Review*, 36(2): 32-36.

411. PCAST (1998). *Teaming with Life: Investing in Science to Understand and Use America's Living Capital*. Washington: President's Committee of Advisers on Science and Technology.

412. Pearson, R.G., Dawson, T.P., Berry, P.M. and Harrison, P.A. (2002). "SPECIES: A Spatial Evaluation of Climate Impact on the Envelope of Species", *Ecological Modelling*, 154(3): 289-300.

413. Peinel, G., Rose, T. and San Jose, R. (2000). "Customised Information Services for Environmental Awareness in Urban Areas", *IST-2000*. Nice, France. http://istevent.cec.eu.int/.

414. Peng, Z. and Tsou, M. (2003). *Internet GIS: Distributed Geographic Information Services for the Internet and Wireless Networks*. New Jersey: John Wiley & Sons.

415. Percy, L. and Rossiter, J.R. (1992). "Advertising Stimulus Effects: A Review", *Journal of Current Issues in Research in Adverstising*, 14(1): 75-90.

416. Perens, B. (1997). "The Open Source Definition", *Open Source Initiative*. http://www.opensource.org/docs/definition.php.

417. Perrings, C., Williamson, M., et al. (2002). "Biological Invasion Risks and the Public Good: An Economic Perspective", *Conservation Ecology*, 6(1): http://www.consecol.org/vol6/iss1/art1/.

418. Peterson, A.T. (2003). "Projected Climate Change Effects on Rocky Mountain and Great Plains Birds: Generalities of Biodiversity Consequences", *Global Change Biology*, 9(5): 647-655.

419. Peterson, A.T., Ball, L.G. and Cohoon, K.C. (2002). "Predicting Distributions of Mexican Birds Using Ecological Niche Modelling Methods", *Ibis*, 144(1): 27-32.

420. Peterson, A.T. and Kluza, D.A. (2003). "New Distributional Modeling Approaches for Gap Analysis", *Animal Conservation*, 6(1): 47-54.

421. Peterson, A.T., Martínez-Meyer, E., González-Salazar, C. and Hall, P.W. (2003). "Modeled Climate Change Effects on Distributions of Canadian Butterfly Species", *Canadian Journal of Zoology*, Submitted: http://pubs.nrc-cnrc.gc.ca/.

422. Peterson, A.T., Navarro-Sigüenza, A.G. and Benitez-Diaz, H. (1998). "The Need for Continued Scientific Collecting: A Geographic Analysis of Mexican Bird Specimens", *Ibis*, 140(2): 288-294.

423. Peterson, A.T., Ortega-Huerta, M.A., et al. (2002). "Future Projections for Mexican Faunas Under Global Climate Change Scenarios", *Nature*, 416(6881): 626-629.

424. Peterson, A.T., Sanchez-Cordero, V., et al. (2001). "Effects of Global Climate Change on Geographic Distributions of Mexican Cracidae", *Ecological Modelling*, 144(1): 21-30.

425. Peterson, A.T., Scachetti-Pereira, R. and Kluza, D.A. (2003). "Assessment of Invasive Potential of Homalodisca Coagulata in Western North America and South America", *Biota Neotropica*, 3(1): http://www.biotaneotropica.org.br/v3n1/en/.

426. Peterson, A.T. and Shaw, J.J. (2003). "*Lutzomyia* Vectors for Cutaneous Leishmaniasis in Southern Brazil: Ecological Niche Models, Predicted Geographic Distributions, and Climate Change Effects", *International Journal of Parasitology*, 33(9): 919-931.

427. Peterson, A.T., Soberon, J. and Sanchez-Cordero, V. (1999). "Conservatism of Ecological Niches in Evolutionary Time", *Science*, 285(5431): 1265-1267.

428. Peterson, A.T., Stockwell, D.R.B. and Kluza, D.A. (2002). "Distributional Prediction Based on Ecological Niche Modeling of Primary Occurrence Data", *Predicting Species Occurrences: Issues of Scale and Accuracy*. J.M. Scott et al. Washington: Island Press. 617-623.

429. Peterson, A.T., Tian, H., et al. (2003). "Modeling Distributional Shifts of Individual Species and Biomes", *Biodiversity and Climate Change*. T.E. Lovejoy and L. Hannah. New Haven: Yale University Press. In print.

430. Peterson, A.T. and Vieglais, D.A. (2001). "Predicting Species Invasions Using Ecological Niche Modeling", *BioScience*, 51(5): 363-371.

431. Peterson, A.T., Vieglais, D.A., Navarro-Sigüenza, A.G. and Silva, M. (2003). "A Global Distributed Biodiversity Information Network: Building the World Museum", *Why Museums Matter: Avian Archives in an Age of Extinction*. N.J. Collar et al: Bulletin of the British Ornithologists' Club. 186-196.

432. Petkova, E., Maurer, C., et al. (2002). *Closing the Gap - Information, Participation, and Justice in Decision-Making for the Environment*. Washington: World Resources Institute.

433. Pfirman, S. and AC-ERE (2003). "Complex Environmental Systems: Synthesis for Earth, Life, and Society in the 21st Century", *National Science Foundation*. http://www.nsf.gov/geo/ere/ereweb/acere_synthesis_rpt.cfm.

434. Phillips, D. (1995). Correspondence Analysis (= Social Research Update, Issue 7). Guildford: University of Surrey. http://www.soc.surrey.ac.uk/sru/sru7.html.

435. Pick, T., Menger, M., Jensen, S. and Lethen, J. (2000). "Access to Environmental Information – Towards a Digital Global Knowledge Marketplace", *14th Symposium Computer Science for Environmental Protection*. Bonn, Germany: Metropolis. http://enviroinfo.isep.at/UI%20200/ui2000-etc-cds-paper2.el.hsp.pdf.

436. Pickering, A., Ed. (1992). *Science as Practice and Culture*. Chicago: University of Chicago Press.

437. Pillmann, W. (2000). "Environmental Communication – Systems Analysis of Environmentally Related Information Flows as a Basis for the Popularization of the Framework for Sustainable Development", *14th Symposium Computer Science for Environmental Protection*. Bonn, Germany: Metropolis. http://enviroinfo.isep.at/UI%20200/PillmannW270700.el.ath.pdf.

438. Pinkerton, J.P. (2003). "Fill 'er Up With Hydrogen", *Los Angeles Times (April 13, 2003)*. http://pqasb.pqarchiver.com/latimes/.

439. Plewe, B. (1997). *GIS Online: Information Retrieval, Mapping, and the Internet*. Clifton Park, NY: OnWord Press/Delmar Learning.

440. Pohl, I. (2003). "Advocacy on the Web: An Evaluation of a Nonprofit Advocacy Organization's Use of the Web to Support its Mission and Goals", *University of North Carolina at Chapel Hill, School of Information and Library Science*. http://ils.unc.edu/MSpapers/2846.pdf.

441. Porter, M.E. and van der Linde, C. (2000). "Green and Competitive – Ending the Stalemate", *Harvard Business Review on Business and the Environment*. Boston: Harvard Business School Press.

442. Potter, R.F. (1999). "Measuring the 'Bells & Whistles' of a New Medium: Using Content Analysis to Describe Structural Features of Cyberspace", *49th Annual Conference of the International Communication Association (ICA-99)*. San Francisco, USA.

443. Preece, J. (2000). *Online Communities: Designing Usability, Supporting Sociability*. Chichester: John Wiley.

444. Preece, J., Rogers, Y. and Sharp, H. (2002). *Interaction Design: Beyond Human-Computer Interaction*. New York: John Wiley & Sons.

445. Price, J. (2000). "Modeling the Potential Impacts of Climate Change on the Summer Distributions of Massachusetts Passerines", *Bird Observer*, 28(4): 224-230.

446. Pustejovsky, J. (1998). *The Generative Lexicon*. Boston: MIT Press.

447. Pyle, D. (1999). *Data Preparation for Data Mining*. San Francisco: Morgan Kaufmann.

448. Rafalaf, A. (2003). "The Next Big Thing. Is It Finally Time to Invest Green?", *Fortune Small Business*, 13(5): 103-106.

449. Rajesh, R. (2003). Personal Communication, DNRM Listserv Manager. New Delhi, TARU Leading Edge.

450. Ray, E.T. (2001). *Learning XML*. Sebastopol: O'Reilly & Associates.

451. Raymond, E.S. (2000). "The Cathedral and the Bazaar". http://catb.org/~esr/writings/cathedral-bazaar/.

452. Rayport, J.F. and Jaworski, B.J. (2001). *Introduction to E-Commerce*. Boston: McGraw-Hill.

453. Retscher, G. (2002). Diskussion der Leistungsmerkmale von Systemen zur Positionsbestimmung mit Mobiltelefonen als Basis für Location Based Services. *Telekartographie und Location Based Services, Geowissenschaftliche Mitteilungen Nr. 58*. Vienna, Technical University Vienna.

454. Rheingold, H. (1993). *The Virtual Community – Homesteading on the Electronic Frontier*. Reading: Addison-Wesley.

455. Richards, W.G. (2002). "Virtual Screening Using Grid Computing: The Screensaver Project", *Nature Reviews Drug Discovery*, 1(7): 551-555.

456. Richardson, D. (2001). "The Practical Reality of Knowledge Management within Development Initiatives", *International Fund for Agricultural Development's Electronic Networking for Rural Asia/Pacific Projects (ENRAP) 2nd Comprehensive Workshop*. Singapore: TeleCommons Development Group. http://www.telecommons.com/.

457. Richardson, D.M. and McMahon, J.P. (1992). "A Bioclimatic Analysis of *Eucalyptus Nitens* to Identify Potential Planting Regions in Southern Africa", *South African Journal of Science*, 88: 380-387.

458. Rose, O.H. and Bridgewater, P. (2003). "New Approaches Needed to Environmental Education and Public Awareness", *Prospects*, 33(3): 263-272.

459. Rousseau, D. (1990). "Assessing Organizational Culture: The Case for Multiple Methods", *Organizational Climate and Culture*. B. Schneider. San Francisco: Jossey-Bass. 153-194.

460. Rozemeijer, N. (2002). "Network Who? The 'Impact' of Networks on the Participation of Communities in Community-Based Natural Resource Management (CBNRM) in Botswana", *9th Biennial Conference of the International Association for the Study of Common Property (IASCP-2002)*. Victoria Falls, Zimbabwe: International Association for the Study of Common Property. http://www.iascp.org/.

461. Rubin, J. (1994). *Handbook of Usability Testing: How to Plan, Design, and Conduct Effective T* New York: John Wiley & Sons.

462. Ruchter, M. and Düpmeier, C. (2003). "Web-based Environmental Information Systems for the Public: Concepts, Potentials, and Applicability Based on a Case Study", *5th International Symposium on Environmental Software Systems (ISESS'03)*. Semmering, Austria: International Federation of Information Processing, IFIP Conference Series. 37-46.

463. Sabharwal, K. and Nicholson, A.J. (1997). "Internet Communication Technologies and Their Effect on the Environmental Professional", *Practice Periodical of Hazardous, Toxic and Radioactive Waste Management*, 1(3): 120-123.

464. Sachs, W., Ed. (2002). *The Jo'burg Memo, Fairness in a Fragile World*. Berlin: Heinrich Boll Foundation.

465. Sahlins, M. (1972). *Stone Age Economics*. London: Aldine de Gruyter.

466. Salmon, G. (2000). *E-moderating: The Key to Teaching and Learning Online*. London: Kogan Page.

467. Såmmé, J. (1997). "Action Space and Participation. Knowledge Creation and Organizational Development", *20th Information System Research Seminar in Scandinavia (IRIS-20)*. Oslo, Norway: University of Oslo. http://iris.informatik.gu.se/conference/iris20/25.htm.

468. Sargeant, A. (2001). "Web-based Fund Raising: Is Anyone Making any Real Money?", *Fund Raising Management*, 32(8): 20-24.

469. Sargent, P. (1999). "Feature Identities, Descriptors and Handles", *Interoperating Geographic Information Systems. 2nd International Conference, INTEROP'99*. Zurich, Switzerland: Springer. 41-53.

470. Savage, C. (2001). "Forrester Identifies Europe's Finance Portal Opportunity", *Forrester Research*. http://www.forrester.com/ER/Press/Release/0,1769,577,00.html.

471. Sawyer, S. (2001). "A Market-Based Perspective on Information Systems Development", *Communications of the ACM*, 44(11): 97-102.

472. Schach, R.S. (1993). *Software Engineering* (2nd ed). Burr Ridge: Irwin & Asken Associates.

473. Schaltegger, S. and Figge, F. (1998). "Environmental Shareholder Value", *WWZ, Sarasin*. http://www.uni-lueneburg.de/umanagement/pdf-dateien/env_shareh_value.pdf.

474. Scharl, A. (2000). *Evolutionary Web Development*. London: Springer.

475. Schatzki, T.R., Knorr Cetina, K. and Savigny, E.v., Eds. (2001). *The Practice Turn in Contemporary Theory*. New York: Routledge.

476. Schauer, T. (2003). "The Sustainable Information Society – Visions and Risks". http://www.global-society-dialogue.org/saskia.pdf.

477. Scheide, W., Strobel, M., et al. (2001). "Flow Cost Accounting in Practice – ERP-Based Solutions of the ECO-Rapid-Project", *Sustainability in the Information Society – 15th International Symposium Informatics for Environmental Protection*. Zurich, Switzerland. 287-296.

478. Schiff, L., Van House, N.A. and Butler, M. (1997). "Understanding Complex Information Environments: A Social Analysis of Watershed Planning", *ACM Digital Libraries Conference (Digital Libraries '97)*. New York: ACM Press. 161-186.

479. Schmidheiny, S., Chase, R. and De Simone, L. (1997). "Signals of Change: Business Progress Towards Sustainable Development", *World Business Council for Sustainable Development*. http://www.wbcsd.ch/.

480. Schmidt-Bleek, F. (1993). *Wieviel Umwelt braucht der Mensch? MIPS – das Maß für ökologisches Wirtschaften*. Berlin: Birkhäuser.

481. Schmidt-Bleek, F. (1998). *Das MIPS-Konzept: Weniger Naturverbrauch – Mehr Lebensqualität durch Faktor 10*. Munich: Droemer Knaur.

482. Schmidt-Bleek, F., Bringezu, S., et al. (1998). *MAIA: Einführung in die Materialintensitätsanalyse nach dem MIPS-Konzept*. Berlin: Birkhauser.

483. Schmidt-Bleek, F., Liedtke, C., et al. (1992-1995). Bauen und Wohnen. Bausteine zum Schließen einer ökologischen Innovationslücke. Wuppertal, Wuppertal Institut für Klima, Umwelt und Energie. http://www.wupperinst.org/.

484. Schmitz, P.-E. and Castiaux, S. (2002). "Pooling Open Source Software – An IDA Feasibility Study", *European Commission (DG Enterprise)*. http://europa.eu.int/ISPO/ida/export/files/en/1115.pdf.

485. Schraml, T. (1997). *Operationalisierung der ökologieorientierten Berichterstattung aus Sicht des Informationsmanagements. Konzeption eines Vorgehensmodells zur formalisierten Explikation von Dokumentstrukturmodellen im Rahmen der Umweltkommunikation von Unternehmen*, Dresden: Technische Universität Dresden.

486. Schrammel, J. and Sefelin, R. (2003). Enviromap – Ergebnisse eines Usability-Tests. Vienna, CURE Center for Usability Research & Engineering: http://www.cure.at/.

487. Schubert, P. and Ginsburg, M. (2000). "Virtual Communities of Transaction: The Role of Personalization in Electronic Commerce", *Electronic Markets*, 10(1): 45-55.

488. Schulz, T.M. (1995). *Ökologieorientierte Berichterstattung von Unternehmen. Ökologieorientierte Berichterstattung in Geschäfts- und Umweltberichten unter Berücksichtigung der Informationsbedürfnisse der Stakeholder, untersucht in der europäischen Chemieindustrie*, St. Gallen: Universität St. Gallen.

489. Schumacher, E.F. (1974). *Small is Beautiful: Economics as if People Mattered*. London: ABACUS.

490. Schwartz, P. and Randall, D. (2003). "How Hydrogen Can Save America", *Wired*, 11(4): http://www.wired.com/wired/archive/11.04/hydrogen.html.

491. SciAm (2003). "Science of War: Nuclear History", *Scientific American*. http://www.sciam.com/special/toc.cfm?issueid=4.

492. Scott, M. (1998). "WordSmith Tools Manual Version 3.0". http://www.lexically.net/wordsmith/version3/manual.pdf.

493. Scott, P. and Jackson, R. (2002). "Environmental, Social and Sustainability Reporting on the Web: Best Practices", *Corporate Environmental Strategy*, 9(2): 193-202.

494. Scott, W. and Gough, S. (2002). "Re-orienting Teacher Education to Address Sustainability: Reporting on a UNESCO Initiative", *Planet*, December 2002 (Special Edition 4): 7-8.

495. Sehmel, H. (2001). "Designing Web Sites to Educate, Induce Action, and Promote Participatory Communication for Decision-Making", *48th Annual STC Conference*. Chicago, USA: Society for Technical Communication. http://www.stc.org/proceedings/ConfProceed/2001/PDFs/STC48-000129.pdf.

496. Selz, D. and Schubert, P. (1997). "Web Assessment: A Model for the Evaluation and Assesment of Successful Electronic Commerce Applications", *International Journal of Electronic Markets*, 7(3): 46-48.

497. Servaes, J., Ed. (2002). *Walking on the Other Side of the Information Highway: Communication, Culture and Development in the 21st Century*. Penang: Southbound.

498. Shapin, S. (1994). *A Social History of Truth: Civility and Science in Seventeenth-Century England*. Chicago: University of Chicago Press.

499. Sharp, J. (1997). *Communities of Practice: A Review of the Literature*. Rock Hill: John Sharp Associates.

500. Short, P.L. (2003). "Global Top 50", *Chemical & Engineering News*, 81(30): 17-19.

501. SIF (2001). "2001 Report on Socially Responsible Investing Trends in the United States", *Social Investment Forum*. http://www.socialinvest.org/areas/research/trends/SRI_Trends_Report_2001.pdf.

502. Sindel, B.M. and Michael, P.W. (1992). "Spread and Potential Distribution of *Senecio Madagascariensis Pior* (Fireweed) in Australia", *Australian Journal of Ecology*, 17(1): 21-26.

503. Singer, E., Groves, R.M. and Corning, A. (1999). "Differential Incentives: Beliefs About Practices, Perceptions of Equity, and Effects on Survey Participation", *Public Opinion Quarterly*, 63(2): 251-260.

504. Siqueira, M.F.d. and Peterson, A.T. (2003). "Consequences of Global Climate Change for Geographic Distributions of Cerrado Tree Species", *Biota Neotropica*, 3(2): http://www.biotaneotropica.org.br/v3n2/en/.

505. Skillius, Å. and Wennberg, U. (1998). *Continuity, Credibility and Comparability. Key Challenges for Corporate Environmental Performance Measurement and Communication*. Sweden: Lund University.

506. Skov, F. (2000). "Potential Plant Distribution Mapping Based on Climatic Similarity", *Taxon*, 49(3): 503-515.

507. Slater, M. and Steed, A. (2002). "Meeting People Virtually: Experiments in Shared Virtual Environments", *Presence and Interaction in Shared Environments*. R. Schroeder. London: Springer. 146-171.

508. Slocum, T.A. (1999). *Thematic Cartography and Visualization*. New Jersey: Prentice Hall.

509. Small, R.V. and Arnone, M.P. (2000). "Evaluating the Effectiveness of Web Sites", *Human Centered Methods in Information Systems: Current Research and Practice*. A. Clarke and B. Lehaney. Hershey: Idea Group Publishing. 91-101.

510. Smith, G. and Middleton, J. (2003). "Interactive Versus Observational Learning of Spatial Visualisation of Geometric Transformation", *Australian Educational Computing*, 18(1): 3-10.

511. Smith, M. (2002). "Tools for Navigating Large Social Cyberspaces", *Communications of the ACM*, 45(4): 51-55.

512. Smith, M.A. (1992). "Voices from the WELL: The Logic of the Virtual Commons", *University of California at Los Angeles, Department of Sociology*. http://www.sscnet.ucla.edu/soc/csoc/papers/voices/.

513. Smith, W.H.F. and Sandwell, D.T. (1997). "Global Sea Floor Topography from Satellite Altimetry and Ship Depth Soundings", *Science*, 277(5334): 1956-1962.

514. Soberón, J., Arriaga, L. and Lara, L. (2002). Issues of Quality Control in Large, Mixed-origin Entomological Databases. *Technical Report 70*, European Environment Agency.

515. SocialFunds (2000). "Investors Prod CEO's to Adopt New Reporting Standards", *SocialFunds*. http://www.socialfunds.com/news/article.cgi/article433.html.

516. Soeftestad, L.T. (1999). "Workshop Report", *The International Workshop on Community-based Natural Resource Management (CBNRM)*. Washington, USA: World Bank. http://www.cbnrm.net/pdf/soeftestad_l_001_98wbreport.pdf.

517. Soeftestad, L.T. (2001). "Aligning Needs and Means: On Culture, ICT and Knowledge in Development Cooperation", *24th Information System Research Seminar in Scandinavia (IRIS-24)*. Ulvik in Hardanger, Norway: University of Bergen. 47-60.

518. Soeftestad, L.T. (2001). "Community-Based Natural Resource Management: Knowledge Management and Knowledge Sharing in the Age of Globalization", *The Community-Based Natural Resource Management Network (CBNRM Net)*. http://www.cbnrm.net/.

519. Soeftestad, L.T. (2002). "CBNRM Net: Knowledge Management and Networking for the Global CBNRM Community of Practice", *9th Biennial Conference of the International Association for the Study of Common Property (IASCP-2002)*. Victoria Falls, Zimbabwe: International Association for the Study of Common Property. http://www.iascp.org/.

520. Soeftestad, L.T. and Sein, M.K. (2004). "ICT and Development: East is East and West is West and the Twain May Yet Meet", *The Digital Challenge: Information Technology in the Development Context*. S. Krishna and S. Madon. Aldershot: Ashgate. In print.

521. Someya, Y. (1998). "e_lemma.txt". http://www.lexically.net/downloads/e_lemma.zip.

522. Spencer, T. (2002). "The Potential of the Internet for Non-profit Organizations", *First Monday*, 7(8): http://www.firstmonday.dk/issues/issue7_8/spencer/index.html.

523. Spencer-Cooke, A. (1995). "Engaging Stakeholders. The Next Challenge in Corporate Environmental Reporting", *IÖW/VÖW-Informationsdienst*, 10(3-4): 4-5.

524. SPREP (1997). "Action Plan for Managing the Environment of the South Pacific Region 1997-2000", *South Pacific Regional Environment Programme*. http://www.sprep.org.ws/.

525. Stainforth, D., Kettleborough, J., et al. (2002). "Distributed Computing for Public-Interest Climate Modeling Research", *Computing in Science and Engineering*, 4(3): 82-89.

526. Stainforth, D., Kettleborough, J., et al. (2002). "Climateprediction.net: Design Principles for Public-Resource Modeling", *14th IASTED International Conference on Parallel and Distributed Computing and Systems*. Cambridge, USA. http://www.iasted.org/conferences/2002/cambridge/pdcs.htm.

527. Stanford, J., Tauber, E.R., Fogg, B.J. and Marable, L. (2002). Experts vs. Online Consumers: A Comparative Credibility Study of Health and Finance Web Sites. New York, Consumer Web Watch. http://www.consumerwebwatch.org/.

Stanoekska-Slaveba, K. and Schmid, B. (2000). "Internet Electronic Product Catalogs: An Ap-·oach Beyond Simple Keywords and Multimedia", *Computer Networks*, 32(6): 701-716.

529. Star, S.L. (1989). "The Structure of Ill-structured Solutions: Boundary Objects and Heterogeneous Distributed Problem Solving", *Distributed Artificial Intelligence* 2. L. Gasser and M. Huhns London: Pitman Publishing. 37-54.

530. Star, S.L. (1995). "Introduction", *Ecologies of Knowledge: Work and Politics in Science and Technology*. S.L. Star. Albany: State University of New York Press. 1-35.

531. Star, S.L. and Griesmer, J.R. (1989). "Institutional Ecology, 'Translations', and Boundary Objects: Amateurs and Professionals in Berkeley's Museum of Vertebrate Zoology, 1907-39", *Social Studies of Science*, 19(3): 387-420.

532. Star, S.L. and Strauss, A. (1999). "Layers of Silence, Arenas of Voice: The Ecology of Visible and Invisible Work", *Computer Supported Cooperative Work*, 8(1-2): 9-39.

533. StatSoft (2000). Correspondence Analysis. Tulsa, StatSoft Inc. http://www.statsoftinc.com/textbook/stcoran.html.

534. Stein, M. (2001). "Environmental Defense: From Brochureware to Actionware", *Benton Foundation*. http://www.benton.org/publibrary/practice/features/environdefense.html.

535. Steven, M., Schwarz, E. and Letmathe, P. (1997). *Umweltberichterstattung und Umwelterklärung nach der EG-Öko-Audit-Verordnung. Grundlagen, Methoden und Anwendungen*. Berlin: Springer.

536. Stewart, S. (1999). Not for Profit. *Industry Standard (December 6, 1999)*: 136. http://www.thestandard.com/.

537. Stillar, G.F. (1998). *Analyzing Everyday Texts – Discourse, Rhetoric, and Social Perspectives*. Thousand Oaks: Sage.

538. Stiller, H. and Liedtke, C. (2000). "Redesigning Commerce, Reframing the Market", *People and Nature: Operationalising Ecological Economics (ISEE-2000)*. Canberra, Australia. http://incres.anu.edu.au/anzsee/ISEE2000/.

539. Stilwell, F. (1994). "Economic Rationalism, Cities and Regions", *Australian Journal of Regional Studies*, 7: 54-65.

540. Stockwell, D.R.B. and Peterson, A.T. (2002). "Controlling Bias in Diodiversity Data", *Predicting Species Occurrences: Issues of Scale and Accuracy*. J.M. Scott et al. Washington: Island Press. 537-546.

541. Stockwell, D.R.B. and Peterson, A.T. (2002). "Effects of Sample Size on Accuracy of Species Distribution Models", *Ecological Modelling*, 148(1): 1-13.

542. Stockwell, D.R.B. and Peterson, A.T. (2003). "Comparison of Resolution of Methods used in Mapping Biodiversity Patterns from Point Occurrence Data", *Ecological Indicators*, 3(3): 213-221.

543. Stokes, J. (2003). *How to do Media and Cultural Studies*. London: Sage.

544. Stone, B.K., Guerard, J.B., Gultekin, M.N. and Adams, G. (2001). "Socially Responsible Investment Screening: Strong Evidence of No Significant Cost for Actively Managed Portfolios". http://www.socialinvest.org/areas/research/Moskowitz/2001_HM_full.pdf.

545. Stone, P.J. (2002). "The General Inquirer". http://www.wjh.harvard.edu/~inquirer/.

546. Stone, P.J., Dunphy, D.C., Smith, M.S. and Ogilvie, D.M. (1966). *The General Inquirer: A Computer Approach to Content Analysis*. Cambridge: MIT Press.

547. Stott, P.A. and Kettleborough, J.A. (2002). "Origins and Estimates of Uncertainty in Predictions of Twenty-First Century Temperature Rise", *Nature*, 416(6882): 723-726.

548. Stott, P.A., Tett, S.F.B., et al. (2000). "External Control of 20th Century Temperature by Natural and Anthropogenic Forcings", *Science*, 290(2): 133-137.

549. Strauss, J. and Frost, R. (2000). *E-Marketing* (2nd ed). Upper Saddle River: Prentice Hall.

550. Suchman, L.A. (2001). "Located Accountabilities in Technology Production", *Department of Sociology, Lancaster University*. http://www.comp.lancs.ac.uk/sociology/soc039ls.html.

551. Surman, M. and Wershler-Henry, D. (2002). *Commonspace: Beyond Virtual Community*. Upper Saddle River: Prentice Hall.

552. SustainAbility (2002). *Good News & Bad – The Media, Corporate Social Responsibility and Sustainable Development*. London: Beacon Press.

553. SustainAbility and UNEP (1999). *Engaging Stakeholders 1999. The Internet Reporting Report*. London: Beacon Press.

554. Swartz, O. (1998). *The Rise of Rhetoric and Its Intersections with Contemporary Critical Thought.* Boulder: Westview.

555. Tatnall, A. and Davey, B. (2002). "Information Systems Curriculum Development as an Ecological Process", *IT Education: Challenges for the 21st Century.* E. Cohen. Hershey: Idea Group Publishing. 206-221.

556. Taylor, M. (1991). "Economic Restructuring and Regional Change in Australia", *Australian Geographical Studies,* 29(2): 255-267.

557. Tehranian, M. (1999). *Global Communication and World Politics - Domination, Development, and Discourse.* Boulder: Lynne Rienner.

558. Thioulouse, J., Chessel, D., Dolédec, S. and Olivier, J.-M. (1997). "ADE-4: A Multivariate Analysis and Graphical Display Software", *Statistics and Computing,* 7(1): 75-83.

559. Thomas, I. and Nicita, J. (2002). "Sustainable Education and Australian Universities", *Environmental Education Research,* 8(4): 475-490.

560. Thomas, J., Grainger, G. and Koomen, K. (1999). "Impact on the Media", *World Information and Communication Report 1999-2000.* M. Tawfik et al. Paris: United Nations Educational, Scientific and Cultural Organization (UNESCO). 116-130.

561. Titscher, S., Wodak, R., Meyer, M. and Vetter, E. (1998). *Methoden der Textanalyse: Leitfaden und Überblick.* Opladen: Westdeutscher Verlag.

562. Tochtermann, K. and Maurer, H. (2000). "Umweltinformatik und Wissensmanagement - Ein Überblick", *14th Symposium Computer Science for Environmental Protection.* Bonn, Germany: Metropolis. http://enviroinfo.isep.at/UI%202000/tochtermann_03072000.sk.pdf.

563. Tonts, M. (1997). "Changing Government Approaches to Rural Development: Perspectives from the Western Australian Wheatbelt", *Space and Polity,* 1(2): 171-190.

564. Travers, A. and Decker, E. (1999). "New Technology and Critical Pedagogy", *Radical Pedagogy,* 1(2): http://radicalpedagogy.icaap.org/content/issue1_2/.

565. Tuldava, J. (1993). "The Statistical Structure of a Text and its Readability", *Quantitative Text Analysis.* G. Altmann and L. Hrebícek. Trier: Wissenschaftlicher Verlag Trier. 215-227.

566. UNECE (1998). "Convention on Access to Information, Public Participation in Decision-making and Access to Justice in Environmental Matters", *United Nations Economic Commission for Europe.* http://www.unece.org/env/pp/.

567. UNEP (1992). "Rio Declaration on Environment and Development", *United Nations Environment Programme.* http://www.unep.org/Documents/Default.asp?DocumentID=78.

568. UNEP (1994). Company Environmental Reporting. A Measure of the Progress of Business and Industry Towards Sustainable Development. *Technical Report 24.* Paris, United Nations Environment Programme. http://www.unep.org/.

569. United Nations (1992). "Agenda 21", *United Nations Division for Sustainable Development.* http://www.un.org/esa/sustdev/documents/agenda21/.

570. United Nations Development Programme (1999). "Human Development in this Age of Globalization", *Human Development Report 1999.* New York: Oxford University Press. 25-56.

571. Vallega, A. (1999). *Fundamentals of Integrated Coastal Management.* Dordrecht: Kluwer Academic.

572. Van Dalen, M. (1997). *Company Environmental Reporting. Conditions for the Optimal Information Structure of Environmental Reports,* Humberside, Flushing: University of Humberside and Zeeland Polytechnic.

573. Van der Heijden, P.G.M., Mooijart, A. and Takane, Y. (1994). "Correspondence Analysis and Contingency Table Models", *Correspondence Analysis in the Social Sciences.* M. Greenacre and J. Blasius. London: Academic Press. 79-111.

574. Van der Velden, M. (2002). "Knowledge Facts, Knowledge Fiction: the Role of ICTs in Knowledge Management for Development", *Journal of International Development,* 14(1): 25-37.

575. Van House, N.A. (1995). "User Needs Assessment and Evaluation for the UC Berkeley Electronic Environmental Library Project", *Digital Libraries '95.* College Station, USA: Hypermedia Research Laboratory. http://www.csdl.tamu.edu/DL95/contents.html.

Van House, N.A. (2002). "Digital Libraries and Practices of Trust: Networked Biodiversity Information", *Social Epistemology,* 16(1): 99-114.

577. Van House, N.A. (2002). "Trust and Epistemic Communities in Biodiversity Data Sharing", *Second ACM/IEEE-CS Joint Conference on Digital Libraries.* New York, USA: ACM Press. 231-249.

578. Van House, N.A. (2003). "Digital Libraries and Collaborative Knowledge Construction", *Digital Library Use: Social Practice in Design and Evaluation.* A.P. Bishop et al. Cambridge: MIT Press. 271-295.

579. Van House, N.A. (2003). "Science and Technology Studies and Information Studies", *Annual Review of Information Science and Technology.* B. Cronin. Washington, DC: American Society for Information Science and Technology. 3-86.

580. Van House, N.A., Butler, M. and Schiff, L. (1998). "Cooperative Knowledge Work and Practices of Trust: Sharing Environmental Planning Data Sets", *The ACM Conference on Computer Supported Cooperative Work (CSCW-1998).* Seattle, USA: ACM Press. 335-343.

581. Van House, N.A., Butler, M.H., Ogle, V. and Schiff, L. (1996). User-Centered Iterative Design for Digital Libraries: the Cypress Experience. *DLIB Magazine (February 1996)*: http://www.dlib.org/dlib/february96/02vanhouse.html.

582. Vanclay, F. and Lawrence, G. (1995). "The Restructuring of Agriculture: Environmental and Social Implications", *The Environmental Imperative: Eco-social Concerns for Australian Agriculture.* F. Vanclay and G. Lawrence. Rockhampton: Central Queensland University Press. 3-19.

583. VBDO (1998). "Environmental Information for Investors", *Dutch Association of Investors for Sustainable Development.* http://www.vbdo.nl/publicaties.asp.

584. Veen, J. (1997). "Hotwired Style: Principles for Building Smart Websites". San Francisco: Hardwired.

585. Veen, J. (2000). *The Art and Science of Web Design.* Indianapolis: New Riders Publishing.

586. Vergano, D., McQuillan, L. and O'Donnell, J. (2003). "Hydrogen-Fueled Cars Face Obstacle-Filled Road", *USA Today (Jan 29, 2003).* http://www.usatoday.com/money/autos/2003-01-29-hydrogen_x.htm.

587. Vidal, J. (1999). Modem Warfare: E-Mail and the Net are Revolutionising the Way Environment, Human Rights and Social Justice Groups Work. *The Guardian (Jan 13, 1999).* http://www.guardian.co.uk/Get_The_Net/Feature/0,3868,322672,00.html.

588. Vieglais, D.A., Wiley, E.O., Robins, C.R. and Peterson, A.T. (2000). "Harnessing Museum Resources for the Census of Marine Life: The FISHNET Project", *Oceanography,* 13(3): 10-13.

589. Visser, U., Stuckenschmidt, H., Wache, H. and Vögele, T. (2001). "Using Environmental Information Efficiently: Sharing Data and Knowledge from Heterogeneous Sources". http://citeseer.nj.nec.com/309536.html.

590. Vodafone (2003). "PASSO Stau-Info", *Vodafone D2 GmbH.* http://www.vodafone.de/kundenbetreuung_services/unterwegs/9052.html.

591. Vogel, D. (2003). "The Official PostNuke Installation & Getting Started Guide". http://download.postnuke.com/.

592. Vollmer, S.A.M. (1995). *EG-Öko-Audit-Verordnung Umwelterklärung: Anforderungen, Hintergründe, Gestaltungsoptionen.* Berlin: Springer.

593. Von Weizsacker, E., Lovins, A. and Lovins, L. (1997). *Factor Four, Doubling Wealth, Halving Resource Use.* London: Earthscan.

594. W3C (2002). "Document Object Model", *World Wide Web Consortium.* http://www.w3.org/DOM/.

595. Wackernagel, M. and Rees, W. (1996). *Our Ecological Footprint – Reducing Human Impact on the Earth.* Garbiola Island: New Society.

596. Wade, R. (2002). "Sustainable Development Education and Curriculum 2000", *Planet,* December 2002 (Special Edition 4): 4-7.

597. Waisbord, S. (2001). "Family Tree of Theories, Methodologies and Strategies in Development Cooperation", *Rockefeller Foundation.* http://www.comminit.com/pdf/familytree.pdf.

598. Walke, B. (1998). *Mobilfunknetze und ihre Protokolle.* Stuttgart: B.G. Teubner.

599. Wallace, M.G., Cortner, H.J., Moote, M.A. and Burke, S. (1996). "Moving Toward Ecosystem Management: Examining a Change in Philosophy for Resource Management", *Journal of Political Ecology,* 3: 18-32.

600. Ward, M. (2003). "Scot Aims for High Office", *BBC News (June 14, 2001).* http://news.bbc.co.uk/1/hi/sci/tech/1387082.stm.

601. Warr, W.A. (2002). "Warr Zone 2001: Science Portals on the Internet", *Wendy Warr & Associates.* http://www.warr.com/wzarc01.html.

602. Warwick, M., Hart, T. and Allen, N. (2002). *Fundraising on the Internet: The ePhilanthropyFoundation.org's Guide to Success Online* (2nd ed). San Francisco: Jossey-Bass.

603. Watson, R.T., Zinyowera, M.C. and Moss, R.H., Eds. (1996). *Climate Change 1995: Impacts, Adaptations and Mitigation of Climate Change: Scientific-Technical Analyses.* Contribution of Working Group II to the Second Assessment Report of the Intergovernmental Panel on Climate Change. Cambridge: Cambridge University Press.

604. Watson, T. (2002). "E-mail Campaigns and E-Newsletters", *Fundraising on the Internet: The ePhilanthropyFoundation.org's Guide to Success Online.* M. Warwick et al. San Francisco: Jossey-Bass. 75-84.

605. WBCSD (2003). *Annual Review.* Geneva: World Business Council for Sustainable Development.

606. Weedman, J. (1998). "The Structure of Incentive: Design and Client Roles in Application-oriented Research", *Science, Technology, and Human Values,* 23(3): 315-354.

607. Weingart, P. (2002). "The Moment of Truth for Science", *EMBO reports,* 3(8): 703-706.

608. Weinreich, N.K. (1999). *Hands-on Social Marketing: A Step-by-Step Guide.* Thousand Oaks: SAGE.

609. Weizenbaum, J. (1966). "Eliza – A Computer Program for the Study of Natural Language Communication Between Man and Machine", *Communications of the ACM,* 9(1): 36-45.

610. Wellman, B. (2001). "Computer Networks as Social Networks", *Science,* 293(5537): 2031-2034.

611. Wellman, B. (2001). "Physical Place and Cyberplace: The Rise of Personalized Networking", *International Journal of Urban and Regional Research,* 25(2): 227-252.

612. Wellman, B. (2002). "Designing the Internet for a Networked Society", *Communications of the ACM,* 45(5): 91-96.

613. Wenger, E. (1998). *Communities of Practice: Learning, Meaning, and Identity.* New York: Cambridge University Press.

614. Wenger, E. (2000). "Communities of Practice and Social Learning Systems", *Organization,* 7(2): 225-246.

615. Wenger, E., McDermott, R. and Snyder, W. (2002). *Cultivating Communities of Practice.* Cambridge: Harvard Business School Press.

616. Wennberg, U. (1996). International Survey of Environmental Reporting 1996. Stockholm, Lund University, KPMG Environmental Advisors.

617. Wernecke, J. (1994). *The Inventor Mentor: Programming Object Oriented Graphics with Open Inventor.* Reading: Addison Wesley.

618. Western Australian Technology and Industry Advisory Council (2000). Drivers and Shapers of Economic Development in Western Australia in the 21st Century. Perth, Western Australian Technology and Industry Advisory Council. http://www.wa.gov.au/tiac/drivers/.

619. Wheeler, D. and Elkington, J. (2001). "The End of the Corporate Environmental Report? Or the Advent of Cybernetic Sustainability Reporting and Communication", *Business Strategy and the Environment,* 10(1): 1-14.

620. Whitten, N.E. and Wolfe, A.W. (1973). "Network Analysis", *Handbook of Social and Cultural Anthropology.* J. Honigmann. Chicago: Rand McNally. 717-746.

621. Wilcox, S.K. (1998). *Web Developer.com – Guide to 3D Avatars.* New York: John Wiley Computer Publishing.

622. Wiley, E.O., McNyset, K., et al. (2003). "Niche Modeling and Geographic Range Predictions in the Marine Environment Using a Machine-Learning Algorithm", *Oceanography,* 16(3): 120-127.

623. Williams, K.D., Senior, C.A. and Mitchell, J.F.B. (2001). "Transient Climate Change in the Hadley Centre Models: The Role of Physical Processes", *Journal of Climate,* 14(12): 2659-2674.

624. Wilson, K.M. (2000). "Communicating Climate Change through the Media – Predictions, Politics and Perceptions of Risk", *Environmental Risks and the Media.* S. Allan et al. London, New York: Routledge. 201-217.

625. Wilson, P. (1983). *Second-hand Knowledge: An Inquiry into Cognitive Authority.* Westport, CT: Greenwood Press.

626. Witherspoon, E.M. (1999). "A Pound of Cure: A Content Analysis of Health Information on Web Sites of Top-Ranked HMOs", *49th Annual Conference of the International Communication Association (ICA-99).* San Francisco, USA.

627. Wöber, K.W., Scharl, A., Natter, M. and Taudes, A. (2002). "Success Factors of European Tourism Web Sites", *9th International Conference on Information Technology and Travel & Tourism (ENTER-2002).* Vienna, Austria: Springer.

628. Woodside, A.G. and Wilson, E.J. (2002). "Respondent Inaccuracy", *Journal of Advertising Research,* 42(5): 7-18.

629. World Bank (2000). Entering the 21st Century. World Development Report 1999/2000. Washington, World Bank. http://www.worldbank.org/wdr/2000.

630. World Bank (2000). What is Knowledge Management? A Background Paper to the World Development Report. Washington, World Bank. http://publications.worldbank.org/.

631. World Bank (2003). Task Manager's ICT Toolkit: A Route Map for ICT Components in World Bank Projects. Washington, World Bank. http://publications.worldbank.org/.

632. World Bank (2003). Task Manager's ICT Toolkit: Good Practice for Planning, Delivering, and Sustaining ICT Projects. Washington, World Bank. http://publications.worldbank.org/.

633. World Commission on Environment and Development (1987). *Our Common Future.* Oxford: Oxford University Press.

634. World Wide Web Consortium (2003). "Scalable Vector Graphics (SVG) 1.1 Specification", *World Wide Web Consortium.* http://www.w3.org/TR/SVG11.

635. Worsley, P. (1997). *Knowledges. What Different People Make of the World.* London: Profile Books.

636. WRI (2003). "World Resources 2002-2004: Decisions for the Earth – Balance, Voice, and Power", *World Resources Institute.* http://pubs.wri.org/pubs_description.cfm?PubID=3764.

637. Wrolstad, J. (2002). "Environment-Friendly Chip Designs in the Works". http://www.newsfactor.com/perl/story/17573.html.

638. Wurbs, J., Nickel, R., Rohn, H. and Liedtke, C. (1996). Materialintensitätenanalysen von Grund-, Werk- und Baustoffen (5): Der Werkstoff Glas. Materialintensität von Behälterglas und Flachglas. Die 0,33l Verpackung aus Aluminium, Weißblech und Glas. Wuppertal, Wuppertal Institut für Klima, Umwelt und Energie (Wuppertal Papers No. 64). http://www.wupperinst.org/.

639. Wycherley, I. (1999). "Greening Supply Chains: The Case of the Body Shop International", *Business Strategy and the Environment,* 8(2): 120-127.

640. Xiang, X., Madey, G., Huang, Y. and Cabaniss, S. (2003). "A Stochastic Simulation of Natural Organic Matter and Microbes in the Environment", *2003 World Conference on Natural Resource Modeling.* Beaufort, USA. http://www.nd.edu/~nom/Papers/papers.html.

641. Zalba, S.M., Sonaglioni, M.I. and Belenguer, C.J. (2000). "Using a Habitat Model to Assess the Risk of Invasion by an Exotic Plant", *Biological Conservation,* 93(2): 203-208.

642. Zhang, X., Keeling, K.B. and Pavur, R.J. (2000). "Information Quality of Commercial Web Site Home Pages: An Explorative Analysis", *21st International Conference on Information Systems (ICIS-2000).* Atlanta, USA: Association for Information Systems. 164-175.

643. Zook, M.A. (2001). "Putting It Together: 'Connected' is a Matter of Geography", *netWorker,* 5(3): 13-17.

Online Resources

644. *ABYZ News Links.* http://www.abyznewslinks.com/.

645. *Agricultural Resources Center & Pesticide Education Project (ARC / PESTed).* http://www.pested.org/.

646. *Air Liquide Group.* http://www.airliquide.com/en/.

647. *Akzo Nobel.* http://www.akzonobel.com/.

648. *Allensbacher Computer- und Telekommunikationsanalyse.* http://www.acta-online.de/.

649. *AltaVista.* http://www.altavista.com/.

650. *Apache Cocoon.* http://cocoon.apache.org.

651. *Apache Tomcat.* http://jakarta.apache.org/tomcat/.

652. *APNEE Take-Up Trials.* http://www.apnee.org/.

653. *APNEE Thessaloniki Regional Server.* http://www.apnee.gr/.

654. *Austrian Federal Ministry of Agriculture, Forestry, Environment, and Water Management.* http://www.lebensministerium.at/en/.

655. *Austrian Society for Environment and Technology (OEGUT).* http://www.oegut.at/.

656. *Baltic Environmental Atlas.* http://maps.grida.no/baltic/.

657. *Bank of English.* http://titania.cobuild.collins.co.uk/boe_info.html.

658. *Barron's Online.* http://www.barrons.com/.

659. *BASF.* http://www.basf.com/.

660. *Bayer.* http://www.bayer.com/.

661. *BBC News World Edition.* http://news.bbc.co.uk/.

662. *Berkeley Open Infrastructure for Network Computing (BOINC).* http://boinc.berkeley.edu/.

663. *British National Corpus.* http://www.natcorp.ox.ac.uk/.

664. *Business Week.* http://www.businessweek.com/.

665. *CalFlora.* http://www.calflora.org/.

666. *CBCRM Resource Center.* http://www.cbcrmlearning.org/.

667. *CBNRM Asia Virtual Resource Centre.* http://www.cbnrmasia.org/.

668. *CBNRM Network.* http://www.cbnrm.net/.

669. *CBNRM Networking.* http://www.cbnrm.org/.

670. *CBNRM Support Programme in Botswana.* http://www.cbnrm.bw/.

671. *CNN International.* http://www.cnn.com/.

672. *CNN Money.* http://money.cnn.com/.

673. *Coastal Resource Inventory.* http://srmwww.gov.bc.ca/dss/coastal/mris/resource.htm.

674. *CoastalManagement.* http://www.coastalmanagement.com/.

675. *CoastBase.* http://www.coastbase.org/.

676. *Coin3D.* http://www.coin3d.org/.

677. *Common Property Resource Management Network (CPRNet).* http://www.cbnrm.net/webhosting/cprnet/.

678. *Cosm Project.* www.mithral.com/projects/cosm.

679. *Dainippon Ink and Chemicals.* http://www.dic.co.jp/eng/.

680. *Darwin Core.* http://speciesanalyst.net/docs/dwc/.

Degussa. http://www.degussa.com/.

ktopGarp. http://www.lifemapper.org/desktopgarp/.

683. *Distributed Generic Information Retrieval (DiGIR).* http://www.digir.net/.
684. *Distributed.net.* http://www.distributed.net/.
685. *Domini Social Investments.* http://www.domini.com/.
686. *Dow Chemical Company.* http://www.dow.com/.
687. *DuPont.* http://www.dupont.com/.
688. *Earth Sciences Portal.* http://esdcd.gsfc.nasa.gov/ESD/portal/.
689. *Earth Simulator Center.* www.es.jamstec.go.jp/.
690. *EarthCharter.* http://www.earthcharter.org/.
691. *EarthTrends.* http://earthtrends.wri.org/.
692. *eBay.* http://www.ebay.com/.
693. *EcoInvest.* http://www.ecoinvest.org/.
694. *Eco-Portal.* http://www.eco-portal.com/.
695. *Ecoradar.* http://www.oekoradar.de/.
696. *ECOresearch Network.* http://www.ecoresearch.net/.
697. *Encyclopaedia Britannica.* http://www.britannica.com/.
698. *EnviroMap Austria.* http://www.enviromap.at/.
699. *Environmental Investors Network.* http://www.environmentalinvestors.com/.
700. *Environmental News Network.* http://www.enn.com/.
701. *Environmental Online Communication.* http://www.ECOresearch.net/springer/.
702. *eSeFDee Marine Sciences Portal.* http://www.dvz.be/Portal/.
703. *EU Water Framework Directive.* http://europa.eu.int/comm/environment/water/.
704. *Fida Distributed Computing.* http://fida.chem.washington.edu/.
705. *FishNet.* http://speciesanalyst.net/fishnet/.
706. *Fortune Magazine.* http://www.fortune.com/.
707. *FRAME.* http://www.frameweb.org/.
708. *GEO Data Portal.* http://geodata.grid.unep.ch/.
709. *Global Eco Database (EcoDB).* http://www.ecodb.com/.
710. *Google.* http://www.google.com/.
711. *Great Internet Mersenne Prime Search (GIMPS).* http://www.mersenne.org/.
712. *Green Century Funds.* http://www.greencentury.com/.
713. *GreenMoney.* http://www.greenmoney.at.
714. *GruenesGeld.* http://www.gruenesgeld.at/.
715. *GuideStar.* http://www.guidestar.org/.
716. *HerpNet.* http://herpnet.org/.
717. *Huntsman.* http://www.huntsman.com/.
718. *IDRC CBNRM Asia.* http://www.idrc.ca/cbnrm/.
719. *Imperial Chemical Industries (ICI).* http://www.ici.com/.
720. *Infrastructure for Spatial Information in Europe.* http://inspire.jrc.it/.
721. *International Council of Chemical Associations.* http://www.icca-chem.org/.
722. *International Herald Tribune.* http://www.iht.com/.
723. *Internet2.* http://www.internet2.edu/.
724. *Invesco Funds.* http://www.invescofunds.com/.
725. *InvestorWeb.* http://www.investorweb.com.au/.
726. *Jakarta Turbine.* http://jakarta.apache.org/turbine/.
727. *Jakarta Velocity.* http://jakarta.apache.org/velocity/.
728. *Java 2 Platform, Enterprise Edition.* http://java.sun.com/j2ee/.

729. *Java 2 Platform, Micro Edition.* http://java.sun.com/j2me/.

730. *Johannesburg 2002 – World Summit on Sustainable Development.* http://www.johannesburgsummit.org/.

731. *Julius Baer Group.* http://www.juliusbaer.com/.

732. *KFUE-WAP.* www.lfu.baden-wuerttemberg.de/wap-kfue/servlet/kfue/.

733. *Kidon Media Link.* http://www.kidon.com/media-link/.

734. *Lifelihoods Connect.* http://www.livelihoods.org/.

735. *Lifemapper.* http://www.lifemapper.org/.

736. *Living Directory Network.* http://www.livingdirectory.net/.

737. *Local Exchange Trading Systems (LETS).* http://www.gmlets.u-net.com/.

738. *Los Angeles Times.* http://www.latimes.com/.

739. *Mammal Networked Information System (MaNIS).* http://elib.cs.berkeley.edu/manis/.

740. *MekongInfo.* http://www.mekonginfo.org/.

741. *MFA Online Questionnaire.* http://www.umweltoekonomie.at/mfa/en/.

742. *Microbiology Information Portal.* http://www.microbes.info/.

743. *Mitsui Chemicals.* http://www.mitsui-chem.co.jp/e/.

744. *MUSE Project.* http://biodiversity.uno.edu/muse/.

745. *MySQL Database Management System.* http://www.mysql.com/.

746. *National Atlas of the United States.* http://nationalatlas.gov/.

747. *National Environmental Directory.* http://www.environmentaldirectory.net/.

748. *NEODAT Project.* http://www.neodat.org/.

749. *NewsLink.* http://newslink.org/.

750. *Newsweek.* http://www.newsweek.com/.

751. *O3-WAP Service.* http://wap.hlug.de/.

752. *Ocean Biogeographic Information System (OBIS).* http://www.iobis.org/.

753. *OneWorld.* http://www.oneworld.net/.

754. *Open Money Project.* http://www.openmoney.org/.

755. *Oracle Collaboration Suite.* http://otn.oracle.com/products/cs/.

756. *Oracle Corporation.* http://www.oracle.com/.

757. *Panchayati Raj and Decentralized Natural Resource Management.* http://www.panchayats.org/.

758. *Partnership for Principle 10.* http://www.pp10.org/.

759. *PlaNetwork.* http://www.planetwork.net/.

760. *Pollution Watch.* http://www.pollutionwatch.org/.

761. *PostgreSQL.* http://www.postgresql.org/.

762. *Postnuke Content Management System.* http://www.postnuke.com/.

763. *Refinery Reform Campaign.* http://www.toxictwostep.org/.

764. *Reuters.* http://www.reuters.com/.

765. *Santa Fe Institute.* http://www.santafe.edu/.

766. *Saudi Basic Industries Corporation (SABIC).* http://www.sabic.com/.

767. *Scientific American.* http://www.sciam.com/.

768. *SETI At Home.* http://setiathome.berkeley.edu/.

769. *Skladka.* http://www.skladka.cz/.

770. *Slashdot.* http://slashdot.org/.

SmartMoney. http://www.smartmoney.com/.

Social Investment Forum. http://www.socialinvest.org/.

773. *SocialFunds.* http://www.socialfunds.com/.
774. *Software for Long Time Scale Simulation of Solids (eOn).* http://eon.chem.washington.edu/.
775. *South Asia Human Development Forum.* http://www.hdfnet.org/.
776. *South Pacific Regional Environment Programme (SPREP).* http://www.sprep.org.ws/.
777. *Species Analyst.* http://speciesanalyst.net/.
778. *Sumitomo Chemical.* http://www.sumitomo-chem.co.jp/.
779. *Sustainable Energy and Economic Development Coalition (SEED).* http://www.seedcoalition.org/.
780. *Time Banks UK.* http://www.timebanks.co.uk/.
781. *Time Online Edition.* http://www.time.com/.
782. *Timedollar Institute.* http://www.timedollar.org/.
783. *Toxics Release Inventory.* http://www.epa.gov/tri/.
784. *Tracker.* http://www.nrmtracker.org/.
785. *UC Berkeley Digital Library Project.* http://elib.cs.berkeley.edu/.
786. *USA Today.* http://www.usatoday.com/.
787. *Wall Street Journal.* http://www.wsj.com/.
788. *Web Services Activity.* http://www.w3.org/2002/ws/.
789. *Web Services for Remote Portlets (WSRP).* http:/www.oasis-open.org/committees/wsrp/.
790. *webLyzard.* http://www.webLyzard.com/.
791. *WildDay.* http://www.wildday.co.uk/.
792. *WordNet.* http://www.cogsci.princeton.edu/~wn/.
793. *Z39.50 Resource Page.* http://www.niso.org/z39.50/z3950.html.

Index

GPSR Compliance
The European Union's (EU) General Product Safety Regulation (GPSR) is a set of rules that requires consumer products to be safe and our obligations to ensure this.

If you have any concerns about our products, you can contact us on

ProductSafety@springernature.com

In case Publisher is established outside the EU, the EU authorized representative is:

Springer Nature Customer Service Center GmbH
Europaplatz 3
69115 Heidelberg, Germany

www.ingramcontent.com/pod-product-compliance
Ingram Content Group UK Ltd.
Pitfield, Milton Keynes, MK11 3LW, UK
UKHW020105200726
13856UKWH00002B/388
9781447137993